Springer Series in Synergetics Editor: Hermann Haken

Synergetics, an interdisciplinary field of research, is concerned with the cooperation of individual parts of a system that produces macroscopic spatial, temporal or functional structures. It deals with deterministic as well as stochastic processes.

J. S. Nicolis

Dynamics of Hierarchical Systems

An Evolutionary Approach

With 145 Figures

Springer-Verlag Berlin Heidelberg GmbH

Professor Dr. John S. Nicolis
Department of Electrical Engineering, University of Patras,
Patras, Greece

Series Editor:
Professor Dr. Dr. h. c. Hermann Haken
Institut für Theoretische Physik der Universität Stuttgart, Pfaffenwaldring 57/IV,
D-7000 Stuttgart 80, Fed. Rep. of Germany

ISBN 978-3-642-69694-7   ISBN 978-3-642-69692-3 (eBook)
DOI 10.1007/978-3-642-69692-3

Library of Congress Cataloging-in-Publication Data. Nicolis, J. (John), 1934- Dynamics of hierarchical systems. (Springer series in synergetics ; 25). Includes index. 1. System analysis. I. Title. II. Series: Springer series in synergetics ; v. 25. T57.6.N53 1986 003 85-26200

© 1984 Springer Science+Business Media New York 1986
Originally published by Springer-Verlag New York Berlin Heidelberg Tokyo in 1986
Softcover reprint of the hardcover 1st edition 1986

2153/3130-543210

Jesus answered "You say that I am a King.
For this I was born, and for this I have
come into the world, to bear witness to
the truth. Every one who is of the truth
hears my voice".
Pilate said to him "What is truth?"

John 18 (37–38)

To the memory of my parents

Preface

The main aim of these lectures is to trigger the interest of the *restless* under-graduate student of physical, mathematical, engineering, or biological sciences in the new and exciting multidisciplinary area of the evolution of "large-scale" dynamical systems. This text grew out of a synthesis of rather heterogeneous material that I presented on various occasions and in different contexts. For example, from lectures given since 1972 to first- and final-year undergraduate and first-year graduate students at the School of Engineering of the University of Patras and from informal seminars offered to an international group of graduate and post-doctoral students and faculty members at the University of Stuttgart in the academic year 1982-1983.

Those who search for rigor or even formality in this book are bound to be rather disappointed. My intention is to start from "scratch" if possible, keeping the reasoning heuristic and tied as closely as possible to physical intuition; I assume as prerequisites just basic knowledge of (classical) physics (at the level of the Berkeley series or the Feynman lectures), calculus, and some elements of probability theory. This does not mean that I intended to write an easy book, but rather to eliminate any difficulty for an eager reader who, in spite of incomplete formalistic training, would like to become acquainted with the physical ideas and concepts underlying the evolution and dynamics of complex systems. After going through these lectures the reader is strongly advised to concentrate on a narrower spectrum of interests and to try to do some original work by digging deeper into a specific subject - after becoming familiar with a broader context. These lectures could be used as the basis for a two-semester senior undergraduate or first-year graduate course on systems science or cybernetics.

Patras, January 1986   *J.S. Nicolis*

Acknowledgements

My thanks go to Professor Haken for encouraging me to put these lectures in written form as a volume in the Springer Series in Synergetics. I must also thank him for inviting me to the "Institute for Theoretical Physics" of the University of Stuttgart as a Visiting Professor during the academic year 1982-1983 while I was on leave of absence from the University of Patras. He provided me with a generous and stimulating hospitality. While at the Institute I enjoyed many animated and fruitful discussions with the graduate students and faculty members, most notably with Professor G. Mahler, Dr. Pjotr Peplowski, Professor T. Nagashima, Mr. G. Haubs, and especially Dr. G. Mayer-Kress.

My thanks also go to my brother Gregoire Nicolis of the Université Libre de Bruxelles for many years of "endless" discussions.

Above any other single individual, however, the main moving force behind my work during the last 27 years has been my wife Lala, who prompted me in the first place to try research as a way of life and who tolerated rather well some "intermittent" behavioral by-products of such a passionate pursuit.

I thank also Mrs. I. Wöhrner for typing the first draft of the main part of the manuscript, Miss E. Effenberg for typing the appendixes, and Mrs. U. Funke for general administrative assistance. I would like to express my special appreciation to Mrs. Niki Sarantoglou of the University of Patras for the impeccable typing of the final draft of the whole manuscript.

Finally, I gratefully acknowledge the help I have received from Dr. H. Lotsch and Miss D. Hollis of Springer-Verlag.

Contents

1. Introduction

1.1 What This Book Is About

"Hierarchy" is a provocative term; it has structural as well as functional conno-
tations. By a hierarchical system we mean, in general, an ensemble of interacting
parts which is composed of (and is analysable or decomposable into) successively
nested sets of interacting subunits.

Each set of interacting components —forming a distinct hierarchical level —
affords a specific state-space description with variables and parameters pertaining to that particular level. The interacting variables (and/or the parameters) at
a higher hierarchical level are "collective properties" (statistical moments or
convolutions) of the dynamics going on at the level underneath. Therefore, as we
move to a higher level, we usually witness a tremendous reduction in the number of
degrees of freedom. The higher level receives selective information "from below",
and in turn it exercises (efferent) feedforward control commands on the dynamics
of the lower level.

The complexity of any system arises from the number of its components and the
way these components are interconnected. This complexity refers to the *hardware* of
the system. The variety of the functional repertoire, on the other hand, determines
the complexity of the *software* of the system.

Structural hierarchical theory (also known as a branch of "complexity theory")
seeks ways of rearranging the components of a system (a) to achieve a compromise
in the conflict between complexity and stability, and (b) to design a working system of a predetermined functional repertoire with as few components as possible.
Specifically, it aims at reducing the quadratically (N^2) rising number of switching
elements as a function of the input-output number N needed to perform a given task
to the theoretical minimum $\sim N \log(N)$. This figure comes simply from the fact that
N! is the minimum number of states that a network should possess in order to handle
N input-output connections. If this is implemented via S switches, the maximum
number of states is 2^S. Hence $2^S \geqslant N!$ and $S \geqslant N \log(N)$.

Functional or dynamical theory, on the other hand, deals with the evolution of
large-scale systems from stable subassemblies, and studies the way these systems
may *change* through the exchange of energy and information with their environment.

(Systems *evolve* along integral lines in state space but they *change* only when de-stabilized, by discontinuously bifurcating and jumping from singularity to singu-larity.) Although two systems do not have to be hierarchically predisposed in order to exchange energy, in order to enter into a *communicative transaction* they do. (We need one hierarchical level to perform a cross-correlation, and another, higher one in order to "cognize" it.) In such a communication transaction, "hierarchical" *cod-ing* and *decoding* are again instrumental in reducing the exponentially (2^N) rising number of decoding steps (as a function of the length N of the transmitted time series) down to an algebraic rate $\sim N$.

In this book we have very little to say about structural hierarchy; we are really interested in the *dynamics of formation and cognition of collective properties*, and not so much in coding-decoding algorithms (which can be devised subsequently by way of implementing the dynamics). We will deal essentially with the problem of mutual *simulation* between two dynamical hierarchical systems, and especially with the prob-lem of the *compressibility* of the information which a given system receives from its partner. We proceed, however, in steps.

First we give a rather extended (albeit, we are afraid, somewhat patchy) histori-cal as well as conceptual account of the sine qua non conditions under which such cognitive systems could emerge at all. In particular, we try to establish the com-patibility between entropy production, progressive differentiation, increasing com-plexity, and efficient self-organization. By self-organization we mean the ability of an open system to *simulate* its environment or even parts (lower hierarchical levels) of itself. We essentially aim at understanding the "hierarchy" —if there is any —of the cosmological, thermodynamic, electromagnetic, and information-processing "arrows" of time. This is an ambitious undertaking, and we should ask for the reader's indulgence at almost every step of the argument. We envisage evolution, or rather change, as a cascade of discontinuous episodes —"bifurcations" —each lead-ing to the appearance of a more complex or more abstract hierarchical level in a given dynamical system.

Since a prerequisite for simulation is the "compressibility" of the information received from the partner system, we examine in detail the *reliability* of the in-formation conveyed via spherical electromagnetic waves propagating in different types of noisy "channels". Simulation of the partner usually proceeds via "rituals" that we call *games*. We describe and develop the dynamics of a number of them, on different hierarchical levels, from the genetic to the psychological, and question their "survival value" or simply their expediency.

Finally, we indulge in a new dynamical "paradigm", namely *dissipative chaos*, which, through the device of "strange attractors", promises in a single stroke to do this: to carry out very complex functional repertoires with very simple "hard-ware" and to behave as a versatile (albeit somewhat uncontrollable) information processor. The new paradigm appears if one realizes that information is not only produced via cascading bifurcations giving rise to broken symmetry, but also via successive iterations giving rise to ever-increasing resolution. Beyond a certain resolution interval, microscopic fluctuations are no longer smeared out; they are amplified and pass from lower to higher hierarchical levels. It is these fluctuations which essentially account for *new* information generated by an evolving hierarchical system.

In short, the purpose of this book is to develop an evolutionary approach for the basic problem of generation, storage and dissipation of information in physical and biological systems via dynamical processes. These processes are associated with branchings-off or cascading bifurcations giving rise to broken symmetry as well as chaotic dissipative dynamics giving rise to decreasing and increasing resolution. For an attractor simulating some aspects of a cognitive system (or a "processor"), there exist two basic requirements: large storage capacity and good compressibility. Stable steady states and stable limit cycles, possessing dimensionality zero and one, respectively, are very poor as information storage units but ideal as information compression devices. Strange attractors on the other hand, via a harmonious combination of expanding and converging trajectories in state space ("positive" λ_+ and "negative" λ_- Lyapounov exponents, respectively), can, in principle, comply with both requirements: They may possess a considerable information dimensionality, which makes them suitable for dynamical information storage, while, being "attractors", ($|\lambda_-| > \lambda_+$) they can serve at the same time as information compressors.

We shall see that (dissipative) chaotic dynamics offers an alternative and perhaps more elegant way toward the formation of hierarchy: Whereas the time-honored statistical-mechanical methods lead to hierarchical structures through the collective properties of a master equation (e.g., moments and cross-moments of the probability density function describing the dynamics at the "microscopic" level), dissipative chaotic systems accomplish the same thing differently: They do it by reducing the number of degrees of freedom in state space —where the dynamical system starting from a set of initial conditions, of dimensionality N sooner or later locks-in at a compact subset of state space (the "attractor"), which has *much lower dimensionality*.

1.2 Statement of the Problem

The general problem we are going to deal with in this book can be stated as follows: Consider two systems in a fluctuating environment as in Fig.1.1. Each system is defined as an ensemble of interacting units whose number is not necessarily large (three interacting components may be enough, as we are going to see, to create very complex behavior for the system involved). The environment fluctuates in the following sense: one of its parameters, instrumental in interfering with the

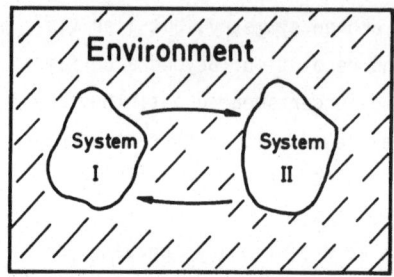

Fig.1.1. Two communicating systems in a fluctuating environment

energetic interactions of systems I and II, changes erratically with time. (For example, if the systems I and II radiate electromagnetic waves to each other, the pertinent parameter is the refractive index of the surrounding medium.)

How can we explain, on the basis of known dynamical laws of physics, the *communication* that takes place between the two systems, i.e., not only the energetic exchanges but also the informational ones? In other words, what is the physical (dynamical) basis of symbolic interactions?

In the first place, we have to clarify what we mean when we say that two systems *communicate*. For a physicist *information* is almost a "metaphysical" word. Indeed, in the current parlance of the mathematical theory of communication (invented by Claude Shannon 37 years ago and not improved considerably since), the solution of the problem of information transfer from place A to place B or from time τ to time $\tau + \Delta\tau$ (in short, from I → II or vice versa) boils down to a variational principle which is an attempt to compromise between two conflicting requirements: speed of transfer and reliability, or degree of errorless reproduction of an abstract pattern sent from I and received by II. In such a formulation of the problem, all physics (dynamics) is left out.

But if that were all, i.e., if the faithful reproduction of patterns were the sole aim of communication theory, then the Xerox machine would be the ideal embodiment of such a process. A Xerox not only reproduces but also actually improves upon the original by enhancing contrast or sharpening the contours, thus making recognition simpler. (This is because the machine uses an inherent nonlinear filtering process of an electrostatic character, a process which produces lateral inhibition.)

The most important aim of the communication process, however, is not copying but *simulation*: it essentially refers to a process of modeling one physical system by another, with the obvious goal of predicting and controlling it. We say that a physical system I simulates another II when it is capable of constructing, out of a finite-length time series that it receives, minimal-length algorithms of *compressed* descriptions of II. These algorithms, when introduced as inputs (programs) in a finite-state machine, will allow it to furnish as an output the behaviour of system II evolving in time —for all times future and past. In short, simulation refers to the construction of "if so, then so" scenarios which somehow fill the role that oracles used to play in ancient times. For system I to be able, in the above sense,

to simulate or model system II (and vice versa), a number of essential prerequisites
are demanded.

In the first place, the system must be hierarchical, i.e., possess at least a
hardware (H) (energetic-structural) level and a software (S) (cognitive) symbolic,
functional level. Information, in the form of discrete time series emanating from
the S-level of system II (S_2), impinges upon level H of system I (H_1), where a set
of convolutions (cross-correlations) takes place between the time series received
and the intrinsic dynamics of system I at level H. The result of such convolutions
is the emergence of a *collective property* possessing far fewer degrees of freedom
than S_2 or H_1. The collective property is now transported at the level S_1 and forms
a single (compound) state there. Figure 1.1 is now transformed into Fig.1.2, where
the above deliberations are sketched.

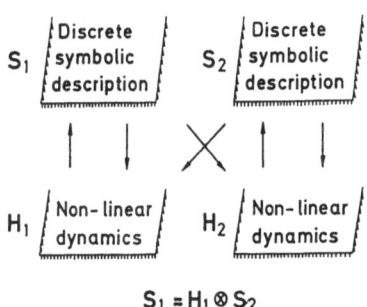

$$S_1 = H_1 \otimes S_2$$
$$S_2 = H_2 \otimes S_1$$

Fig.1.2. Schematic layout of two communicating
hierarchical systems

The communication between the two systems goes on, therefore, as the evolution
in time of a set of pairs of coupled nonlinear integral equations

$$S_1 = H_1 \otimes S_2 \quad,$$

$$S_2 = H_2 \otimes S_1 \quad,$$

(1.2.1)

each pair standing for one statistical moment. For each pair of (1.2.1), the process
clearly amounts to a continuous iteration of S_1 and S_2.

How can such an abstract problem be dealt with quantitatively? We will require
a lot of "homework" before being able to tackle our problem on the basis of the
existing laws of physics.

We shall see, for example, that the "compressibility" of a time series displaying, by a cognitive device, the behavior of a certain system, presumes the existence
of "attractors", both in the dynamical system under observation and the observer
himself. Such attractors come into being as a result of a cascade of instabilities
leading to bifurcation. We shall also see that the degree of compressibility of an
attractor is the difference between the state-space dimensionality of the attractor
and its "fractal" or informational dimensionality. Conclusion: we must learn some-

thing about nonlinear dynamics. Also, we have to remind ourselves that information is transferred by electromagnetic waves. We have therefore to learn how the signal, to be conveyed from the system under observation, is encoded in the degrees of freedom of a dynamical entity such as a spherical wave; how this wave is propagated in fluctuating (stochastic) media; and what amount of information it delivers when it finally impinges on a finite aperture. We will discover that perception is basically ambiguous (many different objects or signals give rise to the same); therefore, we will be forced to evaluate the entropy of an information-carrying spherical wave. Hence the necessity of dealing a little with statistical physics and formal information and coding theory.

Finally, we have to recognize that the two systems I and II involved in mutual modeling actually wage a *game*, or are essentially in conflict, each of them trying to "break" the "code" of the opponent. As a matter of fact, in any real situation "conflict" emerges, since each partner tries to break the code of the opponent and, at the same time, avoid being compressed (or modeled) by the opponent by displaying a seemingly chaotic, i.e., unpredictable, behavior at its own functional level. Hence the necessity of learning something about game theory as well!

In order to approach our problem properly in steps, we require a fair amount of prerequisite knowledge; hence the planning of these lectures according to the following five chapters:

2) Preliminaries on (two-dimensional) nonlinear dynamics and statistical physics.
3) The role of spherical waves as information carriers.
4) Elements of information and coding theory, with applications.
5) Elements of game theory, with applications.
6) Stochasticity due to deterministic dynamics in three- or higher-dimensional space: chaos and strange attractors.

The perceptive reader may notice that the fundamental concept of hierarchy has been somewhat underplayed so far in the introduction. This is because hierarchy — like self-organization (or the ability of a system to compress environmental time series and construct minimal-length algorithms for the simulation of the environment involved) — is an *emergent* property of nonlinear dynamical systems evolving under a discontinuous interplay among the basic four types of physical interactions. It is through a study of such an interplay that one can historically understand the ability to communicate as an evolutionary, emergent property of complex large-scale systems (something which runs against the traditional world view as dictated by a static universe running down to "thermal death"). That is to say, in an expanding universe the interplay of the physical interactions gives rise in a concomitant way to entropy production, progressive differentiation, increasing complexity, and increasing organization. The cosmological, thermodynamic, electromagnetic, and signal-processing "arrows of time" all point in the same direction: the direction of increasing differentiation with time, and the emergence of cognitive systems which can not only simulate the external environment but, in a certain incomplete way, even parts of themselves!

1.3 Some Preliminary Definitions of Complexity and Organization

1.3.1 Complexity

A system may be complex on either a structural or a functional level. *Structural complexity* increases with the number of interacting subunits; the percentage of mutual connectedness among them, in dyadic or plural fashion; and the variance of the probability density function of the strengths of interactions of the individual subunits.

On a *functional level*, complexity increases with the minimum length of the (most compressed) algorithm from which one can retrieve the full behavior of the system. In this sense, as we shall see later, the functional complexity of a system increases with the length of its "phylogenetics tail", namely its evolutionary history marked by a discrete sequence of bifurcation episodes. In that sense, then, a virus or an orange is a more complex system, functionally speaking, than a big star, since in order to understand the function of each one of them, you need as a prerequisite the existence of a supernova.

1.3.2 Organization

A cognitive device capable of compressing and simulating a physical phenomenon possesses a more or less advanced degree of "self"-organization.

In general, self-organization is associated with the spontaneous emergence of long-range spatial and/or temporal coherence among the variables of the (organized) system. Nevertheless the system may be quite "innocent" of this self-organization, namely it may not use it further or may not know how to use it or may not know that it knows how to use it ... The point is that a self-organized system becomes a *cognitive* system only beyond a minimum number (namely two) of hierarchical levels; (this means that one-to-one mapping does not produce cognition). The sine qua non prerequisite for the emergence of a self-organizing *cognitive* system is the ability of this self-organized system to replicate another system (including[1] itself or

---

1 It is sometimes asserted that no self-organizing system can fully simulate itself; this, it is said, could give rise to the "paradox of self-reference" that is to a perpetual nonconverging tautology. For higher cognitive functions this assertion remains undecidable. For genetic self-replication, tautology is transcended via the spontaneous folding of linear polypeptide chains and the resulting formation of stereospecific proteins.

This process of symmetry breaking creates new information in the system and dispels self-reference.

parts of itself). This self-replication ability is in turn *the* indication that the system knows how to *compress and simulate*. Examples: Some chemical reactions constitute dynamical systems which are self-organizing but are not cognitive. The genome of any biological organism on the other hand *is* a *cognitive* system (the simplest one) since via transcription and translation (a process involving two hierarchical levels, the level of nucleic acids and the level of polypeptides) it simulates and eventually reproduces itself. Clearly then, cognition is instrumental in compressing the complexity of another system (including parts of itself) via the formation of collective properties (cross-correlations or statistical crossmoments), thereby dramatically reducing the initial number of the degrees of freedom of the system under observation. Alternatively, as we shall see, an attractor acting on any point of its basin (set of initial conditions) does de facto the same.

Concluding then, we can claim that *communication* is involved in triggering the emergence of new spatial-temporal patterns via a sequence of instabilities producing bifurcations which lead to broken symmetry, and subsequently in creating and maintaining complex attractors in dynamical systems, which attractors compress these patterns. But let us begin from the beginning.

2. Preliminaries from Nonlinear Dynamics and Statistical Physics

2.1 Symmetries and Conservation Principles

What exactly do we mean by the "laws of physics"? Physics is an empirical science. People make lengthy observations of a certain phenomenon and then try to "theorize" about it, namely to *compress* the strings of observation into an essentially simple and brief analytical formula (the "algorithm"). This theorizing eventually is implemented by some sort of convolution between the observed time series and the nonlinear dynamics of the human brain.

By complementing this algorithm with a variety of extrinsic constraints (initial and boundary conditions) and introducing it as an input to a finite-state machine (e.g., a digital computer), people obtain different scenarios of the same phenomenon — depending on the specific choice of initial and boundary conditions. Therefore the "physical laws" are essentially algorithms leading to simulation outcomes or "models" of large classes of dynamical phenomena.

If this is so, then the aim of reducing the communication process to the laws of physics leads to an immediate paradox, namely the effort to explain simulation using simulation! At this moment, however, it is wise to push the paradox under the rug and try to proceed. We shall return to it in Chaps.4-6, where the concepts and examples of hierarchical processing will be helpful in elucidating this paradox of self-reference.

A physical law is always accompanied by a certain number of symmetries. By degree of symmetry we imply the number of equivalent alternative descriptions. For example, a pattern is symmetrical if, after applying to it a certain transformation, it looks the same as it did before. Or to put it in another way, the greater the degree of symmetry of a system (abstract or concrete), the more accurate the prediction that we can make about the evolution of such a system. For instance, we shall see soon that for a closed dynamical system the most probable state is the state of thermodynamic equilibrium, which happens to be the state of fullest possible macroscopic symmetry. No matter which initial condition we start from, no matter what kind of perturbative operation we apply, the above system, with probability one, sooner or later will go to the state of perfect homogeneity. The initial question refers, however, to the symmetry of dynamical laws rather than to patterns (which

come into existence after allowing the laws to operate for given time intervals).
Again, a natural law possesses a symmetry with respect to certain sets of operators
or transformations, if afterwards it appears the same as before. For instance, the
laws obeyed by the nuclear force that binds the atomic nucleus look the same if we
interchange protons and neutrons. We know this because the properties of nuclei
look the same if we apply the above operation (e.g., the properties of the lithium-7
nucleus, which contains three protons and four neutrons, are almost —allowing for
the electromagnetic difference —the same as those of the beryllium-7 nucleus, with
four protons and three neutrons).

Another symmetry possessed by almost all dynamical laws (see below) is time invariance. Symmetries of physical laws are usually referred to as *conservation principles*: they are always *implicitly* contained in the dynamical laws and appear explicitly as (sometimes unexpected) consequences of the workings of the laws. When
unrevealed, they are used in pertinent problems in lieu of the dynamical law itself,
thereby saving the scientist from a lot of unnecessary and redundant "integrations".
The more conservative a system the more predictive it is. So, for instance, Hamiltonian systems are more predictive than dissipative systems far from thermodynamic
equilibrium. The most well-known symmetries in dynamics have to do with symmetries
in space and time. Invariance under translation in time implies energy conservation
and vice versa. Translational invariance in space implies the conservation of linear
momentum, and vice versa (if the momentum is conserved, the underlying law must be
translationally invariant).

Similarly, rotational invariance (i.e., the case where the results of observations do not depend on orientation) implies the conservation of angular momentum,
and vice versa. (For example, Kepler's second law that an orbit under a central
force sweeps out equal areas in equal times is equivalent to angular momentum conservation.)

Now we come to the following crucial question: What is the relation between the
symmetries of a physical law and the symmetries of a pattern generated after allowing this physical law to operate for a certain time interval? Most patterns in nature are highly asymmetrical, yet laws are deduced after laborious studies of large
classes of patterns. The answer to this fundamental inquiry is still unknown. Generally, however, we can argue that in the formation of the degrees of symmetry of a
macroscopic pattern or object, there are some *extrinsic* agents which play significant roles, such as the asymmetry of initial and boundary conditions (which are introduced ad hoc to the solution of our differential equations and determine the
numerical values of the constants of integration) and *also* instabilities which cause
bifurcations, i.e., branchings-off in our solutions, thereby *breaking the symmetry*
of the initial pattern or structure. To those extrinsic factors one has, of course,
to add intrinsic factors arising from the fact that one out of the four basic physical interactions, namely the weak nuclear force, displays a genuine *microscopic*

asymmetry, i.e., mirror-asymmetry (in the decay of the cobalt nuclei). Also charge
— parity invariance (CP) is violated in certain rare decays of the K-meson, thereby breaking the symmetry between matter and antimatter. Since current theory predicts charge-parity-time (CPT) invariance (and there is no experimental evidence
to the contrary), we conclude that the weak interaction is not time invariant.

Finally, there is the third case of perfectly symmetrical laws giving rise to
asymmetrical phenomena, not due to broken symmetries via the intervention of extrinsic agents, but simply due to the fact that the symmetrical solution is *unstable*.
This, however, is the case of *hidden* symmetries, which among other things play a
central role in today's conjectures about the origin of the universe: the story
goes that in the "beginning" two perfectly symmetrical populations of "hyper"-
particles X and $\bar{X}$ had been created. The X's started disintegrating to baryons and
$\bar{X}$'s to antibaryons. The decay process *could* be symmetrical. In such a case the
universe would start and stay as an ensemble of photons in thermodynamic equilibrium. Nevertheless, the present state of the universe encourages the hypothesis
that this homogeneous state was unstable, and the other possibility actually has
been inevitably selected, namely the case of asymmetrical decay. So, for $\sim 10^9$
antibaryons, $\sim 10^9 + 1$ baryons have emerged. This is consistent with two observational
facts: that the observed natural universe contains only baryons and that the ratio
of photons to baryons today is $\sim 10^8$. Later in this chapter, we are going to follow
the "coarse" scenario of how the ensemble of baryons and photons evolved since the
moment of the big bang.

Another accepted example of hidden symmetry worth mentioning is that between
the electromagnetic and the weak force. There are scientists working on "grand-
unification theories" who believe that before long we may end up with one type of
particles and one class of interactions (instead of four), the apparent diversity
of particles and forces which we witness in nature being due to the fundamental
symmetries being hidden.

2.2 Instabilities at the Root of Broken Symmetries, Dissipation, and Irreversibility for Low-Dimensional (Not Statistical) Dynamical Systems

2.2.1 The Role of Gravitation

If the laws of nature are essentially symmetrical with respect to time, how then
does evolution take place? Aside from macroscopic time, how then does evolution
take place? Aside from macroscopic systems containing many degrees of freedom, for
which the answer is provided by the second thermodynamic law and Boltzmann's H-
theorem (and is due, as we are going to see, to the extrinsic asymmetry of the initial conditions), how can we trace the responsibility of change in simple systems,
ignoring completely the influence of initial conditions? In the first place, are

instabilities perhaps built into some of the physical laws themselves, thereby leading, if left unconstrained, to "snowballing" irreversible phenomena? The answer is yes. Among the four basic interactions there is one (gravitation) which displays exactly this kind of behavior.

The potential energy stored in a gravitational system is always negative,and becomes steadily more so as two attracting masses M_1 and M_2 come closer to each other. Unlike all other systems, where the potential energy is always positive and which possesses a point of stable equilibrium at which the potential energy becomes a minimum, even in the simplest gravitational system of two attracting masses there is no equilibrium,since the potential energy goes monotonically to $-\infty$ if the system is left to itself (Fig.2.1).

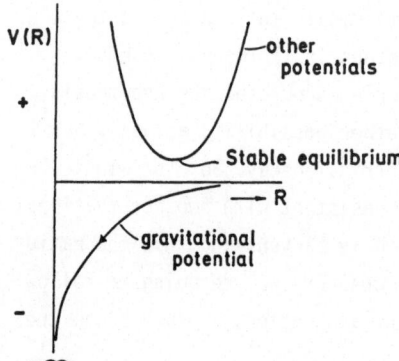

Fig.2.1. Unlike all other potentials, the gravitational potential does not possess equilibria

There is no limit to the kinetic energy the masses M_1 and M_2 can acquire: as M_1 and M_2 approach each other and acquire more and more kinetic energy, their gravitational energy reservoir is depleted and made more negative. Thus if gravitation were the only interaction in physical existence, the universe would collapse irreversibly (but without thermal losses,since viscosity is zero in the absence of electromagnetic interactions) into a mathematical singularity from which it once presumably emerged under the action of the big bang. How long would it take for a homogeneous sphere of total mass M and radius R to collapse? For simplicity,assume that this sphere contains material without internal pressure.

All the individual particles will attract each other and the sphere will contract. Measuring the time from the instant the material is "left" to contract, at time t later the sphere will have a radius r < R. To study how r changes with t, we see that a particle of unit mass at the surface of the sphere will be subjected to a force equal to GM/r^2, so the equation of motion will be $d^2r/dt^2 = -GM/r^2$, which, when integrated, gives $(dr/dt)^2 = 2GM(1/(r - 1/R))$; at r = R, dr/dt = 0 and t = 0, r = R.

The integral of this equation,subject to the above initial conditions,gives

$$t = \left(\frac{R}{2GM}\right)^{\frac{1}{2}} \int_r^R \left(\frac{r}{R - r}\right)^{\frac{1}{2}} dr \quad .$$

The time T for total collapse can be deduced from the above result by putting
r = 0; we get $T = (\pi/2)(R^3/2GM)$, which for a sphere as large as the sun is in the
order of 30 minutes! So if the internal pressure of the sun were suddenly removed,
it would collapse in a matter of half an hour. If is, of course, the cooperation
among all the four basic physical interactions that is responsible for the longevity
of the sun (~10 billion years). Most objects in nature possess, as a result of
such cooperation, sufficient internal pressure to withstand or at least postpone
the catastrophic effects of gravitational stability.

The fact that gravitational systems posses no inherent equilibria also makes them
"nonthermodynamic" systems. To understand what this means let us follow a "thought
experiment" devised by the astrophysicist *Narlikar* [2.1]: We know from experience
that if we place a hot system in contact with a cold one, heat flows from the former
to the latter. The hot body gets progressively colder and the cold body gets hotter
until the two have the same temperature, an instance which signals the establishment
of thermodynamic equilibrium — a most symmetrical state since fluctuations above or
below the final common temperature are a priori equally probable.

Imagine now what would happen if we placed a hot star I in the vicinity of a cold
star II. Heat will flow from I → II. Nevertheless, each star is in equilibrium under
two forces: its self-gravitation and its internal pressure due to γ-rays and neutrinos
emerging outwards from the cascades of nuclear fusion taking place in the core of
each star. The gravitation tends to contract each star and the internal pressure
just cancels this. Now when energy passes from star I, its internal pressure is
reduced and so it begins to shrink. But this shrinking leads to compression of the
gases inside and consequently to a rise in their temperature. In star II the reverse process takes place: it gains heat from star I and therefore increases its
internal pressure. This upsets the equilibrium inside star II and it expands. By
expanding, however, it provokes cooling of its gases. As a net result, then, star
I becomes hotter and star II becomes cooler, again as a result of the negative
energy of gravitation, in apparent contradiction to the second law of thermodynamics. Such an unstable behavior clearly promotes and amplifies any initial small
asymmetry to the status of plastic distortion or broken symmetry.

The tidal effect of gravitation gives another example of an irreversible phenomenon (that of *gravitational capture*) due to dissipation which follows from broken
symmetry. Tidal effects essentially have to do with instantaneous (action at a
distance) energy transport via induction between two *extended* viscous masses, e.g.,
the earth and the moon. If we assume that the two bodies are perfectly spherical
and so viscous that they do not yield to the differential gravitational pull exercised between each other, then gravitational capture is impossible, and the two
bodies will simply undergo elastic bouncings according to the law of conservation
of linear momentum, in a perfectly reversible manner. No coupling between their
spins is going to take place. In fact, any spin of the bodies is irrelevant due

to their perfectly spherical shape, which implies that such a spin is not affected by the interaction. The application of the principle of conservation of angular momentum leads to the conclusion that the final separation of the paths of the bodies must be equal to the original separation; therefore it corroborates the reversible character of the interaction. However, in reality, due to the finite magnitude of viscosity, the two bodies *do* get plastically *deformed* after the gravitational pull. As a result of this nonlinear deformation (broken symmetry), dissipation takes place (dynamical energy is turned into heat through the imperfections of the elastic deformation of the material): the deformation of the initially spherical bodies being nonelastic, it is irreversible. Further, the deviation of their shapes from spherical causes a coupling between their orbital and spin angular momenta. So both *energy* and *angular momentum* may now be exchanged between the rotation of one body and the orbit of the other. A well-known example is the earth-moon system, where tidal forces act in the ocean, the atmosphere, and the solid earth. If we imagine the surface of the earth as covered with oceans at the sides A and A' (Fig.2.2), A is pulled most strongly towards the moon and the A' less strongly. The main mass of the earth which lies in between is pulled with a force intermediate to the two extremes. Thus both parts of the ocean A, A' bulge outwards relative to the center of the earth (Fig.2.2). Near land masses this results in a great inflow and a subsequent outflow of water from the sea. Water moving over land experiences friction and loses energy. This energy comes from the earth's rotational energy, with the result that the rate of rotation of the earth about its axis slows down (by a thousandth of a second per century —so much so since the earth-moon line takes almost a month to circle the earth while the rotation occurs once a day, i.e., much faster). The tides raised on the moon by the earth are much bigger, and the effect of this tidal friction has been to slow down the moon's rotation so that it now always presents the same face to the earth.

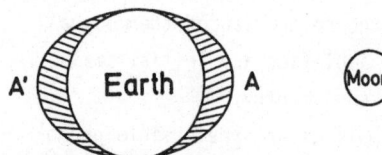

Fig.2.2. Tidal effects on the earth caused by the moon

Applying conservation of angular momentum to the (closed) earth-moon system shows that the angular momentum of the moon's revolution is, in the same direction as the rotation of the earth, increasing: the moon's orbit is growing, as is observable. So the dissipation of tidal friction leads to a spin-orbit coupling, driving the moon away from the earth —an irreversible phenomenon.

Imagine now a body encountering the earth so that at closest approach it is moving in the direction opposite to the daily rotation of the earth. Without dissipative forces it would eventually have to recede to infinity, but now, tidal friction will diminish its angular momentum. If this loss and the associated loss of energy are sufficiently large, they could prevent its escape. Thus the tidal phenomenon can, at least in principle, lead to gravitational capture, an irreversible evolutionary phenomenon impossible without dissipation, i.e., without gravitational pull on an extended viscous object.

Having dealt, for the time being, (rather extensively) with the gravitational interaction per se and the interplay between gravitation and viscosity, let us now turn to other types of instabilities provoked by a sequence of couplings among the four different types of physical interactions, just in order to appreciate in a broader context the role of gravitation as the ultimate controller of evolution in the universe.

2.2.2 Comments on the Role of Coupling Among the Four Basic Interactions in Evolution

We have seen in the previous section that if a gravitational system is left to itself it collapses very rapidly. For the sun this "free-fall" time is about half an hour and for our galaxy ~100 million years.

The ages of the universe in general (~15-20 billion years) and our galactic and solar system in particular (~5-15 billion years) suggest, therefore, a number of "hang-ups" (as $Dyson$ [2.2] calls them), i.e., a number of seemingly accidental obstacles, usually arising from the interplay among the four basic interactions and a number of (unknown) initial conditions associated with the (still unknown) dynamics of the big bang. These hang-ups arrest temporarily the process of the degradation of energy dictated by gravitation alone and are responsible for the longevity of the universe. This longevity is the sine qua non prerequisite for the emergence of more and more complex structures and ultimately of life forms, with an increasing number of hierarchical levels, both structural and cognitive. This is due to evolution being a very slow process.

The relative strengths of the four basic interactions are:

Gravitational	10^{-39}
Electromagnetic	1
Strong nuclear	10^3
Weak nuclear	10^{-6}.

The first two forces are long-range ones, with intensity falling in vacuum as ~$1/R^2$, while the latter two are short-range (they essentially cease to act beyond a distance of the order of a nucleus diameter, 10^{-13}cm) and fall off exponentially. Following $Dyson$ [2.2], let us briefly review the nature of the accidental obstacles, which act, so to speak, as deus ex machina in prolonging dynamical activity in the physical universe.

A first hang-up is the *spin* possessed by many extended objects. The origin of spin is unknown (certainly it is not explicitly included in the dynamical laws), and it constitutes a clearly broken symmetry due to some special initial conditions. Extended objects cannot collapse gravitationally if they spin rapidly. Instead of collapsing, the outer parts of the object settle into stationary orbits around the inner ones (e.g., our solar system with its planets, satellites, and asteroids).

A second hang-up is the so-called *thermonuclear* hang-up. It has to do with the burning of hydrogen to form helium when it is heated and compressed. The thermonuclear burning (fusion reaction between hydrogen nuclei) releases energy in the form of γ-rays and neutrinos which oppose further compression. Thus a star that contains a large portion of hydrogen cannot collapse gravitationally beyond a certain point until all the hydrogen is consumed.

Now the important thing here is the fortunate fact that an isotope of helium with a mass number 2 (the nucleus of which would consist of two protons and no neutrons) does not exist. If helium-2 existed, the proton-proton reaction would yield a helium-2 nucleus plus a photon. The helium-2 nucleus would, in turn, spontaneously disintegrate into a deuteron, a positron, and a neutrino. The first reaction is a strong one so the hydrogen would burn very fast to produce helium-2. The attractive nuclear force between two protons is on the order of 20 million volts, but it just barely fails to produce a bound state of the helium-2 nucleus by a mere margin of half a million volts. So if the strong force were 2.5% stronger, the sun would burn up its hydrogen reservoir in a matter of minutes. What actually happens is this: Ordinary hydrogen can react with itself only by the weak interaction process. In this process two protons fuse to form a deuteron plus a positron and a neutrino. (The positron subsequently interacts with electrons of the surrounding plasma and gives photons in the γ-range. Through Thompson scattering it takes about a million years for these photons to reach the surface of the sun and spread from there into the optical range.) Now since the decay or the relaxation times are inversely proportional to the *intensities* of interactions, the proton-proton reaction proceeds about $(10^{+3}/10^{-6})^2 \sim 10^{18}$ times more slowly than the strong nuclear reaction. Hence the longevity of the stars and our sun in particular.

A third hang-up, instrumental this time for evolution on earth, has to do with the fact that the interior of the earth contains modest amounts of *radioactive materials* uranium and thorium, materials which are not "cooked" in the interior of stars (the fusion process does not go beyond the heaviest stable element, namely iron) but which are "cooked" during the explosions of a supernova. (Such a thermonuclear detonation is due perhaps to the fact that for large stars in their last phase, with masses above the "Chandraseckar limit", the core, composed mainly of carbon and oxygen, burns instantaneously to iron. The star's interior implodes rapidly and the energy produced causes an explosion of the exterior masses.) These nuclei, via a competition between surface tension and electrostatic energy, slowly disintegrate via β-decay by transforming neutrons to protons, electrons (which are emitted), and antineutrinos. The friction of the escaping electrons in the surrounding medium is responsible for the heating of the interior of the earth; hence the volcanic activity and the formation and renewal of the atmosphere, as well as the renewal of the surface of the earth with raw materials.

It is now time to turn to instabilities, and therefore evolutionary phenomena caused not by a physical law per se but rather by an interplay between a physical law and *specific* initial and boundary conditions. What follows is a number of discrete and representative examples referring both to hidden symmetries and broken symmetries.

2.2.3 The Overdamped Nonlinear Oscillator: A Case of Spontaneously Breaking Symmetry

Consider a massive particle m moving in a viscous medium (e.g., undergoing ν shocks per unit time) under the action of a spring of constant K (Fig.2.3).

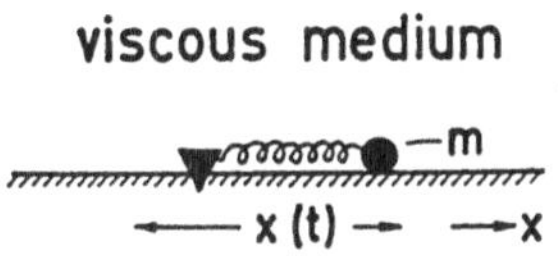

viscous medium

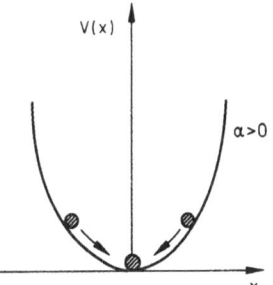

Fig.2.3. A massive particle under the action of a spring in a viscous medium

According to Newton's second law, for constant m the force exercised on the body equals the rate of momentum change, i.e., $F = \partial \bar{P}/\partial t$ or $F = m \, d^2x/dt^2$, $F = (-\nu \, dx/dt) - Kx$; so the equation of motion will be: $m\ddot{x} + \nu\dot{x} + Kx = 0$, or $\ddot{x} + (\nu\dot{x}/m)$ + $(Kx/m) = 0$, or $\ddot{x} + \xi\dot{x} + \omega_0^2 x = 0$, where ω_0 is the eigenfrequency of the (damped) oscillation. For $|\xi| \gg 1$ we can, in the absence of specific initial or boundary conditions advising the contrary, omit the inertial term so that the equation of motion becomes simply $\dot{x} = -\alpha x = F$, where $\alpha = \omega_0^2/\xi$. We can imagine the motion of the particle in a potential well $V(x)$ which is related to the force F, as $F = -\partial V/\partial x$ or

$$V = \frac{1}{2}\alpha x^2 \; . \tag{2.2.1}$$

Clearly there is one steady state $dx/dt = 0$, $x = 0$, and it is stable as shown in Fig.2.4.

$V(x)$

$\alpha > 0$

Fig.2.4.
The potential $V(x) = \alpha x^2/2$ for $\alpha > 0$

x

Suppose now that the restoring force becomes slightly nonlinear, either as a result of finite elongation —where Hook's law does not hold—or as a result of a not-perfectly elastic string. The equation of motion can be written again as

$$\ddot{x} = -\alpha x - \beta x^3 ,$$ (2.2.2)

where $\beta > 0$. The potential well now acquires the form

$$V(x) = \frac{1}{2} \alpha x^2 + \frac{1}{2} \beta x^4 .$$ (2.2.3)

Two cases are worth distinguishing:

a) $\alpha > 0$, $\beta > 0$. In this case the form of the new potential is the same as before, and the only steady state, which is again stable, is the $x = 0$ (this can also be deduced from the fact that the equation $\alpha x + \beta x^3 = 0$, for $\alpha > 0$, $\beta > 0$, possesses only one real value $x = 0$).

b) $\alpha \leqslant 0$, $\beta > 0$. As α crosses the critical value $\alpha_c = 0$, the form of the potential changes, although remaining symmetrical (Fig.2.5).

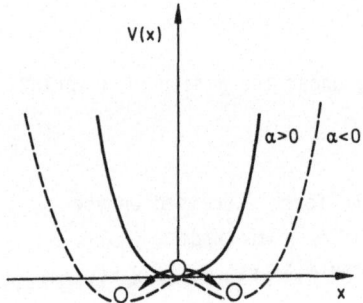

$V(x)$

$\alpha > 0$ $\alpha < 0$

x

<u>Fig.2.5.</u> The potential $V(x) = (\alpha x^2/2 + \beta x^4/4)$ for $\alpha > 0$ (——) and for $\alpha < 0$ (---)

We now have three steady states or, to put it another way, the solutions of the equations $\alpha x + \beta x^3 = 0$, for $\alpha < 0$, $\beta > 0$, are now all three real. However, the former symmetrical stable steady state $x = 0$ has become unstable, and the particle will drift to one of the equally probable states $x = \sqrt{|\alpha|/\beta}$ or $x = -\sqrt{|\alpha|/\beta}$. This example shows that in order to have uncertainty as to what the next state of a system will be, one does not need many degrees of freedom: merely an instability is enough to provoke, through a bifurcation, unpredictable behavior normally attributed to stochasticity.

2.2.4 The Laser: A Case of Broken Symmetry

Consider a material body filling a cylindrical cavity which is bounded by two semitransparent parallel mirrors, as in Fig.2.6d.

We suppose that the atoms of the material can be in two distinct states E_1, E_2, with populations following, at equilibrium, Boltzmann's distribution:

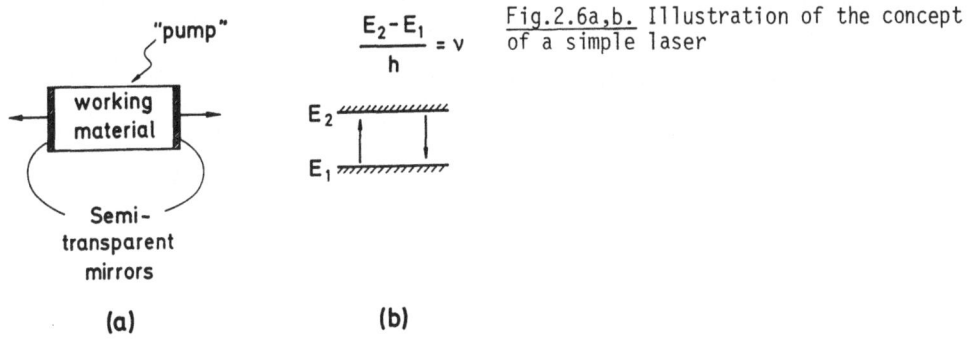

Fig.2.6a,b. Illustration of the concept of a simple laser

$$N_2 \sim N_1 \exp\left(\frac{-E_2 + E_1}{kT}\right) \ .$$

We excite the atoms of the body via an external pump at a frequency $\nu = (E_2 - E_1)/h$ (Fig.2.6b). Atoms from the lower energy state are going to be "transported" to the higher energy state and, from there, spontaneously fall back to the lower state after emitting one photon (of frequency ν). We are interested in the dynamical evolution of the number of photons produced in the cavity, $x(t)$.

Clearly the rate of change dx/dt of the number of photons can be considered at every moment as the difference between a *gain factor* (increasing the number of pro- duced photons as a result of transitions undergone by the excited atoms) and a *loss factor* (decreasing the number of photons in the cavity due to escape from the semi- transparent walls of the cavity).

The gain factor A will be proportional to the number of photons existing in the cavity as well as to the number of atoms N excited per unit time, and hence propor- tional to their product. Thus, $A = GNx$, where G is a proportionality factor giving the degree of susceptibility of the working material and the strength of the pump- ing source. The loss factor B will be simply proportional to the existing number of photons in the cavity $B = \tau x$, where τ is a factor indicating the degree of trans- parence of the reflecting mirrors. So

$$dx/dt = GNx - \tau x \ . \tag{2.2.4}$$

However, the number of the excited atoms decreases with the emission of photons so we can write $N = N_0 - \alpha x$, where N_0 is the number of pumped atoms (per unit time), and α is a coefficient, again related to the susceptibility of the material. Sub- stituting into (2.2.4), we get the equation

$$\dot{x} = G(N_0 - \alpha x)x - \tau x$$

or

$$\dot{x} = -K_1 x - K_2 x^2 \ , \tag{2.2.5}$$

where $K_1 = \tau - GN_0$ and $K_2 = G\alpha$. Here K_2 is always positive. We distinguish two cases:

1) $K_1 > 0$ or $GN_0 < \tau$,   weak pump

2) $K_1 \leq 0$ or $GN_0 \geq \tau$,   strong pump .

We observe that our dynamical equation (2.2.5) is quite similar to that (2.2.2) of the previous example, the difference being that now, our particle is restrained by a restoring force, which instead of a cubic nonlinearity has a quadratic one. Thus we can forget for a moment about the laser (except, of course, that x must be a positive integer) and assume that we have a particle whose instantaneous excursion from the equilibrium position x(t) is determined by a potential well

$$V(x) = \frac{1}{2} K_1 x^2 + \frac{1}{3} K_2 x^3 \quad . \tag{2.2.6}$$

For the case $K_1 > 0$, the potential well has the shape indicated in Fig.2.7, and the only stable steady state is x = 0. In our example, this means that when transients die out, the *steady-state* number of photons (i.e., the number of coherent photons, see below) in the cavity is zero. (The device behaves like a common lamp.) However, when the pumping power exceeds a certain threshold $G_c = \tau/N_0$, then $K_1 < 0$ and the "potential well" takes the form of Fig.2.8, so there exists a stable steady state with x ≠ 0, $x^* = |K_1|/K_2$, where the device behaves as a *coherent* source of radiation. For the sake of completeness, we add in what follows an operational definition of coherence, which will be further used in Chaps.3 and 4.

Originally a "coherent wave" means a wave of the type $\exp[j(\omega t - \varphi)]$, where φ is constant or remains bounded with respect to a given reference (say the phase of the local oscillator of a receiver). In an incoherent wave, the phase φ fluctuates at random.

The essential difference between the two types lies in the way in which the power of the sum of several oscillators depends on the power of the individual summed oscillators. If, for example, two radio (man-made) waves of equal amplitude inter-

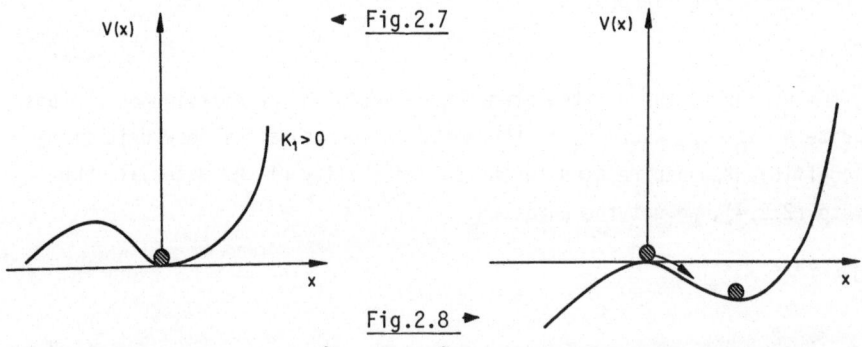

Fig.2.7. The potential $(K_1 x^2/2) + (K_2 x^3/3)$ for $K_1 > 0$

Fig.2.8. The potential $(K_1 x^2/2) + (K_2 x^3/3)$ for $K_1 \leq 0$

fere, the total power density may be anything from zero (if the waves are in phase opposition) to four times the power density of either wave (if the waves are in phase). Two incoherent sources of equal power, however, will always give *double the intensity* (square of the amplitude) of either source: if n independent oscillators of unit amplitude interfere at a point in space, the total power density is propor-tional to

$$W \sim \left| \sum_{i=1}^{n} E_i \right|^2 = \left| \sum_{i=1}^{n} e^{j\varphi_i} \right|^2 \quad ,$$

if we assume unit amplitude for all interfering components. Or,

$$W \sim \left(\sum_{i=1}^{n} \cos\varphi_i \right)^2 + \left(\sum_{i=1}^{n} \sin\varphi_i \right)^2 = n + 2 \sum_{\substack{i+j}}^{n(n-1)/2} \cos(\varphi_i - \varphi_j) \quad ,$$

where the sum is taken over the $n(n-1)/2$ terms with $i \neq j$.

Let $\varphi_i - \varphi_j = \varepsilon_\sigma$, then,

$$W \sim n + \sum_{\sigma=1}^{n(n-1)} \cos\varepsilon_\sigma \quad . \tag{2.2.7}$$

Let us examine the two extreme cases of full coherence and full incoherence:

1) If all φ_i are equal, i.e., all the oscillators are in phase, then $\varepsilon_\sigma = 0$ for all σ and $W = n + (n-1)n = n^2$.

2) Now let the individual phases be random, independent, and *uniformly* distri-buted between $-\pi$ and π. What will be the distribution of $\varepsilon_\sigma = \varphi_i - \varphi_j$? In probability theory we can prove that if we have two uniform distributions of variables x_i, x_j in an interval a, b (Fig.2.9), then the probability density function of their dif-ference follows the so-called Simpson distribution (Fig.2.9b) in the interval 2a, 2b. So, if the φ_i's are distributed uniformly between $-\pi$ and π, one expects that ε_σ will be Simpson distributed between -2π and 2π. However, owing to the periodicity of $\cos\varepsilon_j$, this is completely equivalent to a uniform distribution from $-\pi$ to π as

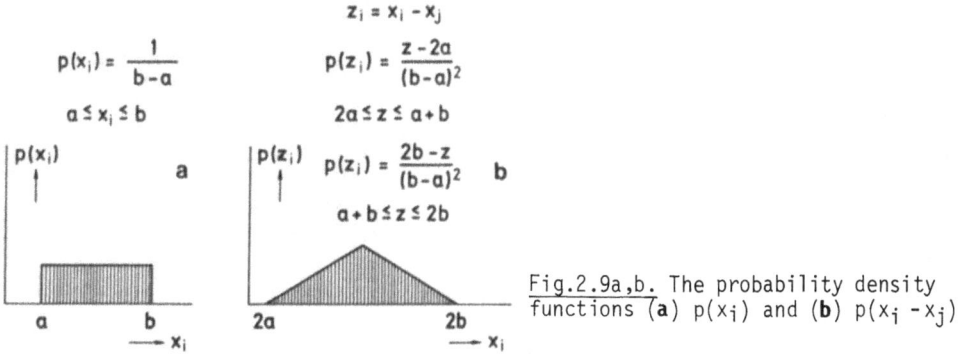

Fig.2.9a,b. The probability density functions (a) $p(x_i)$ and (b) $p(x_i - x_j)$

the probability densities in any other interval from $m\pi$ to $(m+2)\pi$, where m is an integer, will be added to those in the basic interval$(-\pi,\pi)$ or any other chosen basic interval of width 2π. So, taking the average of W in (2.2.7) above, we get

$$<W> \sim \left\langle n + \sum_{\sigma=1}^{n(n-1)} \cos\varepsilon_\sigma \right\rangle$$

$$= n + \frac{n(n-1)}{2\pi} \int_0^{2\pi} \cos\varepsilon \, d\varepsilon = n \quad . \tag{2.2.8}$$

The result is that if the phases of the n oscillators are all equal, the total power density is proportional to the square of their number, n^2 (coherence). If their phases are random and uniformly distributed over an interval 2π, the total power density is proportional to their number, n (incoherence). So, if out of 100 interacting oscillators just 10 of them are coherent, they "behave" as ~100 and can "overtake" the remaining ones which behave just as 90. The most usual case, of course, is that of *partial coherence,* where either the phases of the individual oscillators are bounded via recursive relations $|\varphi_i - \varphi_j| < \Phi_{ij}$, with bounding limits Φ_{ij} not all zero, or the phases of the individual oscillators are random but not uniformly distributed on the basic phase cycle 2π.

The question naturally arises, how can we make the phases of individual oscillators *interact* so that we can control their degree of coherence? In normally coupled linear oscillators, the net outcome of interaction is energy exchange between their amplitudes, whereas their phases or their frequencies remain unaffected. The picture changes, however, with coupled nonlinear oscillators: in such oscillators (see below) there is an *intrinsic* coupling between amplitude and phase, or amplitude and frequency. Thus, if two such nonlinear oscillators get coupled, they start interacting (when the amplitude of their oscillations is still small) by exchanging energy between their amplitudes A_1, A_2 (Fig.2.10). Soon, however, due to the internal coupling $A_1 \rightleftharpoons \varphi_1$, $A_2 \rightleftharpoons \varphi_2$, the energetic interactions between the amplitudes will give rise to coupling between the individual phases φ_1, φ_2; hence the possibility of obtaining coherence (and therefore organization), something impossible with linear oscillators. Finally, before moving to another example, let us examine more closely the above-mentioned "intrinsic" relationship between the amplitude and, say, the frequency of a nonlinear oscillator.

Let us write down the pendulum equation for finite excursions from the stable equilibrium as shown in Fig.2.11. The force acting is the tangential projection of the weight of the oscillating mass, $-mg \sin\theta$. If L is the length of the pendulum, the inertial term $md\theta/dt$ giving the rate of the momentum change can be written as

$$m \frac{d(\omega L)}{dt} = m \frac{L d^2\theta}{dt^2} \quad ,$$

since $\omega = d\theta/dt$. So,

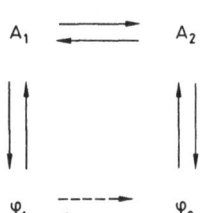

$A_1 \rightleftharpoons A_2$

$\varphi_1 \rightleftharpoons \varphi_2$

Fig.2.10. Amplitude interactions between coupled nonlinear oscillators are
transferred to phase coupling

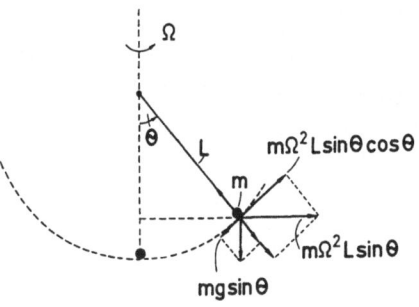

Fig.2.11. The simple pendulum ($\Omega = 0$) and
the rotating nonlinear oscillating pendulum ($\Omega \neq 0$)

$$\frac{d^2\theta}{dt^2} + \frac{g}{L} \sin\theta = 0 \quad \text{or} \quad \frac{d^2\theta}{dt^2} + \omega_0^2 \sin\theta = 0 \qquad (2.2.9)$$

will be the equation for the pendulum, where $\omega_0 = \sqrt{g/L}$ is the eigenfrequency of the
linearized system, i.e., the frequency at which the linear and harmonic pendulum
would oscillate for small deviations from equilibrium.

$$|x| \ll 1 \quad , \quad \sin x \sim x \rightarrow \frac{d^2\theta}{dt^2} + \omega_0^2\theta = 0 \quad .$$

Now in order to deal with the nonlinear oscillation, we expand $\sin\theta$ and keep
only the first nonlinear term. So (2.2.9) becomes

$$\frac{d^2\theta}{dt^2} + \omega_0^2\theta - \frac{\omega_0^2}{6}\theta^3 = 0 \quad . \qquad (2.2.10)$$

Let us try a solution of (2.2.10) of the form

$$\theta = \theta_0 \sin\omega t + \varepsilon\theta_0 \sin 3\omega t \quad , \qquad (2.2.11)$$

where $\varepsilon \ll 1$ is a dimensionless constant, and $\theta_0 \ll 1$ is the amplitude of the linearized
harmonic oscillator. The presence of a term like $\sin 3\omega t$ in (2.2.11) is suggested
by the trigonometric identity

$$\sin 3x \equiv 0.75 \sin x - 0.25 \sin 3x \quad ; \qquad (2.2.12)$$

the θ^3 term in (2.2.10) will generate from the cube of $\sin \omega t$ a term in $\sin 3\omega t$. So,
to satisfy (2.2.10) we have to add in the linearized solution a term such as
$\varepsilon\sin 3\omega t$ —just to cancel the $\sin 3\omega t$ term generated by θ^3. Proceeding further, we
discover that this new $\varepsilon\sin 3\omega t$ term in the trial solution (2.2.11) will generate,
on being cubed, a term in $\varepsilon^3\sin 9\omega t$, and so on. However, by choosing $\varepsilon \ll 1$, the
series may be expected to converge rapidly, since higher powers of ε are involved
as factors in the higher-frequency terms. We now have to determine ε and ω. From

the outset, however, we know that the *nonlinear* (2.2.10) inevitably leads to *non-harmonic* oscillations (albeit periodic). So ω will be just the fundamental frequency in a discrete Fourier spectrum, which theoretically will contain *all* higher harmonics 3^n [in our case this spectrum has been truncated to the third harmonic ($n = 1$)]. Also we are eager to find out how much this fundamental frequency ω will differ from the eigenfrequency ω_0 of the linear (harmonic) oscillator.

We now insert our perturbation solution (2.2.11) into (2.2.10); we get, investigating each term separately:

$$\ddot{\theta} = -\omega^2 \theta_0 \sin\omega t - 9\omega^2 \varepsilon \theta_0 \sin 3\omega t \quad , \tag{2.2.13}$$

$$\theta^3 = \theta_0^3 (\sin^3 \omega t + 3\varepsilon \sin^2 \omega t \sin 3\omega t + \ldots) \quad , \tag{2.2.14}$$

where we have discarded the terms of order ε^2 and ε^3 since our assumption is $\varepsilon \ll 1$. Using the trigonometric identity (2.2.12) we get:

$$\omega_0^2 \theta = \omega_0^2 \theta_0 \sin\omega t + \omega_0^2 \varepsilon \theta_0 \sin 3\omega t \tag{2.2.15}$$

$$-\frac{1}{6} \omega_0^2 \theta^3 = -\frac{3\omega_0^2}{24} \theta_0^3 \sin\omega t + \frac{\omega_0^2}{24} \theta_0^3 \sin 3\omega t \quad . \tag{2.2.16}$$

Adding the members of (2.2.13,15,16) vertically, we obtain the left-hand side equal to zero, using (2.2.10), and the right-hand side in the form $A\theta_0 \sin\omega t + B\omega_0 \sin 3\omega t$ where

$$A = -\omega^2 + \omega_0^2 - \frac{3}{24} \omega_0^2 \theta_0^2 \tag{2.2.17}$$

and

$$B = -9\omega^2 \varepsilon + \omega_0^2 \varepsilon + \frac{\omega_0^2}{24} \theta_0^2 \quad . \tag{2.2.18}$$

Now, from the equation $A\theta_0 \sin\omega t + B\theta_0 \sin 3\omega t = 0$, we observe that in order to hold for all times, $A = 0$ and $B = 0$ simultaneously. So from $A = 0$ we get

$$\omega^2 = \omega_0^2 \left(1 - \frac{\theta_0^2}{8}\right)$$

and for $|\theta_0| \ll 1$ making use of the approximation $\sqrt{1-x} \sim 1 - x/2$, $|x| \ll 1$, we finally get

$$\omega \sim \omega_0 \left(1 - \frac{\theta_0^2}{16}\right) \quad , \tag{2.2.19}$$

which shows that the fundamental frequency of the nonlinear oscillation is *smaller* than the eigenfrequency of the linear approximation.

From the relation $B = 0$ we get the expression for the perturbation parameter: $\varepsilon \sim \theta_0^2/192$ if we put $\omega^2 \sim \omega_0^2$ ($\theta_0 \ll 1$), or, more precisely: $\varepsilon = \theta_0^2/[24(8-9\theta_0^2/8)]$. So our solution reads

$$\theta \cong \theta_0 \sin\omega t + \frac{\theta_0^3}{192} \sin 3\omega t \tag{2.2.20}$$

and we can see that in a nonlinear oscillator, apart from the noharmonicity, we have a coupling between the fundamental frequency and the amplitude of the oscillation; so energetic interactions between amplitudes among nonharmonic oscillators give rise to frequency (or phase) coupling and therefore open up the prospect of phase coherence [2.3].

Let us now move to an even more general example of *branching solutions* promoted by an instability.

2.2.5 The Rotating Pendulum: A Case of Bifurcation Leading to Spontaneous Symmetry Breaking

We take up the pendulum of the preceding section, but now we consider the system (i.e., the oscillating sphere of mass m) "imprisoned" within a solid thin two-dimensional torus of radius L, where it (the sphere) can slide without friction, and the torus rotates as indicated in Fig.2.11 with a uniform angular velocity . In the absence of rotation of the frame, the system has two steady states, one stable ($\theta = 0$) and the other unstable ($\theta = \pm\pi$).

Let us examine how this state of affairs is influenced by the introduction of the new degree of freedom in the motion of the particle, namely the rotation of the frame. In the non-inertial frame of the rotating system, the acting forces are (i) the tangential component of the weight, and (ii) the tangential component of the centrifugal force.

Thus the equation of motion will be

$$mL \frac{d^2\theta}{dt^2} = -mg \sin\theta + m\Omega^2 L \sin\theta \cos\theta \quad ,$$

or

$$\frac{d^2\theta}{dt^2} = \sin\theta\left(-\frac{g}{L} + \Omega^2\cos\theta\right) = \frac{g}{L} \sin\theta\left(\frac{\Omega^2 L}{g} \cos\theta - 1\right) ,$$

or, if we call the composite parameter $\Omega^2 L/g = \mu$, and $g/L = \omega_0^2$,

$$\frac{d^2\theta}{dt^2} = \omega_0^2 \sin\theta(\mu \cos\theta - 1) \quad . \tag{2.2.21}$$

For $|\theta| \ll 1$, $\cos\theta \sim 1$, $\sin\theta \sim \theta$ and (2.2.21) would give

$$\frac{d^2\theta}{dt^2} = \omega_0^2(\mu - 1)\theta \quad ,$$

or

$$\frac{d^2\theta}{dt^2} + \omega^2\theta = 0 \quad , \tag{2.2.22}$$

where $\omega = \omega_0 \sqrt{(1-\mu)}$ is now the eigenfrequency of the linearized pendulum in the rotating frame. So we see as a by-product of our process that the eigenfrequency of a linear pendulum (real, if $\mu < 1$) *slows down* in a rotating frame of reference versus its value in an inertial frame ($\Omega = 0$). In case $\mu > 1$, ($\Omega^2 > g/L$ or $\Omega > \omega_0$) the linear "oscillator" in the rotating frame does not oscillate but executes a damping non-periodic motion.

Let us now turn to the study of the stability of the solutions of (2.2.21). Steady states ($\dot{\theta} = $ const) require simultaneously $\theta = 0$ and $\ddot{\theta} = 0$, i.e.,

$$\sin\theta(\mu\cos\theta - 1) = 0 \quad . \tag{2.2.23}$$

Equation (2.2.23) is satisfied either for $\sin\theta = 0$, $\theta = \pm K\pi$ ($K = 0,1,2,...$), or for $\cos\theta = 1/\mu$, $\mu \geqslant 1$.

The first solution, $\theta = \pm K\pi$, provides the "classical" nonrotating steady states $\theta = 0$ (stable) and $\theta = \pm\pi$ (unstable). The second solution, $\theta = $ arc $\cos\mu^{-1}$, provides *two symmetrical branches* for each value of $\mu > 1$. For $\mu = 1$, three branches merge at $\theta = 0$, as is indicated in Fig.2.12, where the solution of (2.2.21) is plotted as function of the "control parameter" μ. For $\mu \to \infty$, $1/\mu \to 0$, $\theta \to \pi/2$.

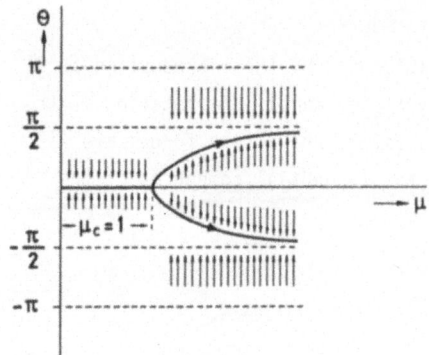

Fig.2.12. Bifurcation diagram of the rotating nonlinear pendulum oscillator, where $\mu = \Omega^2 L/g$

So we see that *beyond* a certain critical value of the angular velocity Ω of the frame, the traditional stable steady state $\theta = 0$ becomes unstable, and the new (symmetrical) stable steady-state solutions are given by $\theta = $ arc $\cos\mu^{-1}$.

Of course, we may as well reduce the preceding analysis to the motion of a heavy particle in a potential well. What is the potential function now? The external force acting on the system is $F = \omega_0^2 \sin(\mu\cos\theta - 1)$. It is related to the potential in any Hamiltonian system, such as the one under consideration, by $F = -\partial V(\theta)/\partial\theta$, from which

$$V(\theta,\mu) = \omega_0^2 \left(\frac{\mu}{2}\cos\theta - 1\right)\cos\theta \quad ; \tag{2.2.24}$$

a *periodic function* with period 2π.

Let us plot this function for $\mu < 1$ and $\mu \geqslant 1$. The curve $V(\theta,\mu)$ is shown in Fig. 2.13, (a) for $\mu < 1$ and (b) for $\mu \geqslant 1$ (both for a single period).

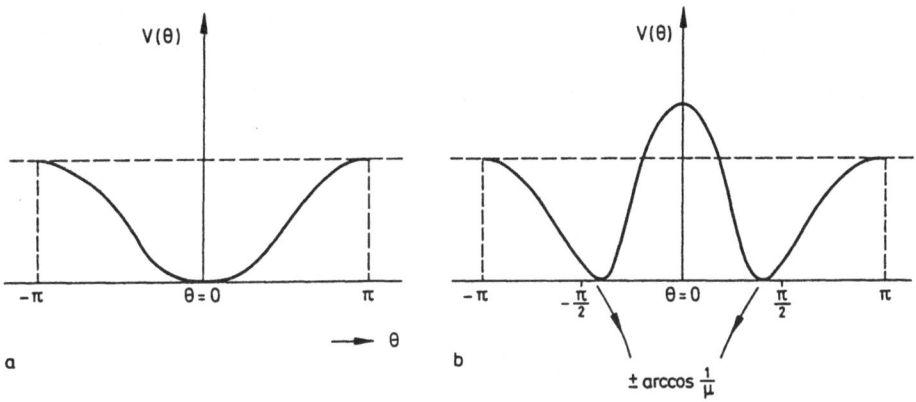

a

b

$\pm \arccos \frac{1}{\mu}$

Fig.2.13a,b. Sketch of the potential function of the pendulum oscillator for (a) $\mu < 1$ and (b) $\mu \geqslant 1$ (see text)

2.2.6 Broken Symmetry Through a Hysteresis-Like Process

We finally offer an example of an altogether different type of state-transition dynamics which takes place not as a result of an instability of the existing state, but is rather based on hysteresis or *plastic modification* occurring in systems possessing a single steady state [2.4].

Consider a one-dimensional system X (say, an overdamped linear oscillator) obeying the differential equation $\dot{x} = -x$. The system obviously possesses a single, "symmetrical", stable steady state $x = 0$. Imagine now that we couple this simple system with an "environment" Y, also one dimensional, whose single variable y is shifted by an environmental control parameter a in the following way: $dt/dt = a$, where $a(t) = \text{const}$ (or zero) in general, but gets activated, so to speak, within a time interval 2T, as shown in Fig.2.14; afterwards it returns to zero. The system X is now coupled to the environment in the simplest possible way, namely $dx/dt = -x + y$.

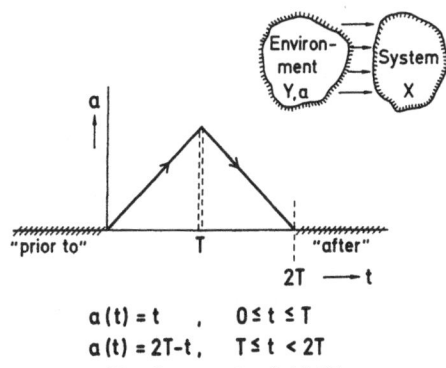

$a(t) = t$, $0 \leq t \leq T$
$a(t) = 2T - t$, $T \leq t < 2T$
$a(t) = 0$, $t < 0, t \geqslant 2T$

Fig.2.14.
The parameter a(t) (see text)

Let us follow what happens within the time interval 2T and afterwards:

a) *Interval 0...T*: $dx/dt = -x + y$, $dy/dt = t$, $da/dt = 1$, so

$$dx/dt = -x + \frac{t^2}{2} \; ,$$

and

$$x(t) = ce^{-t} + \frac{t^2}{2} - t + 1 \qquad\qquad (2.2.25)$$

where c is a constant.

At $t = T$ (middle of the interval),

$$x(t) = ce^{-T} + \frac{T^2}{2} - T + 1$$

and

$$y(T) = \frac{T^2}{2} \; .$$

b) *Interval T...2T*: $dx/dt = -x + y$, $dy/dt = 2T - t$, $da/dt = -1$, so

$$\frac{dx}{dt} = -x + 2Tt - \frac{t^2}{2} + K \; ,$$

where K is a constant whose value is calculated from the fact that $y = 2Tt - (t^2/2) + K$, for $t = T$, must be equal to $T^2/2$; so $K = -T^2$ and

$$x(t) = ce^{-t} - \frac{t^2}{2} + t(2T + 1) - (2T + T^2 + 1)$$

$$= ce^{-t} - \frac{t^2}{2} + t(2T + 1) - (T + 1)^2 \; ; \qquad\qquad (2.2.26)$$

also:

$$y(t) = 2Tt - \frac{t^2}{2} - T^2 \; . \qquad\qquad (2.2.27)$$

At the end of the time interval 2T: $y(2T) = T^2$. At times $t > 2T$, $y(2T)$ is going to stay unchanged. So for times $t > 2T$ the behavior of x, which is always given by the differential equation $\dot{x} = -x + y$, will be described by

$$\frac{dx}{dt} = -x + T^2$$

with a stable steady state which is now $x^* = T^2$—different from the previous one. What happened? How exactly has the former stable steady state disappeared and given way to the new one? We witness here a case of "plastic" change induced in the system X, not via an instability giving rise to a bifurcation, but rather via a process proceeding with a finite reaction rate. What has happened is that the state $x^* = 0$ has been *deleted*, and a quantity of "reactant" $y = T^2$ has been *trapped* permanently in the system X: this is indeed the analogue of a "memory trace" left in the system X as a result of a transient rate change undergone by an external control parameter.

2.2.7 Essentials of Stability Theory

We have occupied ourselves in a number of preceding sections with isolated —but, it is hoped, representative —examples of *evolution* in simple nonlinear systems as an outcome of specific instabilities. It is time now to sketch the steps of a general stability theory —always in a *two-dimensional* state space.

a) General Criterion

Let S be a steady state of a certain two-dimensional dynamical system and ε a small, closed contour around S. The steady state S is said to be *stable* if, for every given ε, we can always find a neighbourhood $\delta(\varepsilon)$ surrounding S so that every possible trajectory emerging from the interior of $\delta(\varepsilon)$ never reaches the envelope ε (Fig. 2.15). If such a neighbourhood $\delta(\varepsilon)$ cannot be found, then the stability of S cannot be guaranteed [2.5].

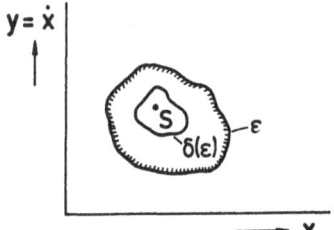

Fig.2.15.
Defining the stability of a steady state

Consider now a dynamical system described by two coupled nonlinear differential equations in autonomous form, i.e., in a form such that time does not appear as a free parameter:

$$\frac{dX_1}{dt} = f_1(X_1, X_2; \mu) \quad , \tag{2.2.28}$$

$$\frac{dX_2}{dt} = f_2(X_1, X_2; \mu) \quad . \tag{2.2.29}$$

The system evolves in a two-dimensional state space of variables X_1, X_2. At each point of a trajectory determined by specific initial conditions $X_1(0) = X_{1_0}$, $X_2(0) = X_{2_0}$, the slope will be given by

$$\frac{dX_2}{dX_1} = \frac{f_2}{f_1} \quad . \tag{2.2.30}$$

The *singular points* (steady states) in the trajectory are therefore those for which $f_2 = 0$ and $f_1 = 0$, i.e., those points where we cannot define the tangent. *Closed trajectories* (see below), on the other hand, corresponding to a periodic regime, will have a fundamental period determined by

$$T = \oint \frac{dX_1}{f_1(X_1, X_2; \mu)} \quad . \tag{2.2.31}$$

For the moment, let us assume that we are dealing with an N-dimensional system. The dynamical equations will be

$$\frac{dX_i}{dt} = f_i(X_1, X_2, \ldots, X_N; \mu) \qquad i \in (1, \ldots, N) \quad . \tag{2.2.32}$$

The steady states will be determined by the real solutions of the system of coupled nonlinear algebraic equations:

$$f_i(X_1, X_2, \ldots, X_N; \mu) = 0 \quad . \tag{2.2.33}$$

Suppose we somehow solve this system and determine the steady states $(X_1^*, X_2^*, \ldots, X_N^*)_\xi$, where $\xi \in (1, \ldots, K)$ is the number of real solutions of the system.

We next proceed in the examination of the stability of those states (the system will occupy one of them depending on the initial and boundary conditions). The stability is to be examined by small perturbations $x_i(t)$ from the equilibria, namely under the restrictions

$$x_i(t) = |X_i - X_i(t)| \ll \varepsilon \quad ,$$

where ε is an arbitrarily small positive number, for all i.

For a given steady state we now expand the second nonlinear members of the dynamical equations in Taylor series around the steady state value(s) of the individual variable(s) and, provided that the f_i's are smooth enough, we keep in the expansion only the linear terms. So we get

$$\frac{dx_i(t)}{dt} = \sum_{j=1}^{N} (X_j - X_j^*) \left(\frac{\partial f_i}{\partial X_j} \right)_{X=X_j^*} \tag{2.2.34}$$

[The constant term in the Taylor expansion

$$f_i(X_1^*, X_2^*, \ldots, X_N^*; \mu) = 0 \quad ,$$

since $dX_i^*/dt = 0$].

We call

$$\left(\frac{\partial f_i}{\partial X_j} \right)_{X=X_j^*} = a_{ij} \tag{2.2.35}$$

the interaction parameter(s) exercised by the variable X_i on the variable X_j (in general, of course, $a_{ij} \neq a_{ji}$). Among other things, $a_{ij} = 0$ may indicate the absence

30

of the variable X_j in the polynomial f_i. The elements a_{ij} constitute the so-called interaction matrix A.

The linearized differential system with the perturbations $x_i(t)$ as variables now has the form

$$\frac{dx_i}{dt} = \sum_{j=1}^{N} a_{ij} x_j \quad ,$$

or

$$\dot{x} = Ax \quad .$$
(2.2.36)

Accepting for the perturbations $x_i(t)$ of the individual variables the form $x_i(t) \sim e^{\lambda_i}$, we get the linear system

$$\lambda_i x_i = \sum_{j=1}^{N} a_{ij} x_j \quad ,$$

or, in matrix form,

$$\lambda x = Ax \quad .$$
(2.2.37)

The request for nontrivial solutions $[x_i(t) \neq 0]$ leads to the characteristic equation

$$\begin{vmatrix} a_{11} - \lambda & a_{12} \cdots\cdots a_{1N} \\ a_{21} & a_{22} - \lambda \cdots a_{2N} \\ \vdots & \vdots \qquad \vdots \\ a_{N1} & a_{N2} \cdots\cdots a_{NN} - \lambda \end{vmatrix} = 0 \quad ,$$
(2.2.38)

from which the eigenvalues λ_i of the interaction matrix can in principle be calculated. To give specific examples, let us now return to our two-dimensional system for which the characteristic equation will have the form

$$\begin{vmatrix} a_{11} - \lambda & a_{12} \\ a_{12} & a_{22} - \lambda \end{vmatrix} = 0 \quad ,$$
(2.2.39)

or

$$(a_{11} - \lambda)(a_{22} - \lambda) - a_{12} a_{21} = 0 \quad ,$$
(2.2.40)

or

$$\lambda^2 - (a_{11} + a_{22})\lambda + (a_{11} a_{22} - a_{12} a_{22}) = 0 \quad ,$$
(2.2.41)

which may be written as

$$\lambda^2 - b\lambda + \gamma = 0 \quad ,$$
(2.2.42)

from which we get the eigenvalues

$$\lambda_{1,2} = \frac{b \pm \sqrt{b^2 - 4\gamma}}{2} \quad . \tag{2.2.43}$$

In general, we will have $\lambda_{1,2} = (\lambda' + j\lambda'')_{1,2}$.

The steady state involved is stable if $\text{Re}\{\lambda_i\} < 0$ for $i = 1$ *and* $i = 2$. If λ_1' λ_2' is positive, the steady state is unstable. For $\lambda_1' = \lambda_2' = 0$ and $\lambda'' \neq 0$ we have a regime of "marginal stability" or neutral stability, in other words, the system performs periodic motion with frequency λ_i'' on a closed trajectory around the steady state whose radius is, of course, small, but does depend on the initial and boundary conditions. Rather than exhausting all possible cases —depending on the parameters b and γ—let us give a specific example, so that things will speak for themselves.

We consider the following process involving a conflict between two populations X_1, X_2 or X and Y. In a "reactor", raw material A mixed with population X increases this population (X grows on A). Population Y feeds on population X and grows. Finally, population Y, in contact with a catalyst or a "metabolite" B, turns to waste (Fig.2.16).

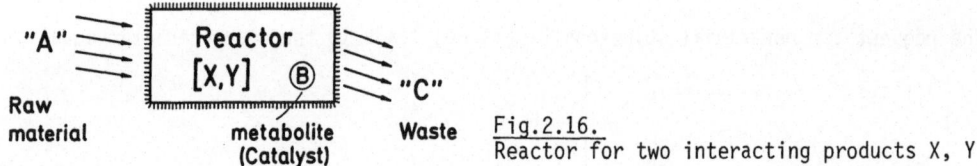

"A"
Raw
material

Reactor
[X,Y] Ⓑ

"C"

metabolite
(Catalyst)

Waste

Fig.2.16.
Reactor for two interacting products X, Y

This scheme may have an application to competitive production processes and also to biological antagonism between two species, one of which preys on the other (e.g., the Volterra-Lotka model of two species of "small" X and "big" Y fish). Now, the reactions should proceed as follows:

$$A + X \xrightarrow{K_1} 2X \quad , \tag{2.2.44}$$

$$X + Y \xrightarrow{K_2} 2Y \quad , \tag{2.2.45}$$

$$Y + B \xrightarrow{K_3} C + B \quad , \tag{2.2.46}$$

where K_1, K_2, K_3 are reaction coefficients.

The corresponding differential equations with respect to X and Y can be deduced immediately on the basis of the "gains" and "losses" that each population is subjected to. For instance, X increases with a rate $K_1 AX$ and decreases with a rate $K_2 XY$; hence

$$\frac{dX}{dt} = K_1 AX - K_2 XY \quad . \tag{2.2.47}$$

On the other hand, Y increases with a rate K_2XY and decreases with a rate K_3BY. So

$$\frac{dY}{dt} = K_2XY - K_3BY \quad . \tag{2.2.48}$$

Our nonlinear second member polynomials in this example are therefore

$$f_1 = K_1AX - K_2XY \quad \text{and} \quad f_2 = K_2XY - K_3BY \quad .$$

The parameters K_1, K_2, K_3, A, B stand for μ in the formal equations (2.2.32).

Let us find the steady states. From the solution of the system $f_1 = 0$, $f_2 = 0$, we get two real solutions, namely,

$$X_I^* = \frac{K_3B}{K_2} \quad , \quad Y_I^* = \frac{K_1A}{K_2} \quad , \tag{2.2.49}$$

and

$$X_{II}^* = 0 \quad , \quad Y_{II}^* = 0 \quad , \tag{2.2.50}$$

which are illustrated in Fig.2.17.

Let us perform for each of them the stability analysis.

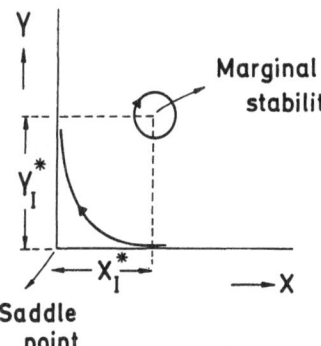

Saddle
point

Fig.2.17. Stable regimes of the reactor model. The steady state X_I^*, Y_I^* is marginally stable and the steady state X_{II}^*, Y_{II}^* is unstable

b) *Specific Analyses*

Steady State I. We have to calculate the eigenvalues of the linear matrix A, which amounts to calculating the coefficients a_{ij}. We have

$$a_{11} = \left(\frac{\partial f_1}{\partial X}\right)_* = 0 \quad , \quad a_{12} = \left(\frac{\partial f_1}{\partial Y}\right)_* = -K_3B \quad , \tag{2.2.51}$$

$$a_{21} = \left(\frac{\partial f_2}{\partial X}\right)_* = K_1A \quad , \quad a_{22} = \left(\frac{\partial f_2}{\partial Y}\right)_* = 0 \quad ,$$

where the asterisk means that the numerical values of the derivatives have to be taken at $X_I^* = K_3B/K_2$ and $Y_I^* = K_1A/K_2$.

So the characteristic equation (2.2.39) gives

$$\lambda = \pm \sqrt{K_1 K_3 AB} \quad . \tag{2.2.52}$$

This means that steady state I is marginally stable, and when the system, under a slight perturbation, leaves this state, it settles into a periodic trajectory equal to the perturbation value and with circular frequency equal to $\sqrt{K_1 K_3 AB}$ (always *clockwise*, in state space, where the values of variables grow *from* the origin).

Steady State II. Following an analysis similar to the one before, but now calculating the numerical values of the derivatives at the points $X_{II}^* = 0$, $Y_{II}^* = 0$, we find

$$a_{11} = K_1 A \quad , \qquad a_{12} = 0 \quad ,$$
$$a_{21} = 0 \quad , \qquad a_{22} = -K_3 B \quad . \tag{2.2.53}$$

The characteristic equation is now

$$(K_1 A - \lambda)(K_3 B + \lambda) = 0 \quad , \tag{2.2.54}$$

from which we find the eigenvalues

$$\lambda_1 = K_1 A > 0 \quad \text{and} \quad \lambda_2 = -K_3 B < 0 \quad .$$

The state 0,0 (fortunately for ecology!) is unstable (a saddle point, see below).

2.2.8 Behavior of a Two-Dimensional Dynamical System in the Vicinity of Singular Points (Steady States)

We now find it necessary to take, so to speak, a magnifying glass and try to see (again to a first approximation) exactly how the system behaves in the vicinity of a singularity and especially an *unstable state*. The reason for such an inquiry is the following. Once a given state, hitherto stable, becomes unstable by shifting one of the control parameters μ beyond some critical value (so that, via the numerical values of the interaction coefficients a_{ij}, which depend on μ, the real part of one eigenvalue becomes positive), the question is, where does the system go from there? To answer this, we now have to solve *again* the system of coupled nonlinear algebraic equations $f_i(X_1, X_2, \ldots, X_N; \mu) = 0$, in order to define the *new* set of real solutions (the new set of steady states), since now μ has changed. In this new set, the previously stable and now unstable steady state figures, of course, as a member, but the rest of the states are new. At first glance we could say that after "taking off" from the now unstable steady state, the system will "land" with equal probability at any one of the available states of the new set, thereby frustrating any specific prediction. This is not always the case, however. In order to pass judgement we have to examine carefully the "cartography" of the state space around each such new state. (We will see though that this is not enough.)

Such a study may prove some states to be more probable (i.e., more accessible) candidates than others for receiving the system which, after being expelled from the former state, requires a "landing". In short, we have to examine the "stream-lines" around the available steady states in state space, in the hope of predicting the likelihood of the next state. Essentially, we do that by performing a "cross section" on the potential curves — for instance, the curves in Fig.2.13 — perpendicu-lar to the paper, and then looking "down" to appreciate the road system of the land-scape. This in turn amounts to tracing the *integral curves* $X_2 = f(X_1)$ of a *conserva-tive system*, when the potential function is known.

Since we are interested in (Hamiltonian) systems with multiple steady states, we should start with a potential function of *more than one minima* (Fig.2.13) and try to understand the way the integral curves, i.e., the "road system" in phase space, derive from it.

Referring to the example of the rotating pendulum (Sect.2.2.5) we can write imme-diately the equations for the set of integral curves in the state-space plane $\dot{\theta}$, θ as the law of conservation of energy per unit mass, namely,

$$\frac{1}{2} \left(\frac{d\theta}{dt}\right)^2 + V(\theta,\mu) = E \quad , \tag{2.2.55}$$

where E is the total energy of the pendulum.

Our aim is to discover the topology of the integral curves in the vicinity of minima (stable steady states) and maxima (unstable steady state). Consequently, let us make a Taylor expansion of $V(\theta,\mu)$ around each of these types of singular points, namely,

$$V(\theta) \cong V_0 + \frac{c}{2} (\theta - \theta_0)^2 \quad , \tag{2.2.56}$$

where $c = [\partial^2 V(\theta)/\partial\theta^2]_0$ is a positive number if $V_0 [=V(\theta_0)]$ is a minimum and is a negative number if V_0 is a maximum. Obviously, $[\partial V(\theta)/\partial\theta]_0 = 0$ in both cases, since we are dealing with extrema.

So, making use of the above expansion, it is easy to investigate the behavior of integral curves in the vicinity of steady states by returning to (2.2.55) and re-placing $V(\theta)$ from (2.2.56). We get

$$\frac{1}{2} \left(\frac{d\theta}{dt}\right)^2 + c(\theta - \theta_0)^2 = 2(E - V_0) \quad , \tag{2.2.57}$$

or

$$\frac{\dot{\theta}^2}{2} + c(\theta - \theta_0)^2 = \text{const} \quad . \tag{2.2.58}$$

Here $E > V_0$ in order to have real solutions for θ.

Now if the steady state θ_0 is stable, i.e., it corresponds to a minimum in $V(\theta)$, (2.2.58) constitutes a family of ellipses centered at θ_0. If, on the other hand, the steady state θ_0 is unstable, i.e., it corresponds to a maximum in $V(\theta)$, (2.2.58)

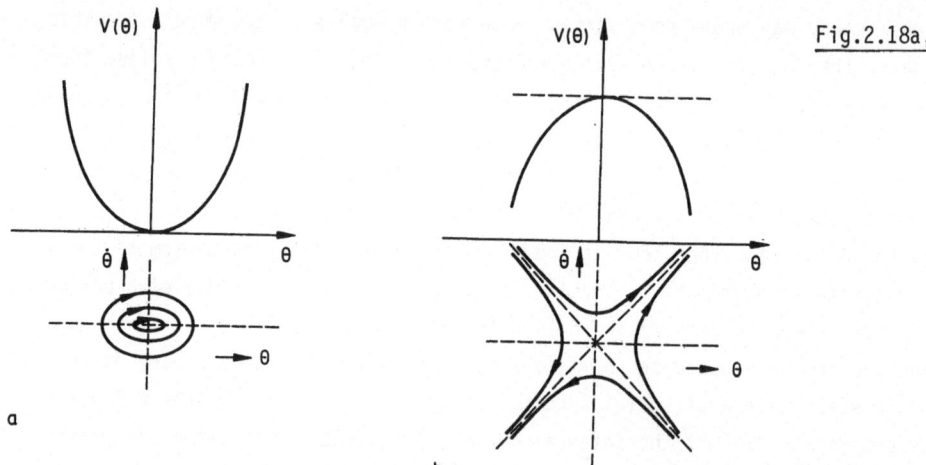

Fig.2.18a,b. In the neighbourhood of (a) a potential minimum the singularity is a focus,
while in the neighborhood of
(b) a potential maximum the
singularity is a saddle point

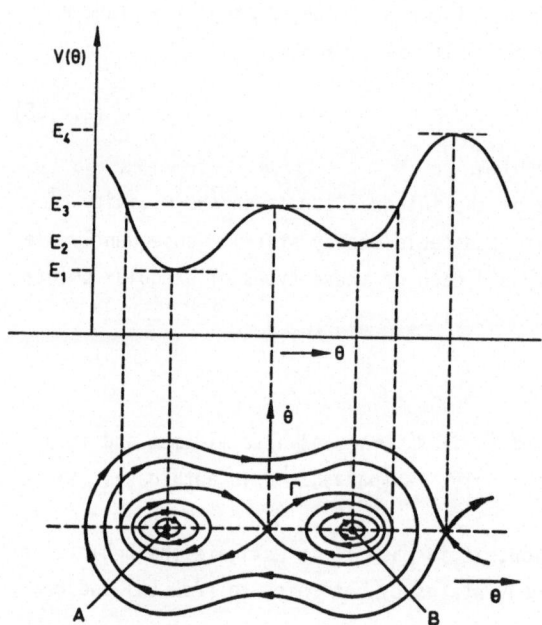

Fig.2.19. A general potential
function and the corresponding
integral curves

gives a family of hyperbolas again centered at θ_0. For a const $= 0$, i.e., for $E = V_0$,
these hyperbolas go to the asymptotes given by $\dot\theta = \pm\sqrt{-c(\theta - \theta_0)}$, which asymptotes are
called *separatrices*, since they divide, so to speak, the different groups of the
hyperbolas. This correspondence between the kinds of steady state and integral
state-space trajectories in their vicinity is shown in Fig.2.18. Combining the two
halves of Fig.2.18 as in a real situation, we get the correspondence illustrated
in Fig.2.19. From the map of the integral lines one would believe that it is easy
in any given situation, having in mind the state of departure, to ascribe different
likelihoods to different successor states (in the example of the rotating pendulum,

symmetry gives, of course, a fifty-fifty chance to each stable steady state). In the example of Fig.2.19, for instance, no integral curve exists near steady state A for $E < E_1$; also steady state B cannot be reached by a system possessing *less* energy than E_2. For systems starting with energies E, such that $E_3 < E < E_4$, there is no steady state. For $E = E_3$ the trajectory is the degenerate curve Γ. Again, there is no steady state.

The conclusion is that by dealing with Hamiltonian systems, we cannot answer the question with which we started this section, namely the mechanism of the succession of states, or the preferential evolution of the dynamical system from state to state via a sequence of instabilities. Hamiltonian systems *do not* reach specific states as a result of evolution, unless they are put there from the very outset, as a result of quite specific initial and boundary conditions. Whenever they are slightly perturbed from a given trajectory, they never return to it. Instead, they follow with the utmost rigidity the new trajectory, which is absolutely specified and dependent on the new initial and boundary conditions (e.g., the total energy of the system).

2.2.9 First Encounter with Nontrivial Dissipative Systems: The Concept of the Attractor in Two Dimensions (Limit Cycle)

With our first specific example, in Sect.2.2.3, we came across a simple dissipative system (the overdamped harmonic and nonlinear oscillator) and we appreciated the fact that the simplest form of an *attractor* is a stable steady state. This is, however, a trivial case, exactly because it is symmetrical and static. We intend in this section, pursuing the problem posed in the previous section, to give examples of two-dimensional dynamical systems capable of being *attracted* to a new dynamical stationary regime, once expelled from a previous one via an instability, in ways which are independent of the initial conditions. We can say that such systems really *change* with time.

Let us start with a dissipative dynamical system where the "viscosity" term is nonlinearly dependent on the dynamical variable itself. In such a system we have, then, a periodic exchange of energy with the environment, an exchange which, under certain conditions, may allow self-sustained dynamical activity (rather than a monotonic damping process, as in the example in Sect.2.2.3). Consider the simple circuit of Fig.2.20. A d.c. source of current charges a capacitor C via an ohmic

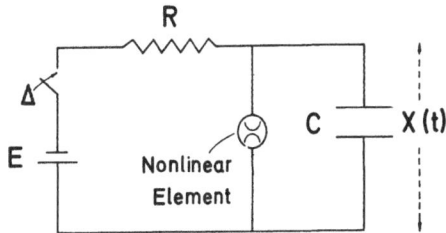

Fig.2.20. Simple electrical circuit generating relaxation oscillations

resistance R. In parallel with the capacitor we interpose a switching element, which opens when the potential difference across the capacitor falls below V_{C_1}, and closes (thereby short-circuiting the capacitor) when the potential across the capacitor goes above V_{C_2}. At a time $t = 0$ we switch on the source, which starts charging the capacitor through the resistance R.

When the potential across C goes above V_{C_2}, the switch closes and the capacitor is discharged through the nonlinear element. When the potential falls below V_{C_1}, the switch opens and the capacitor starts to be charged anew. The shape of this periodic but highly nonlinear process is shown schematically in Fig.2.21. Its fundamental period obviously depends on the parameters R, C, V_{C_1}, V_{C_2}.

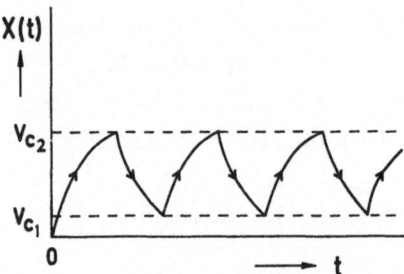

Fig.2.21.
Relaxation oscillations

It can be proved that $x(t)$ obeys the following differential equation

$$\ddot{x} - \varepsilon(1 - x^2)\dot{x} + \omega_0^2 x = 0 \quad , \tag{2.2.59}$$

where ε is the parameter having to do with the strength of the supply E and the response curve of the nonlinear switch. The above process constitutes the famous van der Pol oscillator. All electronic oscillators (transistors, etc.) can, more or less, be faithfully simulated by the above model.

We intend now to analyse the system (2.2.59) from the point of view of stability. The system evolves in a two-dimensional state space $\dot{x}$, x. Let us put, for simplicity, $\omega_0 = 1$, and write (2.2.59) in parametric form:

$$x(t) \sim r(t)\cos t \quad , \tag{2.2.60}$$

$$y(t) = \dot{x}(t) \sim -r(t)\sin t \quad ; \tag{2.2.61}$$

the second relation presumes $\varepsilon \ll 1$, i.e., a slowly evolving amplitude $r(t)$ so that during a full period $\Delta r/r \ll 1$ (Fig.2.22).

From (2.2.60,61), we get $x^2 + y^2 = r^2$ and, differentiating both sides with respect to time,

$$x \frac{dx}{dt} + y \frac{dy}{dt} = r \frac{dr}{dt} \quad , \tag{2.2.62}$$

or, taking (2.2.59) into account, and using (2.2.60,61),

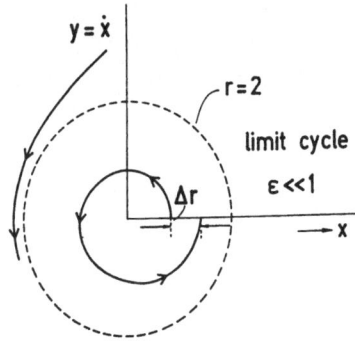

Fig.2.22. The limit cycle

$$xy + y[\varepsilon(1 - r^2\cos^2 t \; y - x] = r \frac{dr}{dt} \quad , \tag{2.2.63}$$

or

$$\frac{dr}{dt} = \varepsilon(1 - r^2\cos^2 t)r \sin^2 t \quad , \tag{2.2.64}$$

or

$$dr = \varepsilon r(1 - r^2\cos^2 t)\sin^2 t \; dt \quad . \tag{2.2.65}$$

The change Δr over a period will then be equal to

$$\Delta r = \int_0^{2\pi} dr = \pi\varepsilon r\left(1 - \frac{r^2}{4}\right) \quad . \tag{2.2.66}$$

We observe that under the approximation $\varepsilon \ll 1$,

$\Delta r < 0 \qquad$ for $\qquad r > 2 \quad ,$

$\Delta r > 0 \qquad$ for $\qquad r < 2 \quad ,$

$\Delta r = 0 \qquad$ for $\qquad r = 2 \quad .$

We therefore deduce that the circle with radius $r = 2$ is *asymptotically stable* and that in its vicinity there is no other cycle. This stable trajectory is a *limit cycle* and belongs to a new category of stationary singularities in state space, different from the steady states.

The result achieved here remains essentially the same for all $\varepsilon \gtrsim 1$. The limit cycle is slightly deformed as ε increases, bringing to mind a sort of hysteresis cycle. It can be proved [2.6] that as ε increases, the second approximation gives for the stationary amplitude

$$x(t) \sim 2 \cos(\omega t + \theta) - \frac{\varepsilon}{4} \sin 3(\omega t + \theta) \quad , \qquad \omega = 1 - \frac{\varepsilon^2}{16} \quad . \tag{2.2.67}$$

The other (trivial) singularity of the system is, of course, the origin, which constitutes an *unstable* steady state. The passage from this steady state to the limit cycle is known as the "Hopf bifurcation". In the above example, the oscillator is "self"-excited.

However, for a stronger nonlinearity in the viscous term (a fifth degree instead of a cubic one), the steady state 0, 0 is stable, and the level of the external excitation or, for that matter, the value of the control parameter ε, has to exceed a certain threshold in order to allow the system to *change* irreversibly from the origin to the limit cycle (Appendix A includes a more extended discussion).

Let us consider in the same context another representative example from Poincaré himself. We are given the following system:

$$\frac{dx_1}{dt} = \alpha x_1 + \beta x_2 - x_1(x_1^2 + x_2^2) \quad , \tag{2.2.68}$$

$$\frac{dx_2}{dt} = -\beta x_1 + \alpha x_2 - x_2(x_1^2 + x_2^2) \quad , \tag{2.2.69}$$

with a trivial steady state $x_1 = x_2 = 0$.

Let us write $x_1 = r\, \cos\varphi$ and $x_2 = r\, \sin\varphi$. Substituting into the original equations, we obtain a system with respect to r, φ, namely,

$$\frac{dr}{dt} = \alpha r - r^3 \tag{2.2.70}$$

$$\frac{d\varphi}{dt} = -\beta \quad . \tag{2.2.71}$$

Equation (2.2.71) gives immediately $\varphi(t) = \varphi_0 - \beta t$.

Let us concentrate on (2.2.70) and determine the steady states; putting $dr/dt = 0$, we have $r_1 = 0$, $r_2 = \sqrt{\alpha}$ ($\alpha > 0$). These two states are represented in the state space by the origin, and a cycle centered on the origin with a radius equal to $\sqrt{\alpha}$ of angular frequency β.

The linearized system is

$$dr/dt = \alpha r \tag{2.2.72}$$

$$d\varphi/dt = -\beta \quad . \tag{2.2.73}$$

It is obvious from (2.2.72) that for $\alpha < 0$, the steady state 0, 0 is stable, and it becomes unstable for $\alpha > 0$ (for $\alpha_c = 0$, the steady state is marginally stable). So let us try to see what happens above the critical value $\alpha_c = 0$, for $\alpha > 0$, i.e., let us try to investigate the stability of the closed trajectory $r = \sqrt{\alpha}$.

Let us slightly perturb this trajectory by putting $r = \sqrt{\alpha} + \delta r$. Substituting into (2.2.70), we get

$$\delta \dot{r} = \alpha(\sqrt{\alpha} + \delta r) - (\alpha^{3/2} + \delta r^3 + 3\alpha\delta r + 3\sqrt{\alpha}\delta r^2)$$

$$\cong \alpha(\sqrt{\alpha} + \delta r) - (\alpha^{3/2} + 3\alpha\delta r) = -2\alpha\delta r \quad . \tag{2.2.74}$$

We deduce therefore that the trajectory $r = \sqrt{\alpha}$ is asymptotically stable, since the radius r decreases for positive r and increases for negative δr. We have again a limit cycle, i.e., a circular, asymptotically stable trajectory which *attracts all*

40

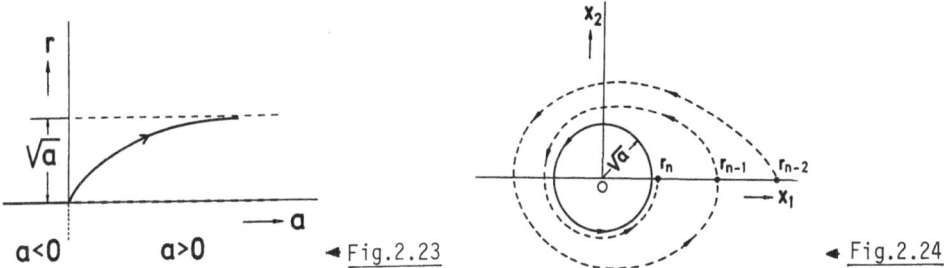

Fig.2.23. Bifurcation diagram for the Poincaré example given in the text

Fig.2.24. The converging "zero-crossings" in the Poincaré example given in the text

trajectories in its neighbourhood. Plotting the diagram, solution versus control parameter, we see that this attractor involves a *broken* symmetry, not only because the radius $\sqrt{a} > 0$, but also because the limit cycle runs in time along only one direction (Fig.2.23). The orbit $r = \sqrt{a}$ is independent of the initial conditions, and its period is intrinsically determined by a system parameter rather than by the initial conditions, as in the case of Hamiltonian oscillators.

Before closing this section, it is of some interest to investigate the recursive relations giving the "zero-crossings" of the limit cycle on the x_1 axis (Fig.2.24), as it shrinks to its asymptotic value $r_\infty = \sqrt{a}$ (for further information see Chap.6). Essentially we are interested in the speed of convergence. From the governing equations in polar coordinates (2.2.70,71) we get, after dividing both sides,

$$\frac{dr}{d\varphi} = \frac{(\alpha r - r^3)}{\beta} \; .$$
(2.2.75)

The distance between two successive crossings is:

$$\int_{r_{n-1}}^{r_n} \frac{dr}{\alpha r - r^3} = -\frac{1}{\beta} \int_0^{2\pi} d\varphi = \frac{2\pi}{|\beta|} \; ,$$
(2.2.76)

or

$$\frac{1}{\alpha} \log \left. \frac{r}{\sqrt{|\alpha - r^2|}} \right|_{r_{n-1}}^{r_n} = \frac{2\pi}{|\beta|} \; ,$$
(2.2.77)

or

$$r_n^2 = \frac{\alpha}{1 + \exp(-4\pi\alpha/|\beta|)[(\alpha/r_{n-1}^2) - 1)]} \; .$$
(2.2.78)

The convergence is exponentially fast; indeed, by successively applying (2.2.78) for $n = 1, \ldots, n$, we get

$$r_n^2 = \frac{\alpha}{1 + \exp(-4\pi\alpha n/|\beta|)[(\alpha/r_0^2) - 1]} \; ,$$
(2.2.79)

where r_0 is the arbitrary radius from which the process started, i.e., the initial condition. We observe that as $n \to \infty$, the above formula, as expected, gives $\lim_{n \to \infty} r_n^2 = \alpha$ *independent* of r_0, i.e., of the initial conditions.

The above example is the last of the representative cases we have presented which are associated with instabilities, bifurcations, and evolution caused by simple one- or two-dimensional nonlinear dynamical systems, both Hamiltonian and dissipative. We have appreciated fully, I hope, the role and significance of non-linearity in provoking complex behavior in structurally simple systems. It is time now to deal with phenomena caused by systems of many degrees of freedom, a group which includes all *large-scale* systems.

2.3 Elements of Statistical Physics and Their Relevance to Evolutionary Phenomena

2.3.1 Some Characteristics of Stochastic Systems

Although in the preceding section we were able to recognize *evolutionary* (irreversible) behavior in simple, nonlinear, dissipative two-dimensional systems (undergoing Hopf bifurcation), we still cannot say that our quest for a preferential switch to one among many possible available "attractors" beyond the point of instability has been completed. The next logical step in our analysis should be to try systems with three degrees of freedom. Nevertheless, for reasons which become clear along the way, and particularly in Chap.6, we intend now to go to the other extreme: by studying systems of many degrees of freedom, we will perhaps be able to spot additional evolutionary characteristics not existing in two dimensions.

Now systems with many degrees of freedom are by necessity *stochastic*. In turn, stochastic systems (macroscopic systems whose dynamics is determined by the interplay of a great number of microscopic units) are de facto *hierarchical* in the sense that they can be described in a complementary way on (at least) two different levels: (1) at the *microscopic* level, where a great number of particles (of the order of magnitude of the Avogadro number) undergo interactions on the basis of Hamiltonian (reversible) dynamics, and (2) at a *macroscopic, phenomenological* level, where the system can be described, for many (but not *all*) practical purposes, by a very few macrovariables (such as volume, pressure, temperature), which macrovariables emerge as *collective properties* of the ongoing dynamics at the microscopic level or as *moments* of a probability density function substituting for the microscopic dynamics.

At this point it is useful, we think, to see more or less explicitly how these averaging processes — via which one *ascends* from the lower (microscopic) to the higher (microscopic) to the higher (macroscopic) level — take place. We represent our system in state space as an N-dimensional vector **x**, which describes a continuous curve, a trajectory, being at a given point or state at a specific moment t. Let

P(**x**;t) be the probability of finding the system at point **x** and at time t. We are interested in specifying the way that this probability density function (p.d.f.) evolves in time. The probability P(**x**;t) will increase due to transitions from other points **x'**, and it will decrease due to the transitions leaving the point (state) **x**, i.e., dP(**x**;t)/dt = "rte in" − "rate out" = I − I' [2.3].

Since the term I consists of all transitions from initial points **x'** →**x**, it is composed of the sum over all initial points **x'**. Each term of it is given by the probability of finding the system at a point **x'**, multiplied by the transition probability per unit time of passing from **x'** →**x**. Thus,

$$I = \sum_{\mathbf{x'}} W(\mathbf{x},\mathbf{x'})P(\mathbf{x'};t) \quad , \tag{2.3.1}$$

where W(**x**,**x'**) is the (transitional) probability for jumping per unit time from **x'** →**x**.

For the "outgoing" transitions (I') we will have

$$I' = P(\mathbf{x};t) \sum_{\mathbf{x'} \neq \mathbf{x}} W(\mathbf{x'},\mathbf{x}) \quad , \tag{2.3.2}$$

where W(**x'**,**x**) is the (transitional) probability for jumping from **x** →**x'** (per unit time).

So our equation, describing the evolution of P(**x**;t) (it is called the "master equation"), reads:

$$\frac{dP(\mathbf{x};t)}{dt} = \sum_{\mathbf{x'}} W(\mathbf{x},\mathbf{x'})P(\mathbf{x};t) - P(\mathbf{x},t) \sum_{\mathbf{x'} \neq \mathbf{x}} W(\mathbf{x'},\mathbf{x}) \tag{2.3.3}$$

[of course $\sum_{\mathbf{x'}} W(\mathbf{x'},\mathbf{x}) = 1$, if the sum includes **x** itself].

For the time being we are not interested in finding ways to solve (2.3.3); what rather interests us here is the way we will pass with the aid of (2.3.3) from the microscopic to the macroscopic description. Multiplying both sides of (2.3.3) by **x** and integrating or summing in the interval of **x**, we get a dynamical equation whose left-hand side describes the time rate of the *median* value <x>, namely,

$$\frac{d}{dt} <x> = f_1\{<x> , < x^2>, ...\} \quad , \tag{2.3.4}$$

where f_1 is, in general, a nonlinear polynomial.

The averaging process above, makes appear on the right-hand side of (2.15) not only the median <x> but higher moments of P(**x**;t) as well. Now why is f_1 nonlinear? [If f_1 were linear, (2.3.4) should read d<x>/dt = f_1(<x>).]

The reason is that in most if not all realistic cases the transitional probability rates W are nonlinear functions of x. In chemical reactions, for instance, these rates are , according to the *law of mass action* ("Fick's law"), proportional to the concentrations of the "reacting species", in other words, to the number of ways a pair or a group of "reactant molecules" can be sorted out from the total population. For example, for the reaction 2X →X' the rate is proportional to

$0.5X(X - 1) = W(X)$, i.e., (2.3.4) for the mean value should, in this particular case, read

$$\frac{d\langle X \rangle}{dt} = -K\langle X(X - 1) \rangle \quad . \tag{2.3.5}$$

Multiplying both sides of (2.3.4) by x^2 and summing up or integrating again, we get a phenomenological equation where the left-hand side gives the rate for the *variance* $\langle \delta x^2 \rangle$ of the distribution $P(\mathbf{x};t)$:

$$\frac{d}{dt} \langle \delta x^2 \rangle = f_2 \{ \langle x \rangle, \langle \delta x^2 \rangle, \ldots \} \quad , \tag{2.3.6}$$

and f_2 stands for another nonlinear polynomial.

Continuing this business, we come up with *coupled* phenomenological equations (2.3.4,6) etc. with respect to the first, second, etc. moment ... and so on, ad infinitum. Thus one is inclined to believe that the averaging processes above, bringing us from the *microscopic* to the *macroscopic* description, do not confer any advantage after all, since we end up again with a system with a great number of degrees of freedom. This conjecture is correct in principle.

In most cases, however, the p.d.f. involved, $P(\mathbf{x};t)$, far from singular points, i.e., far from instabilities, has *one hump*. This means that the median is just the most probable value, and it happens also that all higher moments (variance, skewness, ...) are orders of magnitude below the mean; then the coupling terms $\langle \delta(x^2) \rangle$, ... in the polynomial forms f_1, f_2, ... can be omitted. We end up with the so-called mean-field regime, namely one phenomenological equation which describes the evolution of the mean value

$$\frac{d}{dt} \langle x \rangle = f_1 \{ \langle x \rangle; \mu \} \quad . \tag{2.3.7}$$

This constitutes, for single species systems, the dynamics at the macroscopic level. So in passing from the microscopic to the macroscopic level, we witness indeed a tremendous reduction in the number of degrees of freedom. However, in the vicinity of singular points, as we will see later, this mean-field approximation breaks down, since the p.d.f. $P(\mathbf{x};t)$ develops more than one hump: this signals essentially the prospect of more than one possible state beyond instability, i.e., the appearance of at least a *single* branching ("bi-furcation") in the solution.

Now in a double- or multihumped p.d.f., the median is no longer the most probable value; fluctuations acquire prominence and, at the macroscopic level, due to the coupling of the moment equations (2.3.4,6) ..., the system appears as turbid as at the microscopic level. Nevertheless, when after this period of "turbulence", the system settles for —or tunnels through to —one of the available states, we witness a "shrinking" of the p.d.f. around this new state —and the rehabilitation of the mean-field regime. In conclusion, the benefits of studying a system at the phenomenological rather than at the microscopic level are suspended only in the vicinity of bifurcating points.

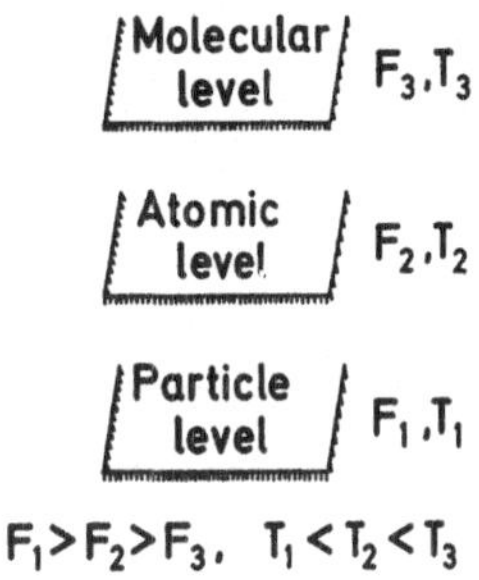

Fig.2.25. The three hierarchical levels of a simple macroscopic body

$$F_1 > F_2 > F_3, \quad T_1 < T_2 < T_3$$

A concomitant benefit that hierarchical systems confer is the so-called "almost decomposability", and this has to do with the ability to study such a system at a given hierarchical level by almost ignoring what takes place "above" and "underneath".

Let us give an example to clarify what this means. Take any piece of solid material (Fig.2.25). It can be "studied", i.e., *acted upon*, by an external observer, depending on the "filter" he is using, at three basic, distinct hierarchical levels: (1) the level of elementary particle interaction (a base level common to all physical systems), (2) the atomic level, and (3) the molecular level.

Now we know that the type of interaction characterizing level 1 is the strong nuclear interaction F_1, characterizing level 2 are the electromagnetic interactions F_2, and characterizing level 3 are the London-van der Waals interactions. We also know the hierarchy of the relative strength of these interactions, namely that $F_2 \sim 10^{-3} F_1$, and that $F_2/F_3 \sim R^{-2}/R^{-7} \sim R^5$, or $F_3 \sim F_2(R^{-5})$, where R is the distance between interacting atoms. The characteristic relaxation times T_i are inversely proportional to the intensities of interactions, so the dynamical processes at level 1 proceed, on average, say 10^6 times faster than those as level 2, and at level 2 some orders of magnitude faster than at level 3.

This means, as we know also from spectroscopy, that for one complete "cycle" at level 2, we have $\sim 10^6$ "cycles" on level 1 and $\sim 10^{-10}$-10^{-4} "cycles" on level 3. So, in order to study matter at the atomic level, it is not necessary to know how matter behaves at the level of nucleon interactions and at the macroscopic level, since concerning level 2 we can comfortably take what is going on underneath, level 1, as an extrinsic boundary condition, and what takes place above, level 3, as a constant. We will see later that this important quality of *decomposability* is one which prompts us, in more complicated cases, to seek or to establish hierarchy in systems as a way of making them more autonomous and more stable.

It is time now to introduce some convenient *macroparameters* which are instrumental in characterizing the collective behavior of systems at phenomenological levels. Needless to say, the best known of such parameters is entropy. Other parameters less known but perhaps more relevant in characterizing the behavior of systems away from equilibrium, like *complexity*, will follow.

2.3.2 Informational Entropy, Physical Entropy, Thermodynamic Entropy

Years of experience in teaching communication theory and statistical physics and thermodynamics have convinced me that the most stimulating way of introducing the concept of *entropy* is by way of introducing the concept of *information*. (The conventional approach starts by defining entropy as a measure of the degree of disorder associated with energy, and there follows a perplexing statement declaring that energy must always "flow" in such a direction that the entropy increases).

What is information? In the absence of any "rigorous" definition, let us for the moment settle for the meaning attached to this word by the layman, and try to find out criteria which discriminate statements bringing forth a lot of information from statements poor in information. Compare, for instance, the two statements or messages:

M_1 = Professor X entered the classroom.
M_2 = Professor X, upon entering the classroom, shot a student.

Everybody agrees that M_1 is trivial and its taking place moves nobody (except perhaps the waiting students), while M_2 causes an international stir (it is broadcast, televised, discussed in the press, and so on); therefore M_2 brings a lot of information —especially if Professor X is well known.

But what exactly is the ingredient responsible for such a difference? One is tempted immediately to say that what makes M_2 more information-prone than M_1 is that M_2 is extremely rare while M_1 happens almost every day (hopefully).

But this conjecture is not correct: Take this message: Tomorrow is Oct. 20. It is rare (it happens once a year). Or consider this message even: Tomorrow is Oct. 20, 1982. This message is unique in the sense that it never happened before and will never happen again. And yet no one takes any information from it. Why? Simply because this message, albeit unique, was expected (perhaps on the basis of the extreme regularity of Newtonian mechanics for short time scales) practically with uncertainty.

So we deduce that the key concept separating messages with small information content from messages with large information content is not the *rareness* of the event but rather its *unexpectedness*. This means that messages which are expected with small a priori probability deliver a great deal of information when actually taking place. Consequently, we can write the following relation between the information I and probability of an event P(M):

$$I = f\left(\frac{1}{P(M)}\right) , \qquad (2.3.8)$$

where f is an unknown function. In order to proceed in our detective search for f, we subject I to only one (obvious) constraint, namely the additive rule: The total information gathered from two incoherent sources 1, 2 should be the sum of individual parts, namely $I = I_1 + I_2$. In that spirit let us try for f the simplest analytical form, namely, $I_1 = 1/P(M_1)$, $I_2 = 1/P(M_2)$. Then we must have

$$I = \frac{1}{P(M_1,M_2)} = \frac{1}{P(M_1)} + \frac{1}{P(M_2)} \ .$$

But this is impossible since $P(M_1,M_2) = P(M_1)P(M_2)$, on the basis of stochastic independence of the sources 1 and 2. So the function $1/P$ won't do. Obviously we need some special function f which has the property

$$f\left(\frac{1}{P(M_1,M_2)}\right) = f\left(\frac{1}{P(M_1)} \cdot \frac{1}{P(M_2)}\right) = f\left(\frac{1}{P(M_1)}\right) + f\left(\frac{1}{P(M_2)}\right) \ . \tag{2.3.9}$$

The logarithm is such a function (we prove below, apropos of the derivation of the expression for the physical entropy of a system, that this function is also unique). So we write

$$I = \log \frac{1}{P}$$

or

$$I = -\log P \ , \tag{2.3.10}$$

which is always non-negative since $0 \leqslant P \leqslant 1$. By choosing as a function f the logarithm with base 2, we also fix the unit of information, the *bit*.

One bit is the amount of information required in order to distinguish between two equiprobable alternatives. So the information received from the outcome of a "head-tail" trial on a fair coin is

$$I = \log_2 2 = 1 \text{ bit} \ .$$

Suppose next that we have a system —which behaves vis-à-vis an observer as a source of information, with a repertoire of Λ possible *discrete* steady states.

The only thing the observer knows about the system, apart from the number Λ of possible steady states, is the set of a priori probabilities P_1, P_2, ..., P_Λ of having the system attracted (via an interplay of initial and boundary conditions) to each one of those states.

Question: What is the uncertainty entertained by the observer —and hence the information acquired after the observation —as to what the state of the system is at a particular moment? The answer is, the average value of the function $\log 1/P_i$. So

$$<I> = - \sum_{i=1}^{\Lambda} P_i \log_2 P_i \text{ bits} \ . \tag{2.3.11}$$

This is called the informational entropy of the system.

The next question is, under what conditions does the informational entropy of the system become maximum? We maximize (2.3.11), subject to the constraint $\sum_{i=1}^{\Lambda} P_i = 1$. Let α be an undetermined Langrangian multiplier. We consider the maximization of

$$F = - \sum_{i=1}^{\Lambda} (P_i \log_2 P_i - \alpha P_i) \quad , \tag{2.3.12}$$

with respect to each P_i. We find the condition

$$\frac{\partial F}{\partial P_i} = -(\log_2 P_i + 1 - \alpha) = 0 \quad ; \tag{2.3.13}$$

hence $\log_2 P_i = \alpha - 1$ for all i. Therefore all P_i are equal. From the normalization condition, we get $P_i = 1/\Lambda$. The maximum entropy then takes place when all states are equally probable (a priori), and its maximum value equals

$$S = S_{max} = - \sum_{i=1}^{\Lambda} \left(\frac{1}{\Lambda} \log \frac{1}{\Lambda} \right) = \log_2 \Lambda \quad , \tag{2.3.14}$$

i.e., simply the logarithm of the number of states.

Let us pass next to the *physical* entropy of a system. Suppose we have a system (a physical body, say) composed of N interacting components which belong to Λ discrete different categories according to a given classification criterion (say a discrete velocity distribution). Let N_i be the population of the ith class $i \in (1, \ldots, \Lambda)$, $\sum_{i=1}^{\Lambda} N_i = N$.

The number of microscopic arrangements or *complexions* responsible for the same macrostate of the system is, of course,

$$W = \frac{N!}{\prod\limits_{i=1}^{\Lambda} N_i!} \quad . \tag{2.3.15}$$

If all complexions are a priori equiprobable, then the probability of having one of them responsible for the given macrostate is $P = W^{-1}$.

Question: What is the uncertainty —and the information to be deduced there from —about which complexion is responsible for the observed macrostate? Think of the system under consideration as divided into two parts 1 and 2. Let W_1 and W_2 be the number of complexions responsible for the state of the first half and the second half respectively; of course, $W = W_1 W_2$.

The degree of ignorance or entropy S as to which microstate is responsible for the observed macrostate is $S(W_1)$ and $S(W_2)$ for the two parts of the system. According to the additive property of information, for the whole system we will have

$$S(W) = S(W_1 W_2) = S(W_1) + S(W_2) \quad . \tag{2.3.16}$$

Differentiating both sides with respect to W_2, we get

$$\frac{\partial S}{\partial W} \frac{\partial W}{\partial W_2} - \frac{dS(W_2)}{dW_2} = 0 \quad ,$$

or

$$W_1 \dot{S} - \dot{S}(W_2) = 0 \quad . \tag{2.3.17}$$

Differentiating (2.3.17) with respect to W_1, we get

$$W_1 \frac{\partial^2 S}{\partial W^2} \frac{dW}{dW_1} + \frac{\partial S}{\partial W} = 0 \quad,$$

or

$$W_2 W_1 \ddot{S}(W) + \dot{S}(W) = 0 \quad,$$

or finally

$$W\ddot{S}(W) + \dot{S}(W) = 0 \quad. \tag{2.3.18}$$

Planck obtained the solution of this equation in the form

$$S = k \ln W + c \quad, \tag{2.3.19}$$

where k is Boltzmann's constant and c a constant determined from the initial conditions. So the physical entropy of the system is

$$S = k \ln W \tag{2.3.20}$$

(how Boltzmann's constant entered the solution of the above equation is explained towards the end of this section).

What is the relationship between the *informational* entropy and the *physical* entropy of a system consisting of N interacting components, which are classified under Λ different categories with a given population per category? Let us put $\Lambda = 2$ to make calculations easier, and start from the expression for the physical entropy (2.3.20), where now $W = N!/N_1!N_2!$ and $N_1 + N_2 = N$. Assuming $N_1, N_2 \gg 1$, we can use Stirling's formula

$$n! = \left(\frac{n}{e}\right)^n \sqrt{2\pi n} \quad. \tag{2.3.21}$$

Taking the logarithm of both sides:

$$\ln(n!) = n \ln(n) - n + \frac{1}{2} \ln(n) + c$$

$$\sim n[\ln(n) - 1] \quad (\text{for } n \gg 1) \quad, \tag{2.3.22}$$

so

$$\ln N_i! \cong N_i(\ln N_i - 1) \quad, \quad \text{for } N, N_1, N_2 \quad.$$

The expression for the physical entropy (2.3.20) then becomes

$$k \ln W = k(\ln N! - \ln N_1! - \ln N_2!)$$

$$\cong k[N(\ln N - 1) - N_1(\ln N_1 - 1) - N_2(\ln N_2 - 1)]$$

$$= k(N \ln N - N_1 \ln N_1 - N_2 \ln N_2)$$

$$= k\left(\ln \frac{N^N}{N_1^{N_1} N_2^{N_2}}\right) = -k \ln\left(\frac{N_1^{N_1} N_2^{N_2}}{N^{N_1} N^{N_2}}\right)$$

$$= -k\left[\ln\left(\frac{N_1}{N}\right)^{N_1}\left(\frac{N_2}{N}\right)^{N_2}\right]$$

$$= -k\left(N_1 \ln \frac{N_1}{N} + N_2 \ln \frac{N_2}{N}\right) , \tag{2.3.23}$$

or, if we divide both members by N, we get the expression for the physical entropy per degree of freedom

$$<S> = \frac{S}{N} = -k\left(\frac{N_1}{N} \ln \frac{N_1}{N} + \frac{N_2}{N} \ln \frac{N_2}{N}\right) . \tag{2.3.24}$$

Now if N, N_1, N_2 are large, we can accept that $N_1/N \sim P_1$, $N_2/N \sim P_2$, where P_1, P_2 are the a priori probabilities of finding the system in category 1 or category 2 [rigorously speaking $P_i = \lim_{N \to \infty} (N_i/N)$, provided that the limit exists].
   Substituting, we get

$$<S> = -k \sum_{i=1}^{2} P_i \ln P_i \tag{2.3.25}$$

or

$$<S> = -k \ln 2 \sum_{i=1}^{2} P_i \log_2 P_i = k \ln 2(S_{inf.}) . \tag{2.3.26}$$

So we see that we pass from the physical entropy per degree of freedom to the informational entropy of a system via the "conversion factor" $k \ln 2$. Now Boltzmann's constant has the numerical value $k = 1.38 \times 10^{-23}$ J/(K bit). Therefore, to discriminate between two equiprobable alternatives (i.e., to acquire one bit of information) we must spend at least $1.38 \times \ln 2 \times 10^{-23}$ J/K. Admittedly information appears cheap, but in very fast computers, we already approach the so-called thermal limit, which severely bounds the capabilities of available hardware [2.7].
   We can now elaborate the conditions under which the entropy of a system increases irreversibly with time, or examine which are the premises under which a disturbed system left alone goes with probability one to a state of maximum disorder or perfect symmetry. Moreover, as we shall see, this state, where all dyadic velocity cross-correlations behave as delta functions, is stable.
   We write for the entropy the common expression

$$S = - \sum_{i=1}^{\Lambda} P_i \ln P_i . \tag{2.3.27}$$

where $P_i = \lim_{N \to \infty} (N_i/N)$. So, for N_i large, we have the equivalent expression

$$S = - \sum_{i=1}^{\Lambda} \frac{N_i}{N} \ln \frac{N_i}{N} = - \sum_{i}^{\Lambda} \frac{N_i}{N} (\ln N_i - \ln N)$$

$$= \sum_{i=1}^{\Lambda} \frac{N_i}{N} \ln N - \sum_{i=1}^{\Lambda} \frac{N_i \ln N_i}{N} = \frac{1}{N}\left(N \ln N - \sum_{i=1}^{\Lambda} N_i \ln N_i\right) \quad . \tag{2.3.28}$$

Differentiating with respect to time and considering a closed system so that the total number N of "particles" or components of the system is constant, we get:

$$\frac{\partial S}{\partial t} = -\frac{1}{N} \sum_{i=1}^{\Lambda} \left(\frac{\partial N_i}{\partial t} \ln N_i + \frac{\partial N_i}{\partial t}\right) = -\frac{1}{N} \sum_{i=1}^{\Lambda} \frac{\partial N_i}{\partial t} \ln N_i \quad , \tag{2.3.29}$$

since

$$\sum_{i=1}^{\Lambda} \frac{\partial N_i}{\partial t} = \frac{\partial N}{\partial t} = 0 \quad . \tag{2.3.30}$$

Let us now investigate the term $\partial N_i/\partial t$. This rate of change of the number of particles N_i in the component i, $i \in (1, \ldots, \Lambda)$ will decrease due to particles leaving i to move to other compartments, and will increase due to the flow of particles into the compartment i from all others.

Let U_{ij} stand for the probability of having one particle jump from compartment i to compartment j per unit time, and U_{ji} for the probability of a jump from j to i per unit time. Then the number of particles leaving compartment i per unit time will be equal to $\sum_{j=1}^{\Lambda} U_{ij} N_i$, and the number of particles entering compartment i per unit time will be equal to $\sum_{j=1}^{\Lambda} U_{ji} N_j$. So,

$$\frac{\partial N_i}{\partial t} = -\sum_{j=1}^{\Lambda} U_{ij} N_i + \sum_{j=1}^{\Lambda} U_{ji} N_j \quad . \tag{2.3.31}$$

Accepting now the initial condition of *molecular chaos* —namely total lack of correlation[1] between the individual positions and the individual velocities of the moving particles from compartment to compartment —amounts to putting $U_{ij} = U_{ji}$. Hence

$$\frac{\partial N_i}{\partial t} = \sum_{j=1}^{\Lambda} U_{ij}(N_j - N_i) \quad . \tag{2.3.32}$$

Substituting this expression for $\partial N_i/\partial t$ in (2.3.29), we get

$$\frac{\partial S}{\partial t} = \frac{1}{N} \sum_{i=1}^{\Lambda} \sum_{j=1}^{\Lambda} U_{ij}(N_i - N_j)\ln N_i \quad . \tag{2.3.33}$$

Interchange of the indices i, j under the accepted chaotic conditions obviously does not change the left-hand side of the above equation, so we can write again

$$\frac{\partial S}{\partial t} = \frac{1}{N} \sum_{i=1}^{\Lambda} \sum_{j=1}^{\Lambda} U_{ji}(N_j - N_i)\ln N_j$$

---

1 This lack of correlation is attributed to cascading collisions among the individual particles —which results in a forgetting of any specific initial condition.

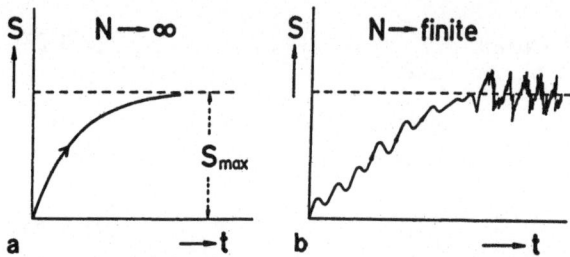

Fig.2.26. (**a**) As the number of interacting particles goes to infinity, the entropy of a perfect gas system increases monotonically. (**b**) For a large yet finite number of interacting "hard spheres", the entropy increases via fluctuations. Once the state of thermodynamic equilibrium is reached, fluctuations above and below are equally probable, due to the perfect symmetry at thermodynamic equilibrium

$$= - \frac{1}{N} \sum_{i=1}^{\Lambda} \sum_{j=1}^{\Lambda} U_{ij}(N_i - N_j) \ln N_j \quad . \tag{2.3.34}$$

Adding (2.3.33,34) we finally obtain

$$\frac{\partial S}{\partial t} = \frac{1}{2N} \sum_{i=1}^{\Lambda} \sum_{j=1}^{\Lambda} U_{ij}(N_i - N_j)(\ln N_i - \ln N_j) \geqslant 0 \quad . \tag{2.3.35}$$

The equality sign holds in cases where $N_i = N_j$ for all i and j, that is, at the state of absolute homogeneity or again at the state of perfect symmetry —which is the state of "thermodynamic" equilibrium.

For $N_i \to \infty$ the function $S = f(t)$ goes monotonically, i.e., irreversibly, to the maximum value of equilibrium, and once there, it stays (Fig.2.26a). For finite, although large, N_i's —which is the realistic case anyway —the entropy goes somewhat as in Fig.2.26b.

Once it reaches the equilibrium state, fluctuations around this state are equally (a priori) probable in both directions. The fluctuations are due to the fact that deviations from the mean values N_i become prominent for N_i finite —or for systems enclosed in finite volumes. In cases where —with the (average) values $N_i \gg 1$ —fluctuations around them nevertheless acquire a variance $\langle \Delta N_i^2 \rangle$ comparable to the N_i's (cases of phase transitions), the deduced relation $\partial S/\partial t \geqslant 0$ needs reconsideration.

Before moving to the calculation of thermodynamic entropy and investigating its relationship with the informational and physical entropies, let us see how we can express the entropy of a *continuous* rather than a discrete distribution of a priori probabilities (Fig.2.27): In order to deduce the entropy, referring to a continuous probability density function $f(x)$, we write $P_i = f_i(x_i)\Delta x_i$, and we substitute the P_i's into our formula for the entropy (2.3.11). We get

$$S = - \sum_{i=1}^{\Lambda} [f(x_i)\Delta x_i]\{\log_2[f(x_i)\Delta x_i]\}$$

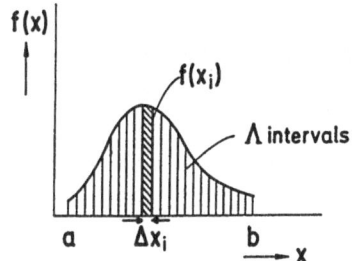

f(x)

f(x_i)

Λ intervals

a   Δx_i   b   x

Fig.2.27. Definitions used in calculating the entropy of a continuous distribution

$$= - \sum_{i=1}^{\Lambda} f(x_i) \log_2 f(x_i) \Delta x_i - \sum_{i=1}^{\Lambda} f(x_i) \Delta x_i \log_2 \Delta x_i \quad . \tag{2.3.36}$$

Let us now pass to the limit allowing $\Delta x_i \to 0$ and $\Lambda \to \infty$. This gives

$$S = \lim_{\substack{\Lambda \to \infty \\ \Delta x_i \to 0}} \left[- \sum_{i=1}^{\Lambda} f(x_i) \log_2 f(x_i) \Delta x_i \right] + \lim_{\substack{\Lambda \to \infty \\ \Delta x_i \to 0}} \left\{ \left[- \sum_{i=1}^{\Lambda} f(x_i) \Delta x_i \right] \log_2 \Delta x_i \right\}$$

or

$$S = - \int f(x) \log_2 f(x) dx + \lim_{\Delta x_i \to 0} (-\log_2 \Delta x_i) \quad , \tag{2.3.37}$$

since

$$\lim_{\substack{\Lambda \to \infty \\ \Delta x_i \to 0}} \left[\sum_{i=1}^{\Lambda} f(x_i) \Delta x_i \right] = \int f(x) dx = 1 \tag{2.3.38}$$

In (2.3.37), the second term of the right-hand side goes to infinity as $\Delta x_i \to 0$, so we conclude that the entropy of a continuous distribution is infinity. There is nothing paradoxical in this conclusion, which, as a matter of fact, should be expected beforehand.

Remember that, in an informational sense, the entropy of a system amounts to the number of "yes-no" questions we have to ask in order to specify completely the state of the system. In a system of an infinite number of states, the complete specification of one of them obviously requires an infinite number of questions and, in turn, should it be carried out, it would furnish an infinite number of bits. Still, by choosing a "quantization steps" $\Delta x_i = 1$ in the above expression, we end up with a finite number of bits.

At this point, at the risk perhaps of moving too quickly ahead towards material properly belonging to Chap.6, it is tempting to introduce the reader to the following example.

Consider within the unit interval $0, \dots, 1$ the very simple (piece-wise linear) recursive relation $X_{n+1} = 2X_n \pmod 1$. Start iterating from an arbitrary initial value X_0 which written in dyadic form is say, $X_0 = 0.110101111010101\dots$ Each iteration moves the dot one position to the right. After ξ iterations the value of the number will be $X_\xi = 2^\xi X_0$.

53

The initial value X_0 can never be known *exactly* —even in theory. Why? Because this would amount to asking an infinite number of questions (and getting an infinite number of answers in the "yes-no" form), that is, it would amount to performing an infinite number of measurements. But we have proved just above that the entropy of a continuous distribution is *infinite*. Given that each measurement requires a finite time (commensurable with the relaxation time of the particle-composed receiver), getting knowledge of any number X_0 with infinite accuracy would require an infinite amount of time.

Now from the above recursive formula, we have $(\Delta X_\xi) = 2^\xi(\Delta X_0)$, which means that the initial uncertainty ΔX_0 (however small, yet finite) is amplified *exponentially* with the number of iterations. If X_0 is known with an accuracy of say $\sim 10^{-31}$ after ~ 100 iterations the initial condition X_0 has been completely forgotten and from then on the recurrent algorithm functions as a "pseudo-random generator" producing an essentially incompressible sequence of digits —although the relationship responsible for such a sequence is quite deterministic. This, in a nutshell, is "chaos": its existence is due to a combination of the finite accuracy of any initial condition and the exponential propagation of this (inevitable) initial error. We end the "parenthesis" here.

Let us see next which particular form of $f(x)$ among those having a fixed variance maximizes S. Let us write

$$S = - \int f(x)\ln f(x)dx \quad . \tag{2.3.39}$$

We must maximize with respect to arbitrary variations of $f(x)$ in the integrand of

$$F(f(x)) = - \int_{-\infty}^{\infty} f(x)\ln f(x)dx - \alpha \int_{-\infty}^{\infty} f(x)dx - \beta \int_{-\infty}^{\infty} x^2 f(x)dx \quad , \tag{2.3.40}$$

where α, β are Lagrangian multipliers under two constraints, namely,

$$\int_{-\infty}^{\infty} f(x)dx = 1 \quad , \quad \int_{-\infty}^{\infty} x^2 f(x)dx = \sigma^2 \quad . \tag{2.3.41}$$

It follows that $-1 - \ln f(x) - \alpha - \beta x^2 = 0$ or $f(x) = \alpha e^{-\beta x^2}$, for the maximum.
Application of the constraint $\int_{-\infty}^{\infty} f(x)dx = 1$ leads to

$$\alpha \int_{-\infty}^{\infty} e^{-\beta x^2}dx = \alpha \left(\frac{\pi}{\beta}\right)^{\frac{1}{2}} = 1 \tag{2.3.42}$$

so
$$\alpha = \sqrt{\frac{\beta}{\pi}} \quad ,$$

and application of the constraint $\int_{-\infty}^{\infty} x^2 f(x)dx = \sigma^2$ leads to

$$\sigma^2 = \alpha \int_{-\infty}^{\infty} x^2 e^{-\beta x^2}dx = \frac{\alpha}{2} \left(\frac{\pi}{\beta^3}\right)^{\frac{1}{2}} = \frac{1}{2\beta} \quad . \tag{2.3.43}$$

Therefore the p.d.f. maximizing the entropy is

$$f(x) = (2\pi\sigma^2)^{-\frac{1}{2}}\exp(-x^2/2\sigma^2) \quad , \tag{2.3.44}$$

i.e., the Gaussian, and the corresponding maximum value of the entropy is equal to

$$S = - \int f(x)\ln f(x)dx = - \int f(x)dx\left[- \frac{\ln(2\pi\sigma^2)}{2} - \frac{x^2}{2\sigma^2}\right]$$

$$= \frac{1}{2}\ln(2\pi\sigma^2) + \frac{1}{2\sigma^2}\int x^2f(x)dx = \frac{1}{2}\ln(2\pi e\sigma^2)$$

$$= \ln(\sqrt{2\pi e}\sigma) \quad . \tag{2.3.45}$$

We move finally to a study of thermodynamic entropy —as defined by Clausius, more than a century ago —and try to relate it to the physical statistical entropy above. The best way is again to start with a specific example, that is, to choose a specific system, in this case an ideal gas in an external field (namely a gravitational field, against which screening is just theoretically impossible —and also negative). A perfect (monatomic) gas at thermodynamic equilibrium is characterized by only one macrovariable, namely its total (average) kinetic energy or its temperature T; the two are related via a transformation factor, Boltzmann's constant, as follows:

$$kT = \frac{1}{3}m\overline{v^2} \quad , \tag{2.3.46}$$

where m is the mass of each molecule of the gas and $\overline{v^2}$ is the variance of the velocity distribution. In the above formula (which actually constitutes a definition of temperature) T is expressed in *kelvin* and the energy in *joules*. The potential energy, i.e., the energy responsible for collisions among the individual molecules of the gas, although not zero, is taken as negligibly small compared to the kinetic energy. The three macrovariables, volume V, pressure p, and temperature T, are related via the equation of state

$$pV = NkT \quad , \tag{2.3.47}$$

where N is the number of molecules of the system concerned. Since gravitational forces act on the individual gas molecules, the gas pressure will not be the same everywhere, but will vary from point to point. For simplicity, we consider the case where the field forces are in a fixed direction, which we choose as the z-axis. We consider two unit areas perpendicular to the z-axis and distance dz apart. If the gas pressures on the two areas are p and p + dp, the pressure difference dp must be equal to the total force on the gas particles in a parallelepiped of unit base and height dz. If F is the force on one molecule (i.e., the number of molecules per unit volume). Hence dp = nF dz. The force on one molecule F is related to the potential energy u(z) of a molecule by F = -du/dz, so

$$dp = -n\,dz\frac{du}{dz} = -n\,du \quad . \tag{2.3.48}$$

From the state equation (2.3.47), with $N/V = n$, we write $p = nkT$; we will assume that the gas is at equilibrium (see the Example below) so that T is the same everywhere. Then $dp = kT\,dn = -n\,du$, from which we get

$$\frac{dn}{n} = d(\log_e n) = -\frac{du}{kT}$$

and

$$n = n_0 \exp(-u/kT) \quad , \tag{2.3.49}$$

where n_0 is a constant — the density of the gas at zero potential.

Equation (2.3.49), relating the variations in density of the gas to the potential energy of its molecules, is called *Boltzmann's formula*. Since the pressure is related to the density by a constant factor kT, an equation similar to the above is valid for the pressure

$$p = p_0 \exp(-u/kT) \quad . \tag{2.3.50}$$

Example. In the field of gravity near the earth's surface, the potential energy of a molecule at height z is $u = mgz$. So

$$p = p_0\, e^{-mgz/kT} = p_0\, e^{-\mu gz/RT} \quad , \tag{2.3.51}$$

where μ is the molecular weight of the gas, and R is the gas constant. (For one mole, the equation of state is $pV = RT$, where $R = kN_0$; N_0 is Avogadro's number). The greater the molecular weight of a gas, the more rapidly its pressure decreases with increasing height: the atmosphere contains an increasing proportion of light gases with increasing height. Keep in mind, however, that the applicability of the barometric formula to the real atmosphere is limited since the atmosphere is not in thermal equilibrium, and, in fact, its temperature varies with height. Let us see this more clearly by trying to apply Boltzmann's formula to all distances from earth. At large distances, the formula for the potential energy of a molecule has to change from the approximate expression $-mgz$ to the rigorous expression $u = -GMm/r$, where G is the gravitational constant and M the earth's mass (r is the distance from the center of the earth).

Then Boltzmann's formula becomes

$$n = n_\infty \exp(GMm/kTr) \quad , \tag{2.3.52}$$

where n_∞ is the density of the gas where $u = 0$ [i.e., at an infinite distance from the earth $(r \to \infty)$].

So $n_\infty = n_E \exp(-GmM/kTr_0)$, where n_E is the density of the gas at the earth's surface, and r_0 is the earth's radius. The above formula tells us that the density of the atmosphere at an infinite distance from the earth should be nonzero. This is absurd, however, since the atmosphere originated — we believe today — from the interior of the planet, and a finite quantity of gas cannot spread over an infinite volume with a density which is nowhere zero.

The solution of the paradox is that the gravitational field *cannot* in fact keep a gas at a constant temperature (in a state of equilibrium), and the atmosphere should therefore be steadily dissipated into space. For the earth this loss is very slow, but in the cases of the moon and mars their atmospheres have been lost more quickly. This happened progressively through collisions which allowed atmospheric molecules to exceed their escape velocity from the corresponding celestial bodies.

Returning to Boltzmann's formula (2.3.49), and replacing the densities by the probabilities of finding a molecule at the specific energy state u, we get

$$P_{(probability)} \sim e^{-u/kT} \quad , \quad \text{or} \quad k \ln P = -\frac{u}{T} \quad , \quad \text{or} \quad k \ln \frac{1}{P} = \frac{u}{T} \quad ;$$

but $1/P = W =$ the number of equiprobable complexions at thermodynamic equilibrium, so, $k \ln W = u/T =$ thermodynamic entropy, after Clausius. Therefore we see that the physical entropy of a system is identical to its thermodynamic entropy.

The above coincidence allows us to easily see the impossibility of getting any information from an "environment" *at thermodynamic equilibrium*, that is, the futility of trying to act like a "Maxwell's demon".

Consider (Fig.2.28) a closed-box system containing, say, two kinds of molecules, the fast (●) and the slow (○) ones, and let an "intelligent" being, holding a tennis racket, be placed in the system near a "trapdoor" separating the two halves of the box. His aim is, whenever he sees a *fast* "ball", to kick it towards the right compartment, and whenever he sees a *slow* "ball" to push into the left compartment. Initially, the two kinds of balls are perfectly intermixed in both compartments, but with the demon working continuously, the system gets progressively differentiated at the expense of *no* external energy.

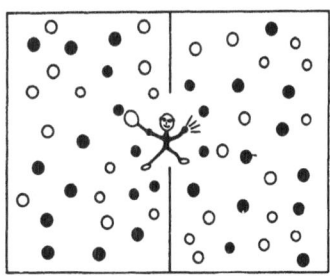

"Maxie" in action

Fig.2.28.
Maxwell's demon at work

How can this thought "experiment" be accommodated within the frame of the second law of thermodynamics? The reader realizes perhaps that a "demon" invented by Maxwell, which puzzled three generations of physicists until finally "exorcised" by Szilard (1929) and Brillouin (1951), constitutes a rather tough problem. Anyway, we start with the implicit assumption that our system is a *thermodynamic* one for which the second law does hold. Then we will assume that the number of molecules of both kinds in the box tends to infinity (otherwise the demon operating his racket at random without seeing anything, namely just by disturbing the equilibrium *long enough*, may, with finite probability, at times separated by a finite amount, indeed produce a notable excursion from the equipartition of molecules in the two halves: see again Fig.2.26b). Given the above conditions, we argue that the demon cannot see and therefore select the individual molecules unless he has at his disposition some sort of energy supply, the use of which transmits photons at a high frequency. The photons scattered from the target molecule give him the information he needs in order to use his racket accordingly.

In the worst case the demon receives the right information but he does not make use of it. Let T_1 be the temperature of the source the demon is using and $T_2 \ll T_1$ the temperature of the box. Let E be the energy the demon spends per observation. The entropy generated by the demon's act equals $S_1 = E/T_1$. The energy E is subsequently dissipated at the ambient temperature T_2, so by the end of the process, the entropy $\Delta S = S_2 - S_1 = E(1/T_2 - 1/T_1) > 0$ will increase in accordance with the second law.

Assume now that the demon does make use of the information he derives from the observation and, moving his racket in the right way, succeeds indeed in increasing (per trial) a little the differentiation of the system, that is, he succeeds in reducing the initial number of complexions W_0, or in decreasing the entropy of the gas in the box. We will see that the increase of the entropy he imparts to the system via his act of observation largely exceeds the (small) decrease of the entropy he obtains when he implements his aim. The entropy *increase* that the demon causes by switching on his light source is

$$\Delta S_1 = \frac{h\nu_1}{T_1} = k\alpha \tag{2.3.53}$$

where k is Boltzmann's constant and $\alpha = h\nu_1/kT_1 \gg 1$. (The larger the value of α, the more unambiguous the discrimination between fast and slow balls.)

Let $S_0 = k \ln W_0$ be the initial entropy in the box prior to the disturbance, and $S_1 = k \ln W_1$ the final value of the entropy after the single action of the racket. Here S_1 will be slightly smaller than S_0; we write $W_1 = W_0 - W$, where W is small.

Then the entropy *decrease* in the box will be equal to

$$\Delta S_2 = S_1 - S_0 = k \ln\left(\frac{W_1}{W_0}\right) = k \ln\left(1 - \frac{W}{W_0}\right) \sim -k \frac{W}{W_0} \quad , \tag{2.3.54}$$

where $W/W_0 \ll 1$.

So the total entropy variation will be

$$\Delta S_1 + \Delta S_2 = K\left(\alpha - \frac{W}{W_0}\right) > 0 \quad , \tag{2.3.55}$$

since $\alpha \gg 1$ and $W/W_0 \ll 1$.

2.3.3 Entropy of a Perfect Gas at Thermodynamic Equilibrium

If an amount of heat ΔQ is exchanged between a body (in our case a perfect gas) and the environment, this quantity of energy will split into two parts, one of which will modify the internal (kinetic) energy of the system by an amount E while the rest will be dissipated in work p V that the system performs versus the environment — or vice versa — depending on the sign of ΔQ. (When the volume increases, energy *leaves* the system.) So we will have,

$$dQ = dE + p\,dV \quad . \tag{2.3.56}$$

From the equation of state for a perfect gas (2.3.46,47), for a mole, we get

$$pV = \frac{1}{3} Nm\overline{v^2} \tag{2.3.57}$$

$$= \frac{2}{3} \left(N \frac{1}{2} m\overline{v^2} \right) = \frac{2}{3} E = RT \quad , \tag{2.3.58}$$

where $E = (N/2)\,vm^2$ is the internal energy of the perfect gas and $\overline{v^2}$ is the variance of the velocity distribution for one molecule. Therefore

$$dE = 3R\,dT^2 \quad . \tag{2.3.59}$$

Also from the same equation $pV = RT$, we get

$$p = RT/V \tag{2.3.60}$$

and

$$pdV = RT\,dV/V \quad . \tag{2.3.61}$$

Substituting into the equation of conservation of energy (2.3.56) we get

$$dQ = R\left(T \frac{dV}{V} + \frac{3}{2} dT\right) \tag{2.3.62}$$

The above quantity is always non-negative. Although in (2.3.62) dQ is expressed in terms of state variables, it is not a state variable itself. The integral on the right side of (2.3.62) depends on the path of integration chosen in the V - T plane. Therefore a value of Q calculated in this manner depends not only on the present state quantities, but on their historical past as well, which boils down to saying that the right-hand side of (2.3.62) is not a total differential. This difficulty is rectified by dividing both sides by T. Provided that T is a positive quantity, $dS = dQ/T$ always remains non-negative. (The variational principle for equilibrium in an open system leads therefore to the condition $dE - TdS \leq 0$ or the minimization of the free energy $F = E - TS$.) This gives

$$dS = R\left(\frac{dV}{V} + \frac{3}{2} \frac{dT}{T}\right) \tag{2.3.63}$$

and the state quantity, the entropy of the gas

$$S = R[\ln V + (3/2)\ln T] + S_0 \tag{2.3.64}$$

or

$$S \sim R \, \ln(VT_m^{3/2}) \; , \tag{2.3.65}$$

where m in T_m stands for "mass".

2.3.4 Entropy of a Photon Gas at Thermodynamic Equilibrium

In the previous section we succeeded in calculating the entropy of a perfect gas as a function of its volume V and its temperature T_m without using the probability density function (p.d.f.) of the velocities, i.e., the Maxwell distribution (something one should perhaps expect from the formula giving the physical entropy of the system). We just made use of the Clausius equivalent formula.

In this section we intend, using macroscopic thermodynamics, to derive the entropy of a box containing *photons* in thermodynamic equilibrium (a perfect relativistic gas) without developing —at this stage— Planck's distribution.

Why make such a calculation? Well, our goal is to see what happens in a closed system where at least two species of particles (one species of baryons and one of photons) coexist. Specifically, we will try to see what happens thermodynamically when the "box" expands or, more correctly, when its metric dilates —as in the case of the universe after the specific moment of the big bang. Of course we are not so foolish as to try to make a mathematical model of the expanding universe. What we would like simply to point out are a number of sine qua non prerequisites that a mixture of gases must incorporate in order to display compatibility between entropy production and progressive differentiation towards more complex patterns, something which at first glance goes against the second law of thermodynamics. Yet this concurrence is exactly what we observe in the universe on many macroscopic scales. For the moment let us proceed in solving our specific problem.

We will assume that a closed box of photons behaves as a *perfect gas*, that is, we will not take into account "light turbulence", a real phenomenon even in "vacuum", which presupposes, however, exceptionally strong light (laser) beams. So, from the picture of a perfect photon gas, we will try to derive some thermodynamic properties of blackbody radiation, e.g., its entropy. From the state equation of a perfect gas we have (2.3.57), which may be rewritten as

$$p = \frac{N}{V} \frac{m \overline{v^2}}{3} = \frac{Nm}{V} \frac{\overline{v^2}}{3} = \frac{1}{3} \rho \overline{v^2} \; , \tag{2.3.66}$$

where ρ is the density of the "fluid".

If we call $u(T)$ the total energy density of the blackbody spectrum, i.e., $u(T) = \int_0^\infty u(\nu,T)d\nu$, where $u(\nu,T)$ is just the coordinate of the Planck spectrum in $J/m^3 Hz$ and ν is the frequency [$u(T)$ is expressed in J/m^3], then we can write that the pressure exercised by the equilibrium radiation on the walls of the enclosure is given by

$$p = u(T)/3 \; . \tag{2.3.67}$$

This result follows directly from the model of a perfect photon gas: for the photon gas all photons have in fact the same velocity c so (2.3.66) becomes

$$p = \rho c^2/3 \quad . \tag{2.3.68}$$

But from Einstein's mass-energy formula, $E = mc^2$, we can write the energy density

$$E/V \sim \rho c^2 \sim u(T) \quad ; \tag{2.3.69}$$

hence (2.3.67). We write down again the fundamental thermodynamic relation of the preceding section

$$dQ = dE + p\,dV \tag{2.3.56}$$

or

$$T\,dS = dE + p\,dV \quad , \tag{2.3.70}$$

where V is the volume of the cavity and T is the radiation temperature. We have

$$E = Vu(T) \quad ,$$

and

$$dE = V\,du(T) + u(T)dV \quad ,$$

or

$$\frac{dE}{T} = \frac{V}{T}\,du + \frac{u(T)}{T}\,dV \quad . \tag{2.3.71}$$

We also have, (2.3.67), $p = (1/3)u(T)$ so

$$p\,dV = u(T)dV/3$$

and

$$\frac{p\,dV}{T} = \frac{u(T)dV}{3T} \quad . \tag{2.3.72}$$

Substituting (2.3.71,72) into the basic thermodynamic formula (2.3.70), we get

$$dS = \frac{4}{3}\frac{u(T)dV}{T} + \frac{V}{T}\frac{du(T)}{dT}\,dT \quad . \tag{2.3.73}$$

Hence

$$\left(\frac{\partial S}{\partial V}\right)_T = \frac{4}{3}\frac{u(T)}{T} \tag{2.3.74}$$

and

$$\left(\frac{\partial S}{\partial T}\right)_V = \frac{V}{T}\frac{du(T)}{dT} \quad . \tag{2.3.75}$$

The second derivative $\partial^2 S/\partial V\partial T$ can be calculated separately from both (2.3.74) and (2.3.75). Equating the two results we get

$$\frac{\partial}{\partial T}\left(\frac{4}{3}\frac{u(T)}{T}\right) = \frac{\partial}{\partial V}\left(\frac{V}{T}\frac{du(T)}{dT}\right) \quad ,$$

or

$$\frac{4}{3}\left(\frac{1}{T}\frac{du(T)}{dT} - \frac{u(T)}{T^2}\right) = \frac{1}{T}\frac{du(T)}{dT} \quad,$$

or

$$\frac{4}{3}\frac{du(T)}{dT} - \frac{4}{3}\frac{u(T)}{T} = \frac{du(T)}{dT} \quad,$$

or

$$\frac{du(T)}{dT} = 4\frac{u(T)}{T} \quad;$$

from which

$$\frac{du(T)}{u(T)} = 4\frac{dT}{T}$$

or

$$\ln[u(T)] = \ln T^4 + \ln\alpha = \ln(\alpha T^4) \quad,$$

where α is a constant, so finally

$$u(T) = \alpha T^4 \tag{2.3.76}$$

(which is, of course, Boltzmann's classic derivation of the Stefan-Boltzmann law).
Substituting in say (2.3.74), we find after integration

$$S = \frac{4}{3T}u(T)V + \text{const.} \tag{2.3.77}$$

The constant of integration can be put equal to zero to make S proportional to the volume, so we get for the entropy of a photon gas at equilibrium

$$S = \frac{4}{3}\alpha T_r^3 V \quad, \tag{2.3.78}$$

where T_r means the temperature of the radiation.

Now before considering jointly the results given by (2.3.65 and 78), we will diverge a little in order to examine (using simple Newtonian dynamics, which is quite adequate for our purpose) the issue of the expansion of the universe. We will meet convincing evidence that the thermodynamic arrow of time is completely subordinate, so to speak, to the "cosmological arrow".

2.3.5 Elements of Newtonian Big Bang Cosmology

If the universe conformed to the Boltzmann view (that of a closed, static universe approach, on the whole, thermodynamic equilibrium), one could not explain the observed progressive differentiation of the physical world, in the sense that even a gigantic fluctuation around the position of equilibrium could only account for an increasing degree of *order*. In addition, the likelihood of such a fluctuation would be extremely small; gigantic fluctuations are extremely rare, so much so that when they occur, they might be justly considered as "miracles". Moreover, their lifetime is very short.

Against the view of a static universe are two other seemingly unrelated phenomena:

a) The epoch-making experimental discovery by E. Hubble in 1929 that the light arriving from distant galaxies is red-shifted in proportion to the distance of the emitting object.

b) An everyday or rather "every-night" phenomenon, namely the darkness of the night sky.

Let us start with the second phenomenon, which was expressed in the form of a paradox by Olbers in 1826. (The issue had been considered even earlier by Halley.)

According to a simple calculation made by Olbers, the night sky should not be dark but in fact immensely bright. The argument goes as follows: Suppose the universe is infinite in extent and static with a uniform and isotropic distribution of stars throughout. Suppose we have n radiating objects per unit volume and that each object has the same luminosity I_0. Consider the volume element dV between two concentric spheres with radii r and $r + dr$, $dV = 4\,r^2 dr$ (Fig.2.29). The luminosity at the point of observation due to the sources in the volume dV, is

$$I = \frac{I_0 n\ dV}{r^2} \quad , \tag{2.3.79}$$

and the total luminosity due to all stars

$$I_t = \int_0^{r_\infty} \frac{I_0 n\ dV}{r^2} = \int_0^{r_\infty} \frac{I_0 n 4\pi r^2 dr}{r^2} = 4\pi I_0 n \int_0^{r_\infty} dr = 4\pi I_0 n r_\infty \quad . \tag{2.3.80}$$

For $r_\infty \to \infty$, $I_t \to \infty$.

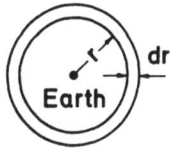

Fig.2.29. Variables used in calculating the intensity of the arriving radiation after the Olbers paradox (see text)

We can, of course, obtain a finite answer if we assume r_∞ to be finite, but in this case the universe would be unstable, that is, it would have a tendency to collapse (as we have seen from the very outset of this book) rapidly towards a center — no matter what. But we have a second possible way out of the paradox $I_t = \infty$, that is, we take into account in an infinite universe the *finite* size of each radiating object.

Suppose each object has a cross-section A normal to the line of sight from the earth. So an object at a distance r would subtend a solid angle $\Omega = A/r^2$; the object would accordingly block radiation from all objects beyond it lying in the same solid angle Ω. Total blockage of radiation will result when the entire solid angle

4π has been covered. This happens when one goes out to a distance R, for which

$$\int_0^R 4\pi r^2 n \frac{A}{r^2} dr = 4\pi$$

or

$$R = 1/An \quad . \tag{2.3.81}$$

The contribution from all sources within this distance R is

$$I_t = 4\pi \int_0^R n I_0 dr = \frac{4\pi I_0}{A} \quad , \tag{2.3.82}$$

which is finite.

Nevertheless, the paradox remains. If the above alternative is accepted, the night sky should be more or less as bright as the day sky. In a static universe at equilibrium there seems to be no way out of the paradox. Even if we assume absorption of distant radiation by intermediate objects, we do not solve the problem since the absorbed radiation heats up the absorbing matter which, in turn, radiates and gives rise to essentially the same outcome as before. The key to the resolution of the paradox comes from the interpretation of Hubble's discovery, namely that in terms of Doppler effect, red-shifting implies that the galaxies are receding with velocities proportional to their distances from earth. The red shift Δv is then expressed as $\Delta v \sim Hr/c$, where r is the distance to the galaxy, H a constant (Hubble constant), and c the velocity of light. The value of H is estimated as $\sim 3 \cdot 10^{-18} s^{-1}$.

The red shift, interpreted as a symptom of an expanding universe at a declining rate as a function of time, solves the problem by modifying the received intensity per source from the value $I_0/4\pi r^2$, which we have being using before, to the value $I_0/4 \ r^2(1+Hr/c)^2$, which when incorporated in the integral for the total received intensity, makes it converge and gives a numerical value consistent with observations. So the answer to the problem "why is the night sky dark?" is, because we are looking away from the sun and because the universe is expanding.

Let us, in the following, examine this expansion in more detail, by trying first to calculate the rate of expansion as a function of time. We assume the universe, at least at the galactic level, to be homogeneous and isotropic (within a few percent); without loss of generality, our galaxy can be taken as the origin for the coordinates of a typical galaxy mass m. The only velocity that any galaxy has with respect to "us" is radial, away from the origin and proportional to r. We write for such a velocity therefore

$$v = Hr \quad , \tag{2.3.83}$$

or

$$dr/dt = Hr \quad ,$$

or, putting $r = r(t)r_0$,

$$dr(t)/dt = Hr(t) \quad . \tag{2.3.84}$$

Here $r(t)$ is called the expansion factor, and again it does not stand for the expanding radius of a sphere in a surrounding preexisting "vacuum", but rather for the expansion of the *metric* of the system itself. The volume of a sphere of radius r about the origin is

$$V = \frac{4}{3}\pi r^3 \quad .$$

(2.3.85)

The mass contained in this sphere equals

$$M = V\rho(t) = \frac{4}{3}\pi r^3(t)\rho(t) \quad ,$$

(2.3.86)

where $\rho(t)$ is the (varying) density. If no matter is created or destroyed during the expansion (or contraction), the mass of this sphere remains constant, and so

$$\rho(t)r^3(t) = \text{const} = \xi \quad .$$

(2.3.87)

To determine $r(t)$, we point out that "galaxy" m is attracted towards the origin by the gravitational force of matter *contained* in the sphere of radius r. (Newtonian theory proves that the gravitational force exerted by any *uniform* spherical cell on a particle *inside* it is zero. Therefore we should not worry about matter existing "outside" the radius r —provided that matter is uniformly distributed about the origin as well.) So we have

$$\frac{GMm}{r^2} = \frac{4}{3}\pi\frac{G\xi m}{r^2} = -\frac{m\,d^2 r}{dt^2}$$

or

$$\frac{d^2 r(t)}{dt^2} = -\frac{4\pi}{3}\frac{\xi G}{r^2(t)} \quad ,$$

(2.3.88)

which after a first integration gives

$$\left(\frac{dr(t)}{dr}\right)^2 = \frac{8\pi\xi G}{3r(t)} + A \quad ,$$

(2.3.89)

where A is an arbitrary constant.

Thus far, our model is a model of a universe filled only with matter. Although at present, radiation contributes less than one per cent of the average energy per unit volume (the entropy distribution between radiation and matter being an altogether different topic), it played a dominant role in the earlier stages of the expansion. We can study the evolution of a universe containing matter and radiation. Let us consider the expansion of the universe to be adiabatic ($dQ = 0$). The first law of thermodynamics states that the change in internal energy of the expanding system equals the work done by the pressure, so

$$dE = -p\,dV \quad .$$

(2.3.90)

Using Einstein's formula $E = Mc^2$, where M includes contributions from both matter and energy, we get

$$E = (\rho_m + \rho_r)Vc^2 = \rho Vc^2 \quad , \tag{2.3.91}$$

where ρ_m is the density of matter and ρ_r is the density of radiation. Since the volume V of the considered element is proportional to $\sim r^3(t)$, we have from (2.3.90)

$$\frac{d}{dt}(\rho r^3) + \frac{p}{c^2}\frac{d}{dt}(r^3) = 0 \quad . \tag{2.3.92}$$

For a universe containing matter only in the present tenuous form, the pressure can be neglected, and we end up with the relation

$$\frac{d}{dt}(\rho_m r^3) = 0 \tag{2.3.93}$$

or

$$\rho_m r^3 = \text{const}$$

(derived already) or

$$\rho_m \sim r^{-3} \quad . \tag{2.3.94}$$

In the present epoch radiation traverses the universe freely with small probability of being scattered by gas or dust; a photon is likely to travel several Hubble distances before being scattered or absorbed. We say that the universe is transparent or optically thin at the present epoch. This was not always so. The typical size of a galaxy is ~ 10 kpc, while the average spacing between galaxies is on the order of ~ 1 Mpc. (Here 1 pc is the distance that light covers in one year).

If we "run" the universe backwards, we come to a period where the galaxies would all have been "touching". Going on backwards, we "see" dust falling back into the stars from which it was blown when these stars exploded as supernovas and, further on, the stars dissolving into the nonluminous gas clouds from which they formed. Complex atoms break down into hydrogen and some helium. We get a fairly uniformly distributed gas of hydrogen and helium spotted here and there with mysterious irregularities (destined to become galaxies). Still further backward, the energy density and the temperature build up; the gas starts to be heated by the photons. The critical moment comes when the temperature of matter reaches ~ 3000 K, beyond which the hydrogen ionizes. This brings us to the age of *coupling* between matter and radiation, due to the Thompson scattering that photons undergo with free electrons: the universe is now "opaque" to radiation and a homogeneous hot plasma as far as we can tell; prior to this era, radiation reigns undisputed.

Let us stop the "movie" at this point and apply the first thermodynamic law to this last era of energetic radiation predominance. We know that (2.3.68)

$$p_r = \frac{\rho_r c^2}{3} \tag{2.3.95}$$

so we get (2.3.92)

$$\frac{d}{dt}(\rho_r r^3) + \frac{1}{3}\rho_r \frac{d}{dt}(r^3) = 0 \quad , \tag{2.3.96}$$

or

$$3\rho_r r^2 \dot{r} + \dot{\rho}_r r^3 + \rho_r r^2 \dot{r} = 0 \quad ,$$

$$4\rho_r r^3 \dot{r} + \dot{\rho}_r r^4 = 0$$

or

$$\frac{d}{dt} (\rho_r r^4) = 0 \quad , \tag{2.3.97}$$

from which

$$\rho_r \sim r^{-4} \tag{2.3.98}$$

in the very early universe — let us say "near" the big bang moment, that is, the first hundred thousand years!

If we now come back to (2.3.89)

$$\left(\frac{dr(t)}{dt}\right)^2 = \frac{8\pi\xi G r(t)}{3r^2(t)} \tag{2.3.99}$$

(putting the constant A = 0), and apply it in the radiation-dominated era, since

$$r(t)\xi = \rho_r r^4 = \xi' = \text{const} \quad , \tag{2.3.100}$$

we get

$$\dot{r}^2 r^2 = \text{const} \quad , \tag{2.3.101}$$

from which it follows that given an early adiabatic expansion, the radiation-domi-nated universe expanded as

$$r(t) \sim t^{\frac{1}{2}} \quad . \tag{2.3.102}$$

2.3.6 Expansion of a Mixture of Matter and Radiation. Differential Cooling and Entropy Production

We now propose to study the way entropy changes in an (ideal) expanding system con-taining two components only: a perfect classical nonrelativistic gas and photons [2.8].

Let us start with a system (a "universe" if you like), containing only matter and, more specifically, a perfect (monatomic) gas at thermodynamic equilibrium. Suppose that the metric in the box starts changing according to a given law $\sim r(t)$. How does this expansion affect the entropy of the gas? In the first place, the expansion does not change the velocity distribution of the gas (whatever this dis-tribution may be — in this case the Maxwellian one) because a rescaling of lengths merely rescales the velocities in the same ratio. For a nonrelativistic gas, the velocity and the momentum distribution coincide, so if the gas is in thermal equi-librium with a Maxwellian distribution, it will remain in equilibrium and the en-tropy will not change. This is demonstrated as follows: In a time dt an atom of the gas with velocity v relative to the expanding substratum will travel a distance v dt. At this new point the substratum will be moving at a rate $\dot{r}(v\,dt)/r$, owing to the expansion, so that the velocity measured by an observer moving with the sub-stratum at this point will be reduced by an amount $dv = -\dot{r}(v\,dt)/r$, from which we have

$$\frac{dv}{v} = -\frac{dr}{r} \tag{2.3.103}$$

or

$$v \sim \frac{1}{r} \quad . \tag{2.3.104}$$

Now since the temperature T_m of an ideal gas is proportional to its average kinetic energy, i.e., $T_m \sim v^2 \sim 1/r^2$, we conclude that

$$T_m \sim \frac{1}{r^2} \tag{2.3.105}$$

and

$$T_m^{3/2} \sim r^{-3} \sim V^{-1} \quad , \tag{2.3.106}$$

where V is the volume of the expanding gas. We see therefore that the entropy of the system

$$S_m \sim R \ln(VT_m^{3/2}) \tag{2.3.65}$$

remains *invariant* during the expansion (or contraction): a "universe" containing only an ideal gas starting from equilibrium would remain in equilibrium forever.

Let us now suppose that our "universe" initially contains only photons (in fact this would be the case if in the big bang, baryons and antibaryons had been produced in equal quantities). Hubble's experimental discovery suggests that as the universe expands, the wavelength of each photon will scale with $r(t)$, i.e., the frequency v will obey the relation

$$v \sim \frac{1}{r(t)} \tag{2.3.107}$$

or, after Planck's law, since $E_r \sim hv$,

$$E \sim \frac{1}{r(t)} \quad , \tag{2.3.108}$$

where E is the energy associated with the specific photon frequency. The expansion clearly does not change the distribution of photons among the various frequencies of the blackbody spectrum because their wavelengths all scale in the same ratio. So the Planck spectrum will be preserved under expansion, although the temperature $T_r \sim E_r$ will decrease in proportion to the average photon energy

$$T_r \sim \frac{1}{r(t)} \tag{2.3.109}$$

or

$$T_r^3 \sim r(t)^{-3} \sim V^{-1} \quad , \tag{2.3.110}$$

where V is the volume of the gas.

Thus the entropy of the expanding photon gas

$$S_r \sim \frac{4}{3} \alpha V T_r^3 \qquad\qquad (2.3.78)$$

(α is Stefan's constant with the numerical value $5.67 \times 10^{-8} W/m^2 K^4$) will remain invariant during the expansion: an expanding (or contracting) system containing *only* photons, starting from thermodynamic equilibrium, would remain at thermodynamic equilibrium for ever.

We now consider the case of a universe homogeneously filled with nonrelativistic matter and electromagnetic radiation together. The expansion (or contraction) causes matter to cool as $T_m \sim 1/r^2(t)$, and radiation to cool as $T_r \sim 1/r(t)$. Matter cools faster than radiation. Therefore a temperature difference will develop between them as the universe expands: energy will tend to flow *from the radiation field to the matter* in an attempt to equalize temperatures. If the expansion rate is smaller than the temperature equalization rate of the mixture, the heat flow will take place under equilibrium conditions, and the entropy of the mixture will remain constant.

If, however, the expansion rate is larger than the temperature equalization rate, the heat flow from radiation to matter will take place under nonequilibrium conditions and the entropy will increase: the universe under such circumstances would "lock itself" far from thermodynamic equilibrium and if ever the phase of expansion were followed by an era of contraction, things would not change.

Figure 2.30 tells the story synoptically. At the earlier phases of the expansion, the rate of expansion is large but the mixture of matter and radiation is so hot that the gradient of temperature equalization exceeds the gradient of expansion: the universe then, starting from thermodynamic equilibrium, remains in that state. As time goes on, the rate of expansion decreases due to the gravitational forces, but as the system cools, the rate of temperature equalization decreases even faster. There comes a critical age T_c, where the two curves cross each other. From then, the system goes (and stays) beyond thermodynamic equilibrium characterized, first of all, by a rate of entropy production

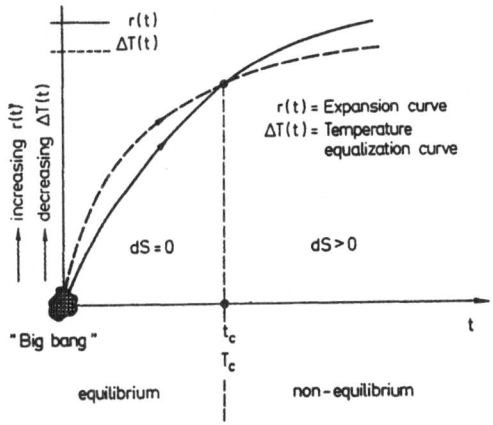

Fig.2.30. When the rate of thermal equalization exceeds the expansion rate, the system is at equilibrium. After a certain critical time T_c, the rate of thermal equalization drops below the rate of expansion: from then on the system moves and stays far from equilibrium

$$\frac{dS}{dt} = \left(\frac{1}{T_m} - \frac{1}{T_r} \right) \frac{dQ}{dt} \quad , \tag{2.3.111}$$

where dQ/dt is the rate of exchange (i.e., of flow) of thermal energy from radiation to matter. Concomitant with *entropy production* is the phenomenon of progressive *differentiation* of the universe from a homogeneous ionized matter-photon (plasma) ball to lumps of matter —galaxies —rushing away from one another (and contracting on an individual basis due to the predominance of gravitational pull on small scales) in a "sea" of long-range propagating radiation.

As galaxies contract, luminous matter starts developing. In the interior of lumps (stars) within given galaxies, once the gravitational pull smashes through the "Coulombic" barrier, nuclear reactions ignite, transforming hydrogen to helium, etc.

Thus we witness a succession of steps or "epochs" in each of which one of the four basic physical interactions is predominant. In the evolution of a big star, the succession goes from gravitation $\longrightarrow$ electromagnetic $\longrightarrow$ strong nuclear (with the help of weak nuclear) $\longrightarrow$ electromagnetic $\longrightarrow$ nuclear $\longrightarrow$ gravitational (total collapse giving rise to neutron stars and black holes or, below the Chandraseckar limit, partial collapse stopping at the white dwarf stage).

During this process of continuous differentiation, we witness also the increase of complexity of matter in the formation of stable elements heavier than primordial hydrogen and helium, manufactured in successive steps up to iron, in the interior of the stars, as well as the whole series of heavier radioactive nuclei produced by still-unknown processes during supernova explosions. Then we witness the formation of molecules out of electromagnetic interactions between atoms which do not necessarily possess a permanent dipole moment but are perfectly symmetrical, atoms like helium, neon, and argon. These forces, $F \sim r^{-7}$ (London-van der Waals interaction), result from interaction between instantaneous dipole moments, i.e., between nonlinear coupled oscillators whose eigenfrequencies depend also on the degree of coupling and persist even at zero temperature. So even in the "inanimate" pre-biological world, we find a systematic tendency to *entropy production*, along with *progressive differentiation* and *complexity increase*.

Let us now briefly examine the energetic (and entropic) exchanges between a common star (our sun) and a "bio-planet" like the earth. Neutrinos and solar wind excluded, we will refer only to the spectrum of electromagnetic radiation. The sun is separated from the earth by 214 sun radii. This means that the electromagnetic (EM) energy hitting the earth scales down between the surface of the sun and the surface of the earth by a factor of $1/(214)^2 \sim 1/46000$. The EM energy which leaves the sun is at thermodynamic equilibrium at a temperature $T_1 \cong 6000$ K; when it hits the earth it has been scaled down by the above factor $\sim 1/46000$, so in the Planck distribution it appears somewhat like the curve bounding the hatched area in Fig.2.31. This spectrum is not in thermodynamic equilibrium as it has been "shifted" to the right of

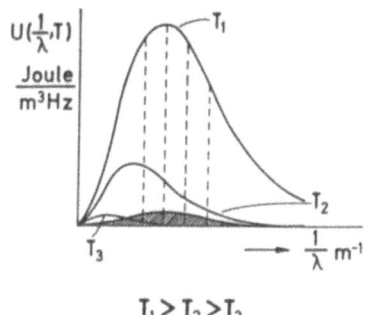

Fig.2.31. The Planck energy distribution of blackbody radiation for three temperatures T_1, T_2, T_3

the corresponding equilibrium curve at the equivalent temperature (the temperature on the surface of the earth, $T_3 \sim 310$ K). Therefore, the electromagnetic energy arriving from the sun possesses less entropy or more order than what one would expect from the Planck distribution at the corresponding temperature; it possesses, as we say, an amount of "negentropy" which can in principle be imparted to and increase the order of a conveniently built sensitive "receiver". The total EM energy from the sun hitting the earth amounts to $\sim 3.2 \times 10^{22}$ J/K year or $\sim 10^{38}$ bits/s in informational terms.

Thus, a rather impressive amount of "information" is available every single moment from the sun, but it is up to the smartness of the "receiver" to appreciate it and make further use of it. For example, a piece of iron does not "understand" the solar information at all; it reflects part of it back (in a more disorganized way than it received it) and uses the rest to heat itself up, i.e., to increase its own entropy.

However, this is not the case for chlorophyl macromolecules: they use the entropy-deficient sunlight in a very efficient way. Instead of leaving their emitted photoelectrons to be thermally dissipated (like the piece of iron), they *use* their energy in order to separate H_2 from O and attach it to C, i.e., to make possible the following reaction:

$$(H_2O)_n + (CO_2)_m + \text{sunlight} \longrightarrow C_mH_{2n}O_n + (O_2)_m \quad . \tag{2.3.112}$$

The resulting molecules of glucose ($C_6H_{12}O_6$), for example, are more complex than the water and carbon dioxide molecules. Moreover, in the reaction above this increase of complexity is accompanied by *entropy production* which exceeds the entropy defect of the solar light, which initiated the process in the first place.

At this hierarchical level of biological complexity, however, we witness for the first time another new, evolutionary by-product, namely *self-organization*. This has to do with the fact that the system under consideration, beyond a certain threshold of complexity, acquires the ability of autonomous self-reproduction, a property which implies that the system uses "self-simulating algorithms". We postpone the discussion on the still quite unknown dynamics of self-organization until Chap.6, but now the time is ripe to deal with the concept of *complexity* —both on a structural level and an operational (algorithmic) level.

2.3.7 The Concept of Complexity

a) Structural Complexity and Its Relationship to the Stability of a System

Intuitively we understand that the structural complexity of a system should increase with the number of interacting variables N. For a given (large) number of variables, the degree of *interconnectedness* among them also contributes to the complexity of the system. Full connectedness implies that all elements a_{ij} in the interaction matrix A of the linearized system (Sect.2.2.7) are different from zero. We usually measure the degree of interconnectedness by a number $0 \leqslant C \leqslant 1$ standing for the fraction of nonzero a_{ij} elements of the matrix A. For given $N \gg 1$ and $C \sim 1$, the a_{ij}'s are essentially random numbers; the complexity increases with the variance σ^2 of a (Guassian) probability density function (p.d.f.) (Fig. 2.32) from which we can sample the individual numbers a_{ij}. As a compact "super-parameter" of the complexity of a randomly connected large network, the nonlinear combination $\eta = \sigma\sqrt{NC}$ has been suggested by McMurtrie [2.8] and others. We have

$$\sigma^2 = \sum_{i=j=1}^{N} \sum^{N} a_{ij}^2 \qquad (2.3.113)$$

for zero mean of the p.d.f. The value of σ characterizes the strength of interaction within the system. How could we approach the problem of stability of such a large system? From what we already know, this "simply" amounts to calculating the eigenvalues of the matrix A.

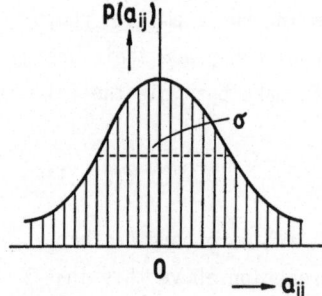

Fig.2.32. The Gaussian p.d.f. of interaction coefficients a_{ij} which are sampled in order to partition the system in hierarchical compartments

But now we must be cautious. The matrix A is a random one and the problem is how we calculate the eigenvalues of a *random* matrix, with real elements. We cannot, however, deduce these eigenvalues as a solution of a deterministic characteristic equation. Attempts have been made to determine in general the theoretical distribution of the eigenvalues of a random real matrix, and they have failed. The failure arises from the fact that the real field is not algebraically closed. For particular cases, computing simulations give the $P(\text{Re}\{\lambda_i\})$ distribution. What we come up with really is not a set of definite numbers $\text{Re}\{\lambda_i\}$, but rather *probabilities* of the $\text{Re}\{\lambda_i\}$ being positive, zero, or negative.

Extensive computing simulation conducted essentially by Ashby, see [2.9], shows that when the parameter $\sigma\sqrt{NC} < 1$, then

$$P(\text{Re}\{\lambda_i\} < 0) \sim 1 \quad \text{for all} \quad i \ . \tag{2.3.115}$$

This means that when the complexity of the randomly connected networks is below a certain value, the system is essentially stable. Nevertheless, the simulation also showed that when the critical value $\sigma\sqrt{NC} = 1$ is exceeded, then for $\sigma\sqrt{NC} \geqslant 1$ there is at least one i for which $P(\text{Re}\{\lambda_i\} > 0) \sim 1$. The system becomes most probably unstable. We see therefore that there is a *conflict between stability and complexity*. If this were so in real systems, one would expect that almost all macroscopic structures of any significance would be inherently unstable. And yet we witness extremely complex systems such as biological organisms along with a great many man-made systems such as computers and airplanes which are extremely stable inspite of their horrendous complexity. (One answer, of course, may be that these large-scale systems are not in fact randomly interconnected. Yet in many cases this conjecture is not supported by existing evolutionary theories: the wiring of a large mammalian brain is a characteristic example.) How does this happen? What type of compromise can be worked out between the seemingly irreconcilable tendencies of stability and complexity? An answer seems to be: "compartmentalization" or "hierarchical decomposition".

Specifically, the existence of any organizational-structural rearranging which can be demonstrated to enhance stability for given degree of complexity could offer guidelines: the approach is to examine the relative probability of stability of matrices with the same N, C, and σ, but with the matrix partitioned in such a way that the nonzero elements a_{ij} occur in blocks.

Consider as an example the very simple matrix shown in Fig.2.33, concerning the connectedness of four variables x_1, x_2, x_3, x_4. If we resample the p.d.f. from which the a_{ij}'s come in such a way as to make the probability of the elements along the BC diagonal very small, then the matrix diagonalized along the AD diagonal gives us a system split into two independent halves.

If, however, we sample our p.d.f. so that we diagonalize the matrix along the BC line (by making the p.d.f. of the elements a_{11}, a_{12}, a_{21}, a_{22}, a_{33}, a_{34}, a_{43}, a_{44} very small) then we obtain the picture in Fig.2.34). We can further arrange the partition of the nonzero elements in such a way as to make a_{13}, a_{31}, a_{24}, a_{42} very strong [i.e., their $P(a_{ij}) \sim 1$], and the elements a_{14}, a_{41}, a_{32}, a_{23} very weak [i.e., their $P(a_{ij}) \sim 0$]. In this way, the original system is split into two compartments, the members of each compartment strongly interacting and the interactions between compartments weak. Computing simulation in such a case gives evidence that for the same degree of complexity, decomposability enhances stability [2.9].

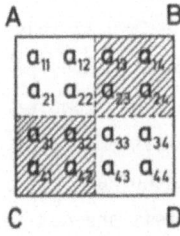
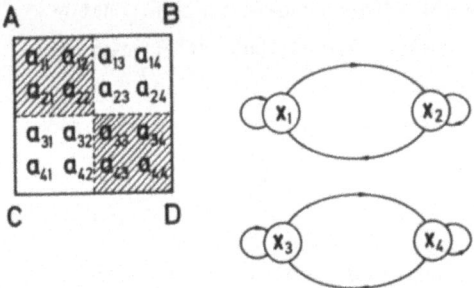

Fig.2.33. (a) Partitioning the random interaction matrix in a way which splits the system into two independent halves. The a_{ij}'s in the areas which are not hatched are very small. (b) Diagram showing the connectedness of the components of the system described by the matrix in (a)

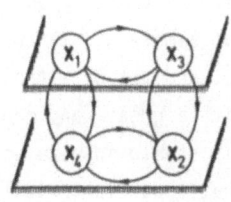

Fig.2.34. (a) Partitioning the random interaction matrix in a way which allows for (loosely) interacting compartments. The a_{ij}'s in the areas which are not hatched are very small. (b) Diagram showing the connectedness of the components of the system described by the matrix in (a)

b) Algorithmic Complexity

The algorithmic complexity of a system is the minimum number of bits (the smallest number of instructions) required by an external observer in order to be able to fully reproduce the system involved. Leaving for Chap.6 the dynamics behind the production of complexity in a system, here we simply intend to elaborate on a definition, largely due to Chaitin [2.10]. The issue is as follows: We receive a time series consisting of 0's and 1's, which time series is produced by just flipping a fair (perfect) coin so that no intrinsic interconnection exists among the individual symbols. How many alternative series of length N can we get? Well, it depends. If we are not interested in fixing the number of 0's and 1's in the series, their total number is 2^N. If we are, their total number is $N!/N_1!N_2!$, where, say, N_1 is the number of 0's and N_2 is the number of 1's ($N_1 + N_2 = N$). Now the problem is, what is the percentage of those 2^N series (each a priori expected, vis-à-vis an external observer, with equal probability 2^{-N}) which are compressible up to, say, K bits? That is to say, how many series can give rise to an algorithm of length N-K, wich, if fed as an input into a finite-state machine, will yield as an output the full series of length N?

With the two symbols 0 and 1, we can construct 2^1 "words" of length 1 bit, 2^2 words of length 2 bits, 2^3 words of length 3 bits, and so on. So the total number of series compressed up to K bits is

$$2^1 + 2^2 + 2^3 + \dots + 2^{N-K-1} = 2^{N-K} - 2 \;, \tag{2.3.116}$$

and their fraction

$$\xi = \frac{2^{N-K} - 2}{2^N} \sim 2^{-K} \quad , \quad \text{for} \quad N \gg 1 \quad . \tag{2.3.117}$$

We discover with some "melancholia" that this fraction rapidly decreases with K, so for $K = 10$, one in a thousand series is compressed up to 10 bits. In general, the relationship between the length of the input alogrithm and the length of the output series is

$$N' \cong N_0 + NS(0,1) \quad , \tag{2.3.118}$$

where N_0 includes a number of instructions intrinsic to the specific machine, and is of the order of ln N, and

$$S(0,1) = -P(0)\log_2 P(0) - P(1)\log_2 P(1) \quad , \tag{2.3.119}$$

is the entropy of the series (Fig.2.35) with $P(0) + P(1) = 1$. So for $P = 1/2$,

$$S(0,1) = 1$$

and N' is in general *greater* than N. As P deviates from 1/2, S(0,1) drops, and for $P = 1$, or $P = 0$,

$$S(0,1) = 0$$

and so

$$N' \ll N \quad .$$

Extreme examples among the 2^N series above are the particular realizations

$$\underbrace{000...0}_{N} \quad \text{or} \quad \underbrace{111...1}_{N} \quad ,$$

where $N' \sim \ln N$. The percentage ξ includes, for arbitrary K, all realizations for which the observer can deduce some stochastic interdependence among the 0's and 1's, i.e., $P \neq 1/2$. What the observer cannot prove, however, is the incompressibility, i.e., the complete randomness of a certain realization. This would go against Gödel's theorem.

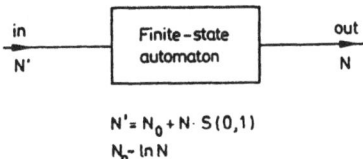

Fig.2.35. A "minimal" program N' is executed in a finite state machine and gives a time series of length N

2.4 Concluding Remarks

In this chapter we have limited the discussion to the dynamics of nonlinear systems in one, two, and many dimensions (dimensionality three in state space has been postponed until Chap.6). Essentially what we saw is that Hamiltonian systems do not change, in the sense of preferentially and irreversibly switching among states, except at the critical points signaling the branching of their solution. Dissipative systems, by contrast, proceed by a succession of broken symmetries (giving rise to irreversibility both in one and two as well as many dimensions). The evolution of dissipative systems, contrary to the Boltzmann description, is characterized by *entropy production* concomitant with *progressive differentiation* and *complexity increase*. We saw that complexity jeopardizes stability unless tempered with hierarchy. A fourth characteristic, *self-organization*, emerges once the complexity of the system exceeds a critical figure.

Self-organization has essentially to do with the ability of a system to simulate (i.e., to compress to minimal-length algorithms) the environment as well as parts of itself. Information theory is needed for further study, since the property to simulate requires the existence of a code or a mapping procedure among hardware and software levels. What we are trying to say is that in order to understand self-organization, a simultaneous treatment on at least two hierarchical levels is needed, so one-to-one mapping does not cause organization.

Chaotic dynamics is needed as well in order to dynamically implement the behaviour of self-compressibility, namely the incorporation of strange attractors in the "cognitive" system with "information dimensionality" smaller than their state-space dimensionality. Chaos is also instrumental in coupling the microscopic with the macroscopic description. For these reasons, analytical discussion on self-organization requires knowledge from Chaps.4 and 6, but prior to that, the *physics* by which information is conveyed in natural systems must be discussed.

Consequently, what we offer next is some material from the dynamics of information carriers, namely topics on radiation, propagation of spherical waves, and information imparted by spherical waves at finite apertures.

3. The Role of Spherical Electromagnetic Waves as Information Carriers

3.1 Radiation from Accelerated Charge in Vacuo. The Concept of "Self"-Force. Thermodynamics of Electromagnetic Radiation

3.1.1 Radiation in Vacuum

Consider an elemental dipole consisting of two charges of opposite-signs -e, e separated by a time-varying distance $l(t)$. The dipole is "elemental" if $l_{max} \ll \lambda$, where λ is the wavelength of the imposed current. We intend to first calculate the energy radiated from this dipole in vacuo and then study the process in a dispersive and lossy medium. To do this, we first have to calculate, at a point R, θ, φ (Fig.3.1), the intensities of the produced electric and magnetic fields or, in more detail, the six components E_r, E_θ, E_φ, H_r, H_θ, H_φ (in spherical coordinates). The source is characterized by coordinates $r'(x',y',z')$, t' and the observation point by $r(x,y,z)$, t, where

$$R = \sqrt{(x - x')^2 + (y - y')^2 + (z - z')^2} \tag{3.1.1}$$

and

$$t' = t - \frac{R}{c} , \tag{3.1.2}$$

c being the velocity of light. The dipole moment equals

$$P(t') = el(t') = el(t - \frac{R}{c}) . \tag{3.1.3}$$

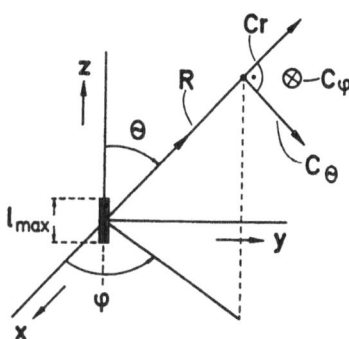

Fig.3.1. Calculating the field from the radiating elementary dipole

77

We start by writing Maxwell's equations in the mks system, that is,

$$\dot{B}(r,t) = - \nabla \times E(r,t) \qquad (3.1.4)$$

(Faraday's law),

$$\frac{1}{c^2} \dot{E}(r,t) + \mu_0 J(r,t) = \nabla \times B(r,t) \qquad (3.1.5)$$

(generalized Ampère's law for time-varying fields),

$$\nabla \cdot E(r,t) = \frac{\rho(r,t)}{\varepsilon_0} \qquad (3.1.6)$$

(Gauss's law),

$$\nabla \cdot B(r,t) = 0 \qquad (3.1.7)$$

(nonexistence of magnetic monopoles),

$$\nabla' \cdot J(r',t') + \frac{\partial \rho(r',t')}{\partial t} = 0 \qquad (3.1.8)$$

(equation of conservation of charge),

where $E(r,t)$ is in V/m, $H(r,t)$ in A turns/m, $D = \varepsilon_0 E$ in C/m^2, $B(r,t) = \mu_0 H(r,t)$ in Wb/m^2, $J(r',t')$ in A/m^2, $\rho(r',t')$ in C/m^3, μ_0 in H/m and ε_0 in F/m. In the general problem as posed above, the intention is to calculate the intensities of the fields E, B, from the densities of the sources (J,ρ) in a medium of permittivity ε_0 and permeability μ_0 (a vacuum). Here

$$c = \frac{1}{\sqrt{\varepsilon_0 \mu_0}} = \text{velocity of light in vacuum}$$

and the symbol

$$\nabla' = \frac{\partial}{\partial x'} + \frac{\partial}{\partial y'} + \frac{\partial}{\partial z'}$$

refers to divergence calculated at the source. The general geometry of the problem for a nonelemental source of finite size, occupying a volume V', is as in Fig.3.2.

It is easiest to arrive at the intensities of the sources by way of calculating the potentials (scalar and vector). Specifically (3.1.7) is satisfied for

$$B = \nabla \times A \qquad (3.1.9)$$

where A is the vector potential.

Substituting B from (3.1.9) into (3.1.4), we get

$$\nabla \times (E + \dot{A}) = 0 \quad , \qquad (3.1.10)$$

from which

$$E + \dot{A} = \pm \nabla \Phi \quad ; \qquad (3.1.11)$$

where $\Phi(r,t)$ is any scalar function (the scalar potential), or, taking into account the fact that in the asymptotic (electrostatic) case $\partial/\partial t \to 0$, $E = -\nabla\Phi$, we get

$$E = - \nabla\Phi - \dot{A} \quad . \qquad (3.1.12)$$

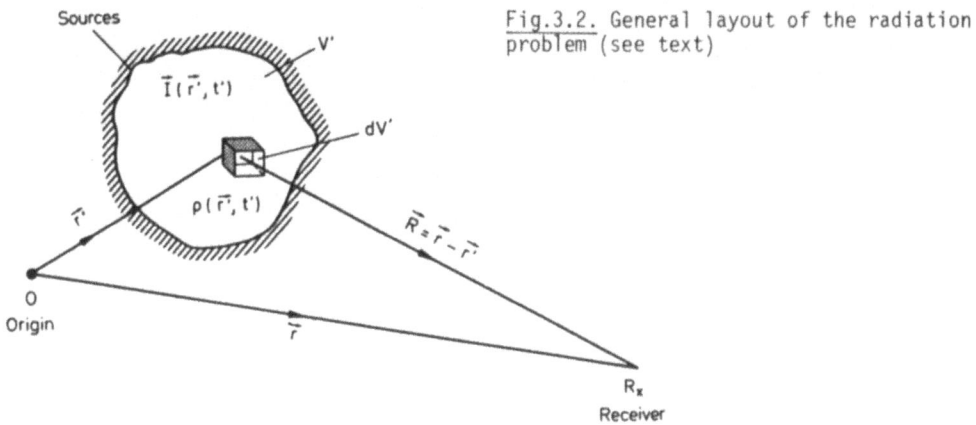

Sources

$\bar{I}(\bar{r}',t')$

V'

dV'

$\rho(\bar{r}',t')$

$\bar{R} = \bar{r} - \bar{r}'$

$\bar{r}$

O
Origin

$\bar{r}$

R_x
Receiver

Fig.3.2. General layout of the radiation problem (see text)

Substituting now **E** and **B** from (3.1.9,12) into (3.1.5), we get

$$\left(\nabla^2 - \frac{1}{c^2}\frac{\partial}{\partial t^2}\right)\mathbf{A} = -\mu_0\mathbf{J} + \nabla\left(\nabla \cdot \mathbf{A} + \frac{1}{c^2}\frac{\partial\Phi}{\partial t}\right) \quad . \qquad (3.1.13)$$

This can be split into two equations, in **A** and Φ respectively, by using the Lorentz gage; we write

$$\nabla^2\mathbf{A} - \frac{1}{c^2}\frac{\partial^2\mathbf{A}}{\partial t^2} = -\mu_0\mathbf{J} \quad , \qquad (3.1.14)$$

then (3.1.13) gives

$$\nabla\left(\nabla \cdot \mathbf{A} + \frac{1}{c^2}\frac{\partial\Phi}{\partial t}\right) = 0 \qquad (3.1.15)$$

or, putting zero as the integration constant,

$$\nabla \cdot \mathbf{A} + \frac{1}{c^2}\frac{\partial\Phi}{\partial t} = 0 \quad . \qquad (3.1.16)$$

Substituting the value of **E** from (3.1.12) into (3.1.6), we get

$$\nabla^2\Phi + \nabla \cdot \dot{\mathbf{A}} = -\frac{\rho}{\varepsilon_0} \qquad (3.1.17)$$

or, taking (3.1.16) into account,

$$\nabla^2\Phi - \frac{1}{c^2}\frac{\partial^2\Phi}{\partial t^2} = -\frac{\rho}{\varepsilon_0} \quad . \qquad (3.1.18)$$

It can be proved that the Lorentz gage (3.1.16) is compatible with (3.1.8), the equation of conservation; the split of (3.1.13) into (3.1.15) and (3.1.16) is therefore correct but still ambiguous. However, since in our problem **A** and Φ are "nonobservables", auxiliary functions, ambiguities in **A**, Φ do not matter. We can always get the same **E** and **B** under the simultaneous transformations

$$A' = A + \nabla\Psi \tag{3.1.19}$$

(where Ψ is *any* scalar function) and

$$\Phi' = \Phi - \frac{\partial\Psi}{\partial t} \quad, \tag{3.1.20}$$

so **E** and **B** remain invariant of the choice of the potentials.

The general solutions of (3.1.14 and 18) give

$$A(r,t) = \frac{\mu_0}{4\pi} \int_{V'} \frac{J(t - R/c)}{R} \, dV' \tag{3.1.21}$$

and

$$\Phi(r,t) = \frac{1}{4\pi\varepsilon_0} \int_{V'} \frac{\rho(t - R/c)}{R} \, dV' \quad. \tag{3.1.22}$$

Let us apply these results now to the example of an elementary dipole. We have

$$p(t') = el(t') \tag{3.1.3}$$

and

$$J = \rho v = \rho \frac{dl}{dt} \tag{3.1.23}$$

so the expression for the vector potential (which is along the $\hat{z}$ direction) becomes

$$A = \frac{\mu_0}{4\pi R} \int_{V'} \rho \frac{dl}{dt} \, dV' \tag{3.1.24}$$

or, since dl/dt is independent of **r**' and $\int_{V'} \rho dV' = e$,

$$A = \frac{\mu_0}{4\pi R} \frac{dp}{dt} \tag{3.1.25}$$

(we point out that since $\partial R/\partial t = 0$, $\partial/\partial t' = \partial/\partial t$).

Let us define a "Hertz vector" Π through the transformation

$$\Pi = \frac{1}{4\pi R} p \quad, \tag{3.1.26}$$

then

$$A = \mu_0 \frac{\partial\Pi}{\partial t} \quad. \tag{3.1.27}$$

From (3.1.27) and Lorentz's condition (3.1.16), we get for the scalar potential

$$\Phi = - \frac{\nabla\Pi}{\varepsilon_0} \quad. \tag{3.1.28}$$

From (3.1.9,12) it is now straightforward to obtain **E** and **B** (or **H**). Taking into account that Π is along the $\hat{z}$ direction, the result is

$$E_\varphi = H_r = H_\theta \equiv 0 \quad, \tag{3.1.29}$$

$$H_\varphi = - \frac{\sin\theta}{4\pi R} \left(\frac{\partial\dot{p}}{\partial R} - \frac{\dot{p}}{R} \right) \quad, \tag{3.1.30}$$

$$E_r = \frac{\cos\theta}{4\pi\epsilon_0 R}\left(\frac{\partial^2 p}{\partial R^2} - \frac{2}{R}\frac{\partial p}{\partial R} + \frac{2p}{R^2} - \frac{\ddot{p}}{c^2}\right) \; , \tag{3.1.31}$$

and

$$E_\theta = -\frac{\sin\theta}{4\pi\epsilon_0 R}\left(\frac{1}{R}\frac{\partial p}{\partial R} - \frac{p}{R^2} - \frac{\ddot{p}}{c^2}\right) \; , \tag{3.1.32}$$

where p is the magnitude of the dipole moment.

The above expressions can be simplified further if we observe that

$$\frac{\partial p}{\partial R} = \frac{\partial p}{\partial t'}\frac{dt'}{dR} = -\frac{1}{c}\frac{\partial p}{\partial t'} = -\frac{1}{c}\frac{\partial p}{\partial t} = -\frac{\dot{p}}{c} \; ,$$

and

$$\frac{\partial^2 p}{\partial R^2} = -\frac{1}{c}\frac{\partial^2 p}{\partial t'^2}\frac{dt'}{dR} = \frac{1}{c^2}\frac{\partial^2 p}{\partial t'^2} = \frac{1}{c^2}\frac{\partial^2 p}{\partial t^2} = \frac{\ddot{p}}{c^2} \; ;$$

so we finally obtain

$$H_\varphi = \frac{\sin\theta}{4\pi R}\left(\frac{\ddot{p}}{c} + \frac{\dot{p}}{R}\right) \; , \tag{3.1.33}$$

$$E_r = \frac{\cos\theta}{2\pi\epsilon_0 R}\left(\frac{\dot{p}}{Rc} + \frac{p}{R^2}\right) \; , \tag{3.1.34}$$

$$E_\theta = \frac{\sin\theta}{4\pi\epsilon_0 R}\left(\frac{\dot{p}}{Rc} + \frac{p}{R^2} + \frac{\ddot{p}}{c^2}\right) \; . \tag{3.1.35}$$

The above expressions are exact provided $l_{max} \ll R$, i.e., they are valid for the elemental dipole, far from the source.

The Poynting vector expressing the energy flowing per second, per unit area (W/m^2) is

$$S = E \times H = \hat{\theta}E_r H_\varphi + \hat{r}E_\theta H_\varphi = S_\theta + S_r \; , \tag{3.1.36}$$

so in general it has a radial component and a component which, if different from zero, would give rise to *radiation of angular momentum* as well.

Let us see what is the case for the elementary dipole. To fix our ideas, let us choose a harmonic oscillator, namely $p = p_0 \cos\omega t$. Substitution into (3.1.33-36) gives the instantaneous terms of the Poynting vector. In order now to discriminate between terms corresponding to dynamical energy stored in the medium (which fluctuates from the source to the observer and back with frequency ω) and terms really corresponding to irreversible one-way travel of energy *from the source* to *the observer* (radiation), we integrate both sides of (3.1.36) with respect to time for an integral number of periods. Obviously the terms responsible for radiation will be the ones which survive the averaging, i.e., give a nonzero result. It turns out that this involves only *one* term, namely the part of

$$E_\theta \sim \frac{\sin\theta}{4\pi\epsilon_0 R}\frac{\ddot{p}}{c^2}$$

and the part of

$$H_\varphi \sim \frac{\sin\theta}{4\pi R} \frac{\ddot{p}}{c} \quad,$$

so, in the case of the dipole, no angular momentum is radiated. (This is not the case, however, for *multipolar* sources: hence the principle of working of any electric motor!)

The density of radiated power in W/m^2 is therefore

$$S_r = \frac{\ddot{p}^2 \sin^2\theta}{16\pi^2 \epsilon_0 R^2 c^3} = \frac{e^2 \dot{v}^2 \sin^2\theta}{16\pi^2 \epsilon_0 R^2 c^3} \quad, \tag{3.1.37}$$

and the total radiated power is calculated by integrating S_r over a sphere of radius R, taking θ from 0 to π (Fig.3.3). We get

$$S = \int_0^\pi \frac{e^2 \dot{v}^2 \sin^2\theta}{16\pi^2 \epsilon_0 R^2 c^3} 2\pi R^2 \sin\theta \, d\theta \quad, \tag{3.1.38a}$$

and since

$$\int_0^\pi \sin^3\theta \, d\theta = \frac{4}{3} \quad,$$

$$S = \frac{e^2 \dot{v}^2}{6\pi \epsilon_0 c^3} \text{ watts} \quad. \tag{3.1.38b}$$

This is the energy radiated (in joules per second) by an accelerated electron in vacuo.

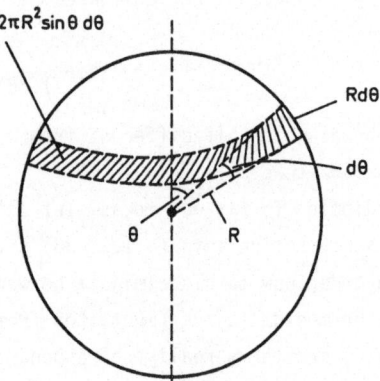

Fig.3.3. Integrating over the surface of a sphere of radius R

3.1.2 The Concept of Self-Force

Let us now examine carefully the forces under which this electron moves. (The concept of a dipole can now be dropped; the positive ion +e is considered as "nailed" at the origin.) The energy loss due to radiation must be compensated by a loss of kinetic energy of the moving electron; hence it decelerates as if an external force F_s were acting against it, thereby causing a damping.

We can calculate this force as follows: the work performed by it within a time interval $\Delta t = t_2 - t_1$ equals in absolute value the energy radiated during the same interval. We get

$$\int_{t_1}^{t_2} F_s \cdot v \, dt = - \int_{t_1}^{t_2} \frac{e^2}{6\pi\epsilon_0 c^3} \dot{v}^2 \, dt \qquad (3.1.39a)$$

or, taking into account the relation

$$\dot{v}^2 = \frac{d}{dt}(v \cdot \dot{v}) - v \cdot \ddot{v} \quad ,$$

we can substitute in the right-hand side and get

$$\int_{t_1}^{t_2} F_s \cdot v \, dt = \frac{e^2}{6\pi\epsilon_0 c^3} \int_{t_1}^{t_2} \ddot{v} \cdot v \, dt - \frac{e^2 (\dot{v} \cdot v)}{6\pi\epsilon_0 c^3} \Big|_{t_1}^{t_2} \quad . \qquad (3.1.39b)$$

Without loss of generality we can choose the times t_1, t_2 so that the (kinetic) energy of the electron is the same at t_1 and t_2. Accordingly, for t_1 and t_2 we will have $\dot{v} \perp v$, i.e., the acting force can be considered to be perpendicular to the instantaneous velocity, so we get

$$\int_{t_1}^{t_2} \left(F_s - \frac{e^2}{6\pi\epsilon_0 c^3} \ddot{v} \right) \cdot v \, dt = 0 \quad . \qquad (3.1.40)$$

In order that (3.1.40) holds for any dt, we must have

$$F_s = \frac{e^2}{6\pi\epsilon_0 c^3} \ddot{v} \qquad (3.1.41)$$

(for $v/c \ll 1$) or

$$F_s = m\tau \dddot{x} \quad , \qquad (3.1.42)$$

where $\tau = e^2/6\pi\epsilon_0 mc^2 \sim 6.3 \times 10^{-24}$ s is a characteristic time constant (see below).

The mass of the electron is m, and $x(t)$ is the instantaneous position of the particle, so the motion of the electron will be given by

$$\ddot{x} + \omega_0^2 x = \tau \dddot{x} \quad , \qquad (3.1.43)$$

if the electron is oscillating around a center with eigenfrequency ω_0, or by

$$\ddot{x} = \tau \dddot{x} \qquad (3.1.44)$$

for an unbound electron. Before dealing with the intricacies of the "self-force" F_s, let us try to find for (3.1.43) solutions of the form $x \sim e^{j\Omega t}$. Substituting in (3.1.43) we get

$$\Omega^2 = \omega_0^2 + j\tau\Omega^3 \sim \omega_0^2 + j\tau\omega_0^3 \quad , \tag{3.1.45}$$

because $\Omega \sim \omega_0$, since the coefficient τ is indeed very small relative to the eigen-period $T = 2\pi/\omega_0$, or because $\omega_0\tau \ll 1$. For most of the electromagnetic spectrum we can say that the condition $\omega_0\tau \ll 1$ is valid. It is violated for cosmic and γ-rays, which cannot be treated by Maxwell's equations. We therefore obtain

$$\Omega \cong \pm \left(\omega_0 + \frac{1}{2}j\tau\omega_0^2\right) \tag{3.1.46}$$

and

$$x(t) \sim e^{-\tau\omega_0^2 t/2}\, e^{j\omega_0 t} \quad \text{for} \quad t \geq 0 \quad ,$$
and
$$x(t) = 0 \qquad \qquad \text{for} \quad t < 0 \quad . \tag{3.1.47}$$

The Fourier spectrum of $x(t)$ is

$$x(\omega) \cong \frac{1}{2}\int_0^\infty e^{-\tau\omega_0^2 t/2}\, e^{j(\omega_0-\omega)t}\, dt$$

$$= \frac{1}{2\pi}\,\frac{1}{j(\omega_0 - \omega) - (\tau\omega_0^2/2)} \quad , \tag{3.1.48}$$

and the power spectrum gives

$$|x(\omega)|^2 \sim \frac{1}{4\pi^2}\,\frac{1}{(\omega - \omega_0)^2 + (\tau^2\omega_0^4/4)} \quad . \tag{3.1.49}$$

This spectrum is plotted in Fig.3.4. The energy radiated by the oscillating electron is *never* monochromatic but, due to the self-force, possesses a natural line width $\Delta\omega$ at half maximum intensity.

The ratio $\Delta\omega/\omega_0^2 \sim \tau \sim 10^{-24}$ s. The physical meaning of τ is the time an EM wave takes to pass through a distance equal to the electron radius. (This result, of

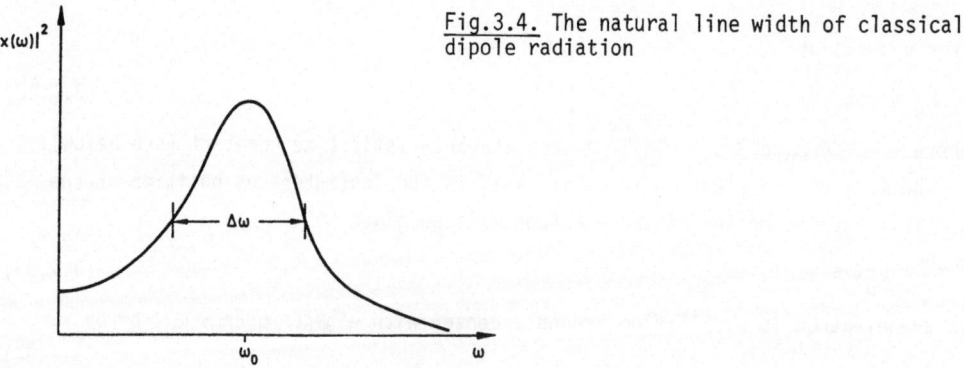

Fig.3.4. The natural line width of classical dipole radiation

course, holds for the classical electron. The quantum mechanical treatment slightly modifies the shape of the above spectrum making it asymmetrical, due to the electron spin.)

Let us now investigate more carefully the nature of the self-force (3.1.41). We have to admit on close inspection that this "force" is indeed a sui generis, first of all because it is simultaneous with the acceleration of the particle, which presumably is damped by way of reaction. Unlike other forces in physics, it depends on the *second* derivative of the velocity, i.e., on the rate of change of the acceleration. In a "normal" (Newtonian) mechanical problem which involves a second-order (in time) differential equation, it is sufficient to specify the position and the velocity of a particle at one instant in order to arrive at a unique solution. In the present case however, the differential equation involved (3.1.43) is *third* order in time, so that to determine uniquely the trajectory of the radiating particle, it is necessary to specify also the *initial acceleration*.

All right then, let us consider the free-electron equation

$$\ddot{x} = \tau \dddot{x}$$

or

$$\dot{v} = \tau \ddot{v} \quad , \tag{3.1.50}$$

and try to solve it by specifying an *arbitrary* initial acceleration $\dot{v}_0 = \dot{v}(0)$. Equation (3.1.50) admits *two* solutions, namely $\dot{v} = 0$ and $\dot{v} = \dot{v}(0) \, e^{t/\tau}$. The second solution, known as the "runaway" solution, is strange, i.e., devoid of any physical meaning, because it represents an unlimited "self"-acceleration, which if allowed would render all radiating particles unstable. What can we do to avoid runaway solutions? Let us write first the general solution of (3.1.43) as

$$\dot{v} = e^{t/\tau} \left[\dot{v}(0) - \frac{1}{\tau} \int_0^t e^{-t'/\tau} \, f_{ext}(t') dt' \right] \quad , \tag{3.1.51}$$

where f_{ext} takes into account not only the restoring force $w_0^2 x$, but also any other external force acting on the particle.

The above solution has been constructed by adding to the general solution of the homogeneous equation (3.1.50) a particular integral of the inhomogeneous equation (3.1.43). Since an *arbitrary* initial acceleration $\dot{v}(0)$ results in runaway solutions, Dirac proposed many years ago that such solutions could be avoided by fixing a *special* value for the initial acceleration rather than leaving it arbitrary. By choosing

$$\dot{v}(0) = \frac{1}{\tau} \int_0^\infty e^{-t'/\tau} \, f_{ext}(t') dt' \quad , \tag{3.1.52}$$

v remains finite as $t \to \infty$. With (3.1.52), the general solution of (3.1.43,51), becomes

$$\dot{v} = \int_0^\infty e^{-\rho} \, f_{ext}(t + \tau\rho)d\rho \quad . \tag{3.1.53}$$

The particle then obeys an integrodifferential equation of motion; its behavior at any time t depends consequently upon the external force acting at times *later* than t.

For a sharp impulse given to the particle, e.g., $f_{ext}(t) \sim A\delta(t)$,

$$\dot{v}(t) = A \, e^{t/\tau} \qquad t < 0$$

and $\hspace{9cm}$ (3.1.54)

$$\dot{v}(t) = 0 \qquad\quad t > 0 \quad .$$

Thus the particle accelerates *before* the force is applied, in anticipation, so to speak, or against what we call the notion of causality.

In conclusion, the runaway solutions seem to be eliminated only at the expense of introducing acausal behavior. The fact that the smallness of the parameter makes the time scale of such a behavior of more or less "academic" interest does not change the serious challenge that such behavior inflicts upon our most cherished notion of physics.

3.1.3 Thermodynamics of Electromagnetic Radiation

An ingenious proposal made by *Wheeler* and *Feynman* in 1945 [3.1] seems to solve the (classical) problem by relating *electrodynamics* to *statistical mechanics*. In broad outline this theory is as follows: Maxwell's equations, like all other interactions in physics (save the weak nuclear interaction), admit symmetrical solutions, i.e., time-advanced and time-delayed ones. And yet observable radiation proceeds *asymmetrically* in time, that is, it transfers energy irreversibly from the radiating source to the observer (where the radiation, carried by coherent spherical waves, is presumably dissipated in heat at the expense of producing information).

Let us adopt the mentality of someone who has never observed radiation phenomena, and ask whether the calculated irreversibility displayed by the Poynting vector is an inevitable (and trivial) outcome of our (prejudiced) omission of the time-advanced components in the solution of Maxwell's equations, or whether it would formally appear even if we had taken into account in our calculations the advanced solutions? If it turns out that it would appear anyway, then we will have to attribute the irreversibility of Poynting vector not to the (fabricated) asymmetry of the elemental dynamical laws but rather to an *extrinsic* asymmetry displayed by the boundary conditions (as in the case of statistical mechanics).

Indeed, a finite closed system of interacting particles has much in common with a finite closed system of *interfering* waves. The wave interference is analogous to the collisions of gas molecules. The progression from a simple symmetrical wave pattern (emanating, say, from the center of a lake upon throwing a stone) to a complex distribution of wavelets reflected incoherently from the irregular boundary of the pond reminds one of the passage of a gas from a highly differentiated distribution (low entropy) to full homogenization (maximum entropy), as a result of collisions among the molecules and their reflections from the walls of the cavity; reflections which quickly destroy any initial cross-correlations or coherence of the individual velocities. To carry the analysis further, use is made of a hypothesis made by the German physicist Tetrode back in 1924, and adopted by *Feynman* and *Wheeler* [3.1]: In order to explain the origin of self-force —which is simultaneous with the moment of acceleration —one has to abandon the concept of electromagnetic radiation as an elementary process. Instead one must try to derive it as a *collective property* of a great number of particle-particle interactions between the source under consideration and the constituent particles of an "absorber", i.e., a closed box of finite dimensions enclosing the source. Specifically the proposal postulates the following:

1) An accelerating point charge in otherwise charge-free space does *not* radiate electromagnetic energy.
2) The fields which act on a given particle arise only from other particles.
3) These fields are represented by one-half the retard plus one-half the advanced solutions of Maxwell's equations.

We will assume then that the radiating source (particle) is enclosed in a perfect absorber, i.e., a box with absolutely opaque walls, walls nevertheless made out of particles similar to the radiating one. At a given moment t, a wave leaves the radiating source-particle. After some time it hits the particles of the absorber, whereupon each one of them goes into (nonrelativistic) motion and reacts by transmitting a wave, half-advanced and half-retarded. Finally, to make calculations more tractable, we will assume that the particles of the absorber are far from one another, so that each "sees" any other as an "elemental" source. We begin by considering the reaction set up between the source +e and a typical charge in the absorber when the particle of the source possesses an acceleration $\dot{v}$ due to collision with a third particle, or otherwise.

Let R_K be the distance between the source and the particle of the absorber. The generalized disturbance by the source traverses this distance and provokes a reaction R_K/c seconds later. The electric field acting on the particle of the absorber will be

$$\sim - \frac{e\dot{v}}{R_K c^2} \sin\theta_K \quad , \tag{3.1.55}$$

where θ_K is the angle between $\dot{v}$ and R_K.

Let e_K and m_K be the charge and the mass of the particle of the absorber. It will acquire in the field (3.1.55) an acceleration $\dot{v}_K$ equal to

$$\sim \frac{e_K}{m_K}\left(-\frac{e\dot{v}}{R_K c^2}\sin\theta_K\right) \sim -\frac{ee_K\dot{v}}{R_K m_K c^2}\sin\theta_K \quad . \tag{3.1.56}$$

Its subsequent motion will generate a field which will be *half-advanced* and *half-retarded*. The advanced part of this field will exert on the source a force *simultaneous* with the original acceleration. The far-field component of this reactive force along the direction of acceleration will be

$$-e\left(\frac{e_K\dot{v}_K}{2R_K c^2}\right)\sin\theta_K \tag{3.1.57}$$

or, substituting $\dot{v}_K$ from (3.1.56),

$$\frac{\dot{v}e^2}{2c^4}\left(\frac{e_K^2}{m_K R_K^2}\right)\sin^2\theta_K \quad . \tag{3.1.58}$$

From (3.1.58) giving the reactive force on the source due to one particle of the absorber, we can now evaluate the total effect due to many particles. If N is the number of the particles of the absorber per unit volume, the number of particles on a spherical shell (surrounding the source) of thickness dr_K will be $4\pi N r_K^2 dr_K$. We now have to integrate (3.1.58) over the volume of a sphere of radius R_K and elemental volume $2\pi R_K^2 \sin\theta_K \, d\theta_K dR_K$. The result is

$$\int_0^{R_K}\int_0^{\pi} \frac{\dot{v}e^2}{2c^4}\frac{e_K^2}{m_K R_K^2}\sin^2\theta_K \ N2\pi R_K^2 \sin\theta_K \ dR_K d\theta_K$$

$$= \left(\frac{2\dot{v}e^2}{3c^3}\right)\left(\frac{2\pi Ne_K^2}{m_K c}\right)R_K \quad , \tag{3.1.59}$$

which as it stands does *not* agree with the self-force expression derived in (3.1.41)!

However, what we failed to appreciate in the course of the above calculation are the *phase relationships* among the individual elementary advanced reactions on the source. There is indeed a phase lag between the outgoing disturbance from the source and the returning reactions —which we have not taken into account.

The advanced force acting on the source due to the motions of a typical particle of the absorber is an *elementary* interaction between *two* charges, and propagates with the velocity of light in vacuum. On the other hand, the disturbance which travels outward from the source and determines the motion of the particle in question is made up not only of the proper field of the originally accelerated charge but also of the secondary fields generated in the material of the absorber. The *elementary* interactions propagate, of course, with the velocity of light; but the combined, collective disturbance travels at a phase velocity c/n, where n is the refractive index —a collective, macroscopic property of the medium. (For the derivation of n see Sect.3.2 where the problem of propagation in dispersive and lossy media is taken up.)

In order to speak of the change in phase velocity of the disturbance, it is necessary to consider a single Fourier component of the acceleration. It is legitimate to decompose the acceleration into spectral components and, having solved the problem for one of them, to recompose the corresponding Fourier components of the radiative reaction (the connection between acceleration and reaction being linear).

Therefore, let the primary (that is the source's) acceleration vary as $\dot{v} = v_0 e^{-j\omega t}$. A disturbance of this frequency will experience in a medium of low density a refractive index given by

$$n^2 \cong 1 - \frac{4\pi Ne_K^2}{m_K \omega^2} .$$ (3.1.60)

Thus, the phase of the radiative reaction which reaches the source from a distance R_K in the absorber will *lag* behind that of the acceleration by

$$\omega\left(\frac{R_K}{c} - \frac{nR_K}{c}\right) = \frac{2\pi Ne_K^2}{m_K c\omega} R_K = \Phi_K .$$ (3.1.61)

We apply this particle's phase correction to the contribution of the absorber (3.1.59) in the range R_K to $R_K + dR_K$, and sum up over all depths in the medium to obtain the total reactive force

$$\frac{2e^2}{3c^3} \dot{v} \int_0^\infty \left(\frac{2\pi Ne_K^2}{m_K c}\right) \exp[-jR_K(2\pi Ne_K^2/m_K c\omega)]dR_K .$$ (3.1.62)

This integral will converge at the upper limit if we allow for the existence of a small but finite absorption in the medium. The result gives the total reaction on the source of the advanced fields from all the particles of the absorber as

$$\sim \frac{2e^2}{3c^3}(-j\omega\dot{v}) = \frac{2e^2}{3c^3}\ddot{v} ,$$ (3.1.63)

which is exactly the result giving the self-force, in Gaussian units. We conclude that the self-force arises not from the direct action of the source upon itself but from a *collective property* of the advanced action upon the source caused by the "future" motion of the particles of the absorber.

We have so far summed up the elementary advanced fields of the particles of the absorbing medium at the source. It remains now to perform a similar calculation at points *outside* the source. The process is too involved so we give directly the result obtained in the Feynman-Wheeler theory. It turns out that the sum of the advanced fields from all the absorber's particles at any point outside the source gives a field which exactly equals (1/2) (retarded field of the source) -(1/2) (advanced field of the source), so the advanced field of the source is *completely* compensated by the advanced field of the absorber (all advanced fields are canceled by destructive interference) and the retarded fields add up to give at any point in space the fully *retarded macroscopic wave* —in accordance with observations.

In conclusion, we see that the effects of the advanced fields show up only in the force of radiative reaction. The electromagnetic arrow of time appears therefore to have a similar origin to the thermodynamic arrow of time, namely the *extrinsic* asymmetry of the boundary conditions. The obvious question: Is the physical *expanding universe* a perfect absorber? This cannot be answered yet. In the absence of a definite answer, one is tempted to think that waves from the "future" should not perhaps be excluded a priori!

3.2 Electromagnetic Wave Propagation in Dispersive Media and Lossy Media

We write down the equation of motion of a charged particle subjected to an incident EM wave, together with Maxwell's equations:

$$\ddot{x} + \nu\dot{x} + \omega_0^2 x = \frac{e}{m} (E + v \times B) \quad , \tag{3.2.1}$$

$$\mu_0 \dot{H} = - \nabla \times E \quad , \tag{3.2.2}$$

$$\varepsilon_0 \dot{E} + \dot{P} = \nabla \times H \quad , \tag{3.2.3}$$

where $u = \dot{x}$ the velocity of the particle, moving under the action of the Lorentz force [right-hand side of (3.2.1), $(B = \mu_0 H)$] and subjected also to a restoring force $\omega_0^2 x$ (if the medium is not ionized) and a dissipative force $\nu\dot{x}$ due to collisions the particle undergoes with its neighboring particles. The polarization vector is $P = Nex$ (its value equals the number of induced dipole moments per unit volume), and N is the number of particles per unit volume of the dispersive medium. The medium is considered homogeneous and isotropic. We can omit the nonlinear term $v \times B$ in the Lorentz force if we appreciate that the intensities of the electromagnetic field involved concern the far field so that

$$H_\varphi = \sqrt{\frac{\varepsilon_0}{\mu_0}} \, \hat{r} \times E_\theta \quad ,$$

$$B_\varphi = \frac{1}{c} E_\theta$$

and

$$|v \times B_\varphi| \leqslant |\frac{v}{c} E_\theta| \quad .$$

Obviously the term vE_θ/c can be neglected compared to the term E_θ, when $v/c \ll 1$, which is the case here. We rewrite the linear equation (3.2.1) in terms of P, namely

$$\left(\frac{\partial^2}{\partial t^2} + \nu \frac{\partial}{\partial t} + \omega_0^2\right)P = \frac{Ne^2}{m} E \quad . \tag{3.2.4}$$

Multiplying both sides of (3.2.2) by $\nabla\times$ and both sides of (3.2.3) by $\mu_0\partial/\partial t$ and adding the results together, we eliminate H, that is, we obtain

$$\varepsilon_0\mu_0\ddot{E} + \mu_0\ddot{P} = -\nabla\times\nabla\times E = \nabla^2 E \qquad (3.2.5)$$

(since $\nabla\cdot E = 0$ in the absence of sources in the medium).

In order next to get an equation in terms of only E, we apply the operator $-\mu_0\partial^2/\partial t^2$ to both sides of (3.2.4) and the operator $(\partial^2/\partial t^2 + \nu\partial/\partial t + \omega_0^2)$ to both sides of (3.2.5). Adding these together we obtain

$$\left(\frac{\partial^2}{\partial t^2} + \nu\frac{\partial}{\partial t} + \omega_0^2\right)\left(\frac{1}{c^2}\ddot{E} - \nabla^2 E\right) + \frac{\mu_0 N e^2}{m}\ddot{E} = 0 \quad . \qquad (3.2.6)$$

We now ask for harmonic solutions $E \sim A\,\exp[j(Kx - \omega t)]$, i.e., propagating plane harmonic waves along say the x direction, where K is the wave number.

Substituting into (3.2.6) we finally get a relationship between ω and K, known as the dispersion relation,

$$(-\omega^2 + \omega_0^2 - j\nu\omega)\left(-\frac{\omega^2}{c^2} + K^2\right) = \frac{\mu_0 N e^2}{m}\omega^2 \quad ,$$

or

$$K^2 = \frac{\omega^2}{c^2}\left(1 + \frac{\mu_0 c^2(Ne^2/m)}{\omega_0^2 - \omega^2 - j\nu\omega}\right) , \qquad (3.2.7)$$

or in terms of the refractive index $n = cK/\omega$,

$$n^2 = 1 + \frac{Ne^2/m\varepsilon_0}{\omega_0^2 - \omega^2 - j\nu\omega} \quad . \qquad (3.2.8)$$

For a fully ionized medium (i.e., a medium of free charges) without losses, $\omega_0 = 0$, $\nu = 0$ and the refractive index becomes

$$n^2 = 1 - \frac{Ne^2/m\varepsilon_0}{\omega^2} = 1 - \frac{\omega_e^2}{\omega^2} \quad , \qquad (3.2.9)$$

where ω_e = plasma frequency, i.e., the frequency of longitudinal waves excited in the ionized fluid. Equation (3.2.9) is the one we used in the Feynman-Wheeler theory (in Gaussian units) in Sect.3.1.3.

In a dispersive medium characterized by the nonlinear function $\omega = f(K)$ (if this function is linear, the medium is just vacuum) the velocity of energy propagation (the group velocity) differs from the velocity of phase propagation. To see this, consider a polychromatic wave (a pulse)

$$\Psi = \int A(K)\,e^{jKx - j\omega t}\,dK \quad , \qquad (3.2.10)$$

where $A(K)$ is the amplitude of the spectrum at the wave number K, propagating in the medium. Let K_0 be the average value of the K distribution in the spectrum of the wave packet. We expand the dispersion relation $\omega = f(K)$ around the reference point ω_0, K_0 so that

$$\omega = \omega_0 + \left(\frac{\partial \omega}{\partial K}\right)_{K_0} \cdot K' + \frac{1}{2} \left(\frac{\partial^2 \omega}{\partial K^2}\right)_{K_0} \cdot K'^2 + \dots + \, , \tag{3.2.11}$$

where $K' = K - K_0$ and $\omega_0 = f(K_0)$. Substituting the above expansion into (3.2.10), we get

$$\Psi = \exp[j(K_0 x - \omega_0 t)]M(x,t) \quad , \tag{3.2.12}$$

where

$$M(x,t) = \int A(K) \exp\left\{jK'\left[x - \left(\frac{\partial \omega}{\partial K}\right)_{K_0} t - \frac{1}{2}\left(\frac{\partial^2 \omega}{\partial K^2}\right)_{K_0} K't - \dots\right]\right\}dK \quad . \tag{3.2.13}$$

Here higher-order derivatives ($\partial^2 \omega / \partial K^2$ and beyond) can be omitted if (as is usually the case) the dispersion relation is a smooth function. We thus obtain

$$M(x,t) \cong \int A(K) \exp\left\{jK'\left[x - \left(\frac{\partial \omega}{\partial K}\right)_{K_0} t\right]\right\}dK \quad , \tag{3.2.14}$$

which means that the amplitude or the energy of the wave packet propagates with a speed equal to

$$v_{gr} = \left(\frac{\partial \omega}{\partial K}\right)_{K_0} \, , \tag{3.2.15}$$

whereas the corresponding phase velocity

$$v_{ph} = \left(\frac{\omega}{K}\right)_{K_0} \quad . \tag{3.2.16}$$

From Fig.3.5 we understand that the v_{gr} (which, unlike v_{ph}, cannot exceed the velocity of light c) may vary widely with respect to v_{ph}.

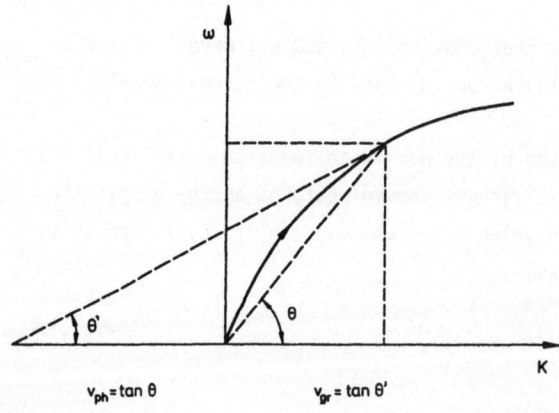

Fig.3.5. The phase velocity (v_{ph}) and the group velocity (v_{gr}) found from the dispersion curve of a slightly dispersive lossless medium

In the interesting (but rare) case where the dispersion relation has a hump followed by a branch of negative slope, the group velocity becomes "negative". What is this supposed to mean? It means that v_{gr} and v_{ph} go in opposite directions. Nevertheless, since the energy, as we have already seen, *always* goes *from the* source *to* the external absorber, a medium with such a dispersion branch implies that the phase fronts, instead of emerging from the source, collapse at the source as time goes on. There is nothing "unphysical" in such behavior since the phase velocity has merely a geometrical character.

The approximate expression (3.2.15) we derived above for the group velocity breaks down not only for non-smooth dispersion relations but also in the case where the medium is lossy, i.e., possesses a finite conductivity in the form of free charges. (A conducting medium has to be differentiated from a lossy dielectric: in the case of a dielectric, the losses are due to collisions between bounded atoms and not to collisions between bounded atoms and free charges.) We will not be interested in the form the group velocity acquires in conducting media. Our attention will rather be drawn towards some new (and unsuspected) characteristics displayed by a wave propagating in a lossy (conducting) medium; these characteristics will be the most suitable introduction to what will come next, namely the spectral expansion of a spherical wave.

Let us investigate, then, the character of the propagation process in a conduction medium, where the intensity of the propagating electric field E gives rise to a current density J [A/m^2], $J = \sigma E$, where σ is the conductivity of the medium.

We start by writing the full set of Maxwell's equations:

$$\nabla \times E = -\frac{\partial B}{\partial t} \quad , \tag{3.2.17}$$

$$\nabla \times H = J + \frac{\partial D}{\partial t} \quad , \tag{3.2.18}$$

$$D = \varepsilon E \quad , \tag{3.2.19}$$

$$B = \mu H \quad , \tag{3.2.20}$$

$$\nabla \cdot D = \rho \quad , \tag{3.2.21}$$

$$\nabla \cdot B = 0 \quad . \tag{3.2.22}$$

From (3.2.17,18), and using (3.2.19,20),

$$\nabla \times \nabla \times E = \mu \frac{\partial}{\partial t}\left(J + \varepsilon \frac{\partial E}{\partial t}\right) \tag{3.2.23}$$

or, taking into account that $\nabla \times \nabla \times E = \nabla(\nabla \cdot E) - \nabla^2 E$, that $\nabla \cdot E = \rho/\varepsilon$, and also that $\varepsilon\mu = (1/v_{ph})^2$,

$$\nabla^2 E - \frac{1}{v_{ph}^2}\frac{\partial^2 E}{\partial t^2} = \mu \frac{\partial J}{\partial t} + \nabla\left(\frac{\rho}{\varepsilon}\right) \quad . \tag{3.2.24}$$

In a medium free of sources ($\rho = 0$) but conducting ($\mathbf{J} = \sigma\mathbf{E}$), the above equation becomes

$$\nabla^2 \mathbf{E} - \frac{1}{v_{ph}^2} \frac{\partial^2 \mathbf{E}}{\partial t^2} = \mu \frac{\partial \mathbf{J}}{\partial t} \quad , \tag{3.2.25}$$

and for harmonic, linear excitation $\mathbf{E} \sim e^{-j\omega t}$; consequently $\mathbf{J} \sim e^{-j\omega t}$. Therefore (3.2.25) becomes

$$\nabla^2 \mathbf{E} + K_0^2 \mathbf{E} = -j\mu\omega\mathbf{J} \tag{3.2.26}$$

or, since $\mathbf{J} = \sigma\mathbf{E}$ and $K_0 = \omega/v_{ph} = \omega\sqrt{\varepsilon\mu}$,

$$\nabla^2 \mathbf{E} + \left(1 + j\frac{\sigma}{\varepsilon\omega}\right) K_0^2 \mathbf{E} = 0 \quad , \tag{3.2.27}$$

or

$$\nabla^2 \mathbf{E} + K^2 \mathbf{E} = 0 \quad , \tag{3.2.28}$$

where

$$K = K_0 \sqrt{1 + j\frac{\sigma}{\varepsilon\omega}} = K_0 \sqrt{1 + js} \quad , \tag{3.2.29}$$

where $s = \sigma/\varepsilon\omega$. We see therefore that a conductive medium is also *by necessity* dispersive.

The solution of (3.2.27 or 28) is therefore of the form

$$E(x,y,z,t) = E(\mathbf{r},t) = E_0 \exp[j(\mathbf{K} \cdot \mathbf{r} - \omega t)]$$

$$= E_0 \exp[j(K_x x + K_y y + K_z z - \omega t)] \quad , \tag{3.2.30}$$

where $K^2 = K_x^2 + K_y^2 + K_z^2$.
We put $K_x = K_x' + jK_x''$, $K_y = K_y' + jK_y''$, $K_z = K_z' + jK_z''$, and

$$K = K' + jK'' = K_0 \sqrt{1 + js} \quad . \tag{3.2.31}$$

From this last relation we calculate the values of K', K'' by solving the system

$$K'^2 - K''^2 = K_0^2 \quad , \tag{3.2.32}$$

$$K'K'' = \frac{K_0^2 s}{2} \quad ,$$

or

$$K' = \frac{K_0}{\sqrt{2}} (1 + \sqrt{1 + s^2})^{\frac{1}{2}} \tag{3.2.33}$$

and

$$K'' = \frac{K_0}{\sqrt{2}} (-1 + \sqrt{1 + s^2})^{\frac{1}{2}} \quad . \tag{3.2.34}$$

Now if anyone is about to identify K' and K'' as the wave number and the absorption coefficient of the wave in the medium, we warn that this conjecture is wrong! The reason why is as follows.

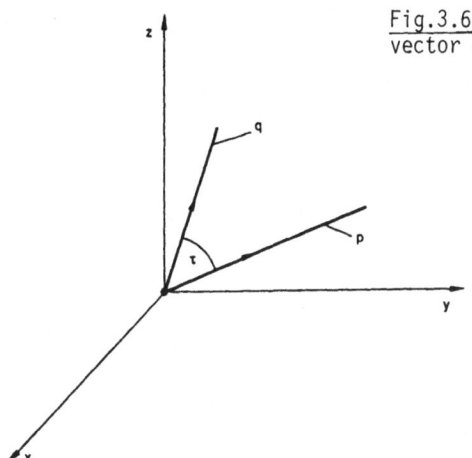

Fig.3.6. The phase vector (**q**) and the attenuation vector (**p**) in a dissipative (conducting) medium

Let us define two vectors $\mathbf{q}(K'_x, K'_y, K'_z)$ and $\mathbf{p}(K''_x, K''_y, K''_z)$, whose coordinates are the real and imaginary parts, respectively, of K_x, K_y, K_z. If α', β', γ' stand for the direction cosines of **q**, and α'', β'', γ'' for the direction cosines of **p** (Fig.3.6), we have

$$K'_x = q\alpha' \qquad\qquad K''_x = p\alpha''$$

$$K'_y = q\beta' \quad \text{and} \quad K''_y = p\beta''$$

$$K'_z = q\gamma' \qquad\qquad K''_z = p\gamma'' \quad .$$

We then have

$$K^2 = (K' + jK'')^2 = K'^2 - K''^2 + 2jK'K''$$

$$= K'^2_x - K''^2_x + K'^2_y - K''^2_y + K'^2_z - K''^2_z + 2j(K'_x K''_x + K'_y K''_y + K'_z K''_z)$$

$$= q^2 - p^2 + 2jpq(\alpha'\alpha'' + \beta'\beta'' + \gamma'\gamma'')$$

$$= q^2 - p^2 + 2jpq \cos\tau \quad , \tag{3.2.35}$$

where τ is the angle between the vectors **p** and **q** ; from these relationships we deduce

$$q^2 - p^2 = K'^2 - K''^2$$

$$\tag{3.2.36}$$

$$pq \cos\tau = K'K'' \quad .$$

The solution of (3.2.36) gives for $\tau \neq \pi/2$

$$q = \frac{K_0}{\sqrt{2}} \left[1 + \sqrt{1 + \frac{s^2}{\cos^2\tau}} \right]^{\frac{1}{2}}$$

and $\tag{3.2.37}$

$$p = \frac{K_0}{\sqrt{2}} \left[-1 + \sqrt{1 + \frac{s^2}{\cos^2\tau}} \right]^{\frac{1}{2}}$$

Here q coincides with K' and p coincides with K" *only* when $\tau = 0$. The solution of (3.2.28) now reads

$$E(x,y,z,t) = E_0 \exp[j(K_x x + K_y y + K_z z) - j\omega t]$$
$$= E_0 \exp[jq(\alpha'x + \beta'y + \gamma'z)]\exp[-p(\alpha''x + \beta''y + \gamma''z)]\exp[(-j\omega t)] \quad ,$$

or

$$E(r,t) = \exp[j(q \cdot r - \omega t)]\exp[-p \cdot r] \quad . \tag{3.2.38}$$

The meaning of $\mathbf{q}$ and $\mathbf{p}$ is now clear: $|\mathbf{q}|$ is the wave number and $|\mathbf{p}|$ is the absorption coefficient of the wave in the conducting medium. The important thing, however, is that the vector $\mathbf{q}$, along which we witness the fastest phase change, does *not* coincide with the direction of the vector $\mathbf{p}$, along which we witness the fastest rate of amplitude drop. What is the angle τ? We need another equation so that from (3.2.37) and this equation the values q, p, τ are unambiguously determined. The third equation will come from the boundary conditions, namely the way the single-ray plane wave enters from vacuum the conducting medium and *splits into two rays* $\mathbf{q}$ and $\mathbf{p}$ (Fig.3.7).

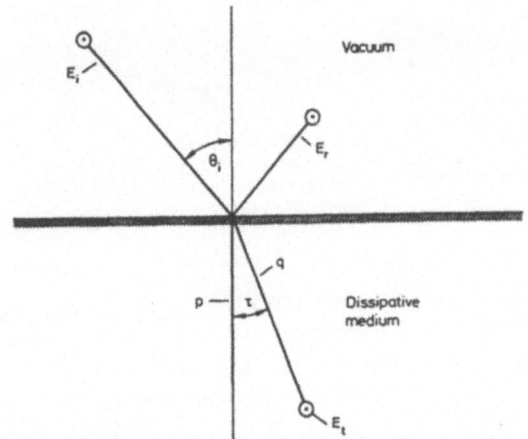

Fig.3.7. Passage of a plane wave from vacuum into a conducting medium

The expressions for the incident, reflected, and transmitted waves are

$$E_i \sim E_{0_i} \exp[jK_0(\alpha_i x + \beta_i y + \gamma_i z)] \quad , \quad z > 0 \quad , \tag{3.2.39}$$

$$E_r \sim E_{0_r} \exp[jK_0(\alpha_r x + \beta_r y + \gamma_r z)] \quad , \quad z > 0 \quad , \tag{3.2.40}$$

and

$$E_t \sim E_{0_t} \exp[jq(\alpha_t'x + \beta_t'y + \gamma_t'z)]\exp[-p(\alpha_t''x + \beta_t''y + \gamma_t''z)] \quad , \quad z < 0 \quad , \tag{3.2.41}$$

where $(\alpha,\beta,\gamma)_{i,r,t}$ are the direction cosines for the incident, reflected, and transmitted waves respectively; the common factor $e^{-j\omega t}$ has been dropped from all the above expressions.

Now the continuity condition for the phase of the incident and transmitted waves at the boundary $z = 0$ for every x and y gives

$$K_0 \alpha_i = q\alpha_t' + jp\alpha_t'' \tag{3.2.42}$$

and

$$K_0 \beta_i = q\beta_t' + jp\beta_t'' \quad , \tag{3.2.43}$$

i.e.,

$$\alpha_t'' = \beta_t'' = 0 \quad , \tag{3.2.44}$$

so that

$$\gamma_t'' = \pm 1$$

or

$$\gamma_t'' = -1 \tag{3.2.45}$$

since for $z \to -\infty$, $E_t \to 0$.

The above phase relations (3.2.42,43) then give

$$K_0 \alpha_i = q\alpha_t'$$

and

$$K_0 \beta_i = q\beta_t'$$

or

$$\frac{\beta_i}{\alpha_i} = \frac{\beta_t'}{\alpha_t'} \quad , \tag{3.2.46}$$

and due to the symmetry of the incident plane wave, we must have $\beta_i = \beta_t' = \beta_r = 0$, so (3.2.39-41) become

$$E_i = E_{0_i} \exp[jK_0(\alpha_i x + \gamma_i z)] \quad , \quad z > 0 \quad , \tag{3.2.47}$$

$$E_r = E_{0_r} \exp[jK_0(\alpha_i x - \gamma_i z)] \quad , \quad z > 0 \quad , \tag{3.2.48}$$

$$E_t = E_{0_t} \exp[jq(\alpha_t' x + \gamma_t' z)]\exp(pz) \quad , \quad z < 0 \quad . \tag{3.2.49}$$

Therefore, **p** will always be directed along the $-z$ axis and

$$\cos\tau = \frac{\mathbf{q} \cdot \mathbf{p}}{qp} = -\frac{q\gamma_t' p}{qp} = -\gamma_t'$$

and, since $\alpha_t'^2 + \gamma_t'^2 = 1$,

$$\gamma_t' = -\sqrt{1 - \alpha_t'^2} \quad ,$$

or since $q\alpha_t' = K_0 \alpha_i = -K_0 \cos\theta$ and $\alpha_t' = -K_0 \sin\theta/q$, it turns out that

$$\cos\tau = \sqrt{1 - \left(\frac{K_0 \sin\theta}{q}\right)^2} \quad , \tag{3.2.50}$$

where $\sin\theta = (q \sin\theta_t)/K_0$, after Snell's law of refraction.

The direction of **q** is then the direction of the refracted ray. Equations (3.2.37, 50) completely solve the problem of the propagation of a wave in a conducting medium.

Now we come to the most important part of our calculations. Let us allow $\sigma \to 0$. Then the system (3.2.36) becomes

$$q^2 - p^2 = K_0^2$$

$$qp \cos\tau = 0 \quad , \tag{3.2.51}$$

that is, we return to a vacuum and, of course, we expect again to find the usual type of wave there, namely the plane wave where $q = K_0$ and $p = 0$. But wait a minute: the solution $q = K_0$, $p = 0$ is *not* the *only* solution of (3.2.51). There is another set, namely $q \neq 0$, $p \neq 0$, $\tau = \pi/2$.

This new (and unsuspected) set of solutions in a vacuum refers to waves which propagate without attenuation in a plane containing the vector **q** and decay exponentially along the half z-axis (the one for which pz < 0).

The entities in this new set are called *evanescent* waves and they need *two* vectors for their determination. The set of plane waves corresponds to K_x, K_y, K_z all real. The set of evanescent waves corresponds to, say, K_x, K_y real and K_z imaginary.

At first glance one would be tempted to dismiss the set of evanescent waves, not considering them as physical realities, but, as we shall see immediately below, these waves are as real as the single-ray plane waves —and indispensable for the analysis of a spherical wave in terms of "ray" components.

3.3 Analysis of a Spherical Wave in Terms of Elemental "Rays". The Mode Theory of Wave Propagation. Excitable Modes (Degrees of Freedom) in a Closed Cavity

3.3.1 Spectral Decomposition of a Spherical Wave

Waves transmitted from all types of electromagnetic sources are spherical, i.e., they have an amplitude which, save for some constants and the harmonic phasors, varies as ~exp(jKR)/R in free space, where R is the distance from the origin, and K is the wave number (Fig.3.8). The characteristic par excellence of a spherical wave is that it diverges at the origin —something expected from the *linear* character of Maxwell's equations.

In a great number of problems having to do with the interaction of spherical waves with *plane* boundaries (finite receiving appertures, reflectors, interfaces, etc.) there is an asymmetry between the shape of the incident wave and the shape of the boundary. Therefore, one wishes to have a spherical wave decomposed into ray components, for each of which the elemental Fresnel formulas of reflection and refraction hold. On geometrical grounds, one is tempted to split the spherical wave into an angular spectrum of plane waves for $0 \leqslant \theta \leqslant \pi$. Such a decomposition, however,

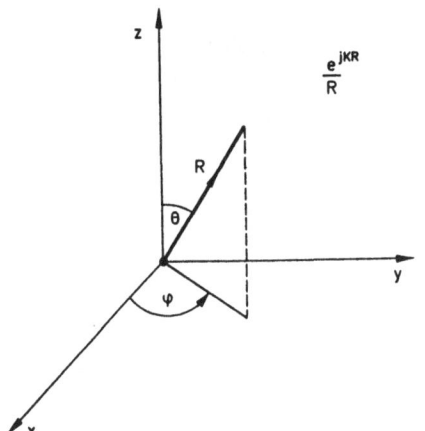

$\frac{e^{jKR}}{R}$

does not work, since any linear combination of plane waves behaves normally at any point in space, including the origin, whereas a spherical wave diverges at the origin. We should therefore abandon intuition in this respect and proceed more formally [3.2].

Our spherical wave has the form $\exp(jKR)/R$, where $R = \sqrt{x^2 + y^2 + z^2}$ (Fig.3.5). On the xy-plane ($z = 0$) the wave has the form $\exp(jKr)/r$ where $r = \sqrt{x^2 + y^2}$. We first analyse on the xy-plane the (circular) wave in terms of a double spatial Fourier spectrum, as follows:

$$\frac{e^{jKr}}{r} = \int\limits_{-\infty}^{\infty} \int\limits_{-\infty}^{\infty} A(K_x, K_y) \exp[j(K_x x + K_y y)] dK_x \, dK_y \quad , \qquad (3.3.1)$$

where $A(K_x, K_y)$ stands for the amplitude of the ray components which propagate on the xy-plane with wave numbers K_x, K_y.

The inverse transform gives

$$A(K_x, K_y) = \frac{1}{(2\pi)^2} \int\limits_{-\infty}^{\infty} \int\limits_{-\infty}^{\infty} \frac{e^{jKr}}{r} \exp[-j(K_x x + K_y y)] dx \, dy \quad . \qquad (3.3.2)$$

We make the transformations:

$$K_x = q \cos\psi \quad K_y = q \sin\psi$$

$$x = r \cos\varphi \quad y = r \sin\varphi \qquad (3.3.3)$$

$(q = \sqrt{K_x^2 + K_y^2}, \, dx \, dy = r \, d\varphi dr)$, whereupon the amplitude $A(K_x, K_y)$ becomes

$$A(K_x, K_y) = \frac{1}{(2\pi)^2} \int\limits_{0}^{2\pi} d\varphi \int\limits_{0}^{\infty} \exp\{jr[K - q \cos(\psi - \varphi)]\} dr$$

$$= \frac{1}{(2\pi)^2} \int\limits_{0}^{2\pi} d\varphi \, \frac{\exp\{jr[K - q \cos(\psi - \varphi)]\}}{j[K - q \cos(\psi - \varphi)]} \Big|_{0}^{\infty} \quad , \quad \text{for} \quad K \text{ real} ; \qquad (3.3.4)$$

this result oscillates without converging.

If, however, we consider even a minute amount of absorption, then the result of the integration with respect to r in (3.3.4) converges in the upper limit and we get the value $j/[K - q \cos(\psi - \varphi)]$; hence,

$$A(K_x, K_y) = \frac{j}{(2\pi)^2} \int_0^{2\pi} \frac{d\varphi}{K - q \cos(\psi - \varphi)} \tag{3.3.5}$$

$$= \frac{j}{(2\pi)^2 K} \int_0^{2\pi} \frac{d\delta}{1 - (q/K)\cos\delta}$$

$$= \frac{j}{2\pi} \frac{1}{\sqrt{K^2 - q^2}} = \frac{j}{2\pi \sqrt{K^2 - K_x^2 - K_y^2}} \quad , \tag{3.3.6}$$

and finally, from (3.3.1),

$$\frac{e^{jKr}}{r} = \frac{j}{2\pi} \int_{-\infty}^{\infty} \int_{-\infty}^{\infty} \frac{\exp[j(K_x x + K_y y)]}{\sqrt{K^2 - K_x^2 - K_y^2}} dK_x \, dK_y \quad . \tag{3.3.7}$$

The above expression can be immediately extrapolated along the ±z-axis, if we keep in mind that (since K is fixed) the independent variables are always *two*, namely K_x, K_y; the only thing which has therefore to change in the above expression is an additive phase shift term $\pm jK_z z$, whereupon we get the result

$$\frac{e^{jKR}}{R} = \frac{j}{2\pi} \int_{-\infty}^{\infty} \int_{-\infty}^{\infty} \frac{\exp[j(K_x x + K_y y + K_z|z|)]}{K_z} dK_x \, dK_y \quad . \tag{3.3.8}$$

The spherical symmetry suggests that we turn to spherical coordinates, namely,

$$K_x = K \sin\theta \cos\varphi \quad ,$$

$$K_y = K \sin\theta \sin\varphi \quad , \tag{3.3.9}$$

$$K_z = K \cos\theta \quad ,$$

and

$$\frac{dK_x \, dK_y}{K_z} = K \sin\theta \, d\theta \, d\varphi \quad .$$

The integration with respect to φ does not present us with any difficulties. Things become interesting, however, as we turn to the integration with respect to θ. Independently gliding along the real axis from $-\infty$ to ∞ are K_x, K_y. The expression in (3.3.8) has to take into account two distinct groups of K_x, K_y. *Group I* contains the K_x, K_y obeying the relation $K_x^2 + K_y^2 \leqslant K^2$; then K_z is real and θ is a real angle changing from zero (K_x, K_y both zero, $K_z = K$) to $\pi/2$ ($K_x^2 + K_y^2 = K^2$, $K_z = 0$), i.e., group I corresponds to propagating waves. *Group II* contains those K_x, K_y for which $K_x^2 + K_y^2 > K^2$; hence K_z is imaginary. Moreover, θ is now a complex angle changing along the negative imaginary axis of the complex plane $\theta = \theta' + j\theta''$ from $\pi/2$ to

$(\pi/2)-j\infty$; i.e., group II corresponds to evanescent waves (see Fig.3.7). The integral in (3.3.8) is thus transformed to

$$\frac{e^{jKR}}{R} = \frac{jK}{2\pi} \int_0^{(\pi/2)-j\infty} \int_0^{2\pi} \exp[j(K_x x + K_y y + K_z|z|)]\sin\theta \, d\theta \, d\varphi \quad . \tag{3.3.10}$$

For the evanescent group of waves we will have, since $\theta = (\pi/2) - j\alpha$, $0 < \alpha < \infty$,

$$K_x = K \cos\varphi \cosh\alpha \quad ,$$

$$K_y = K \sin\varphi \cosh\alpha \quad , \tag{3.3.11}$$

$$K_z = j \sinh\alpha \quad ,$$

as $\alpha \to \infty$, $K_x \to \infty \cos\varphi$, $K_y \to \infty \sin\varphi$, and $K_z \to j\infty$, i.e., we have components which travel along, say, the xy-plane with wavelengths tending to zero and simultaneously decaying along the half z-axis with an attenuation coefficient tending to infinity. The expression (3.3.10) for the spherical wave can be further transformed, if we put

$$x = r \cos\varphi_1 \quad , \tag{3.3.12}$$

$$y = r \sin\varphi_1 \quad ,$$

and perform the integration with respect to φ, to

$$\frac{e^{jKR}}{R} = jK \int_0^{(\pi/2)-j\infty} J_0(u)e^{jK|z| \cos\theta} \sin\theta \, d\theta \quad , \tag{3.3.13}$$

where

$$J_0(u) = \frac{1}{2\pi} \int_0^{2\pi} \exp[jKr \sin\theta \cos(\varphi - \varphi_1)]d\varphi \tag{3.3.14}$$

is the Bessel function of zero order and $u = Kr \sin\theta$. Since

$$J_0(u) = \frac{1}{2} \left[H_0^{(1)}(u) + H_0^{(2)}(u) \right] \quad , \tag{3.3.15}$$

where $H_0^{(1)}(u)$, $H_0^{(2)}(u)$ are the Hankel functions of zero order and first and second kind respectively, i.e.,

$$H_0^{(1)}(u) = \frac{1}{\pi} \int_0^\pi e^{ju \cos\delta} \, d\delta \tag{3.3.16}$$

and

$$H_0^{(2)}(-u) = -H_0^{(1)}(u) \quad , \tag{3.3.17}$$

we can finally write

$$\frac{e^{jKR}}{R} = \frac{jK}{2} \int_{(-\pi/2)+j\infty}^{(\pi/2)-j\infty} H_0^{(1)}(Kr \sin\theta)e^{jK|z|\cos\theta} \sin\theta \, d\theta \quad , \tag{3.3.18}$$

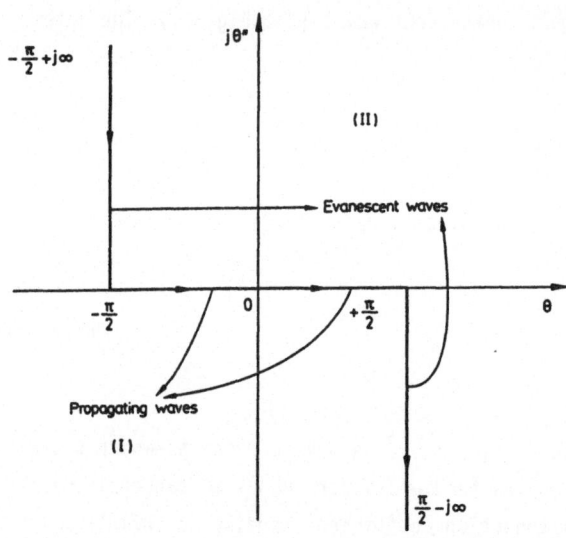

Fig.3.9. The symmetrical integration contour used in the decomposition of a spherical wave
(see text)

where now the contour of integration in the complex θ-plane becomes symmetrical
(Fig.3.9). [Let us note at this point (we will elaborate further at the end of this
chapter and in Chap.4) that if we use a spherical wave as an information carrier,
the object of information will "code" itself, presumably among the ray components
of the carrier. But evanescent waves do not propagate, so the information stored
in them is irretrievably lost. We can therefore appreciate, even at this stage, the
incompleteness of our information carriers and therefore the ensuing ambiguity of
the perceived object.]

We demonstrated above an efficient mathematical way of decomposing a spherical
wave into rays: real and complex. We now intend to consider a spherical wave not
in infinite space but enclosed in a cavity —thereby preparing the way to the treatment of the Planck distribution.

We will, however, proceed in steps. In the first step we consider *one-dimensional
confinement*, namely a spherical wave propagating between two parallel, perfectly
reflecting interfaces with a distance h between them. We will see how this confinement creates a "macroscopic" (classical) "quantization" of the spherical wave, i.e.,
a passage from the continuous spectrum (infinite number of degrees of freedom)
treated thus far to a discrete spectrum (finite number of degrees of freedom) of
"normal modes".

"Quantization", coming from "quantum theory", properly refers to the *discreteness
of the energy levels of each* of these discrete macroscopic normal modes, i.e., to
the numbers of photons that can be accommodated in each normal mode.

These numbers of photons are subject to fluctuations, however. It is exactly the
uncertainty caused by these fluctuations which creates the *entropy* of electromagnetic radiation (and is responsible for the information which is delivered when observation through a *finite* aperture instrument —interfacing the incoming radiation —

takes place). Additional ambiguity comes, of course, from the fluctuations of the propagation medium.

3.3.2 The Wave-Guide Mode Theory of Wave Propagation

Let us consider (Fig.3.10) a spherical wave $\exp(jKR)/R$ transmitted from a centre $(0,z_0)$ within the one-dimensional "wave guide" mentioned above. The problem is to calculate the energy received at another point within the wave guide (r,z).

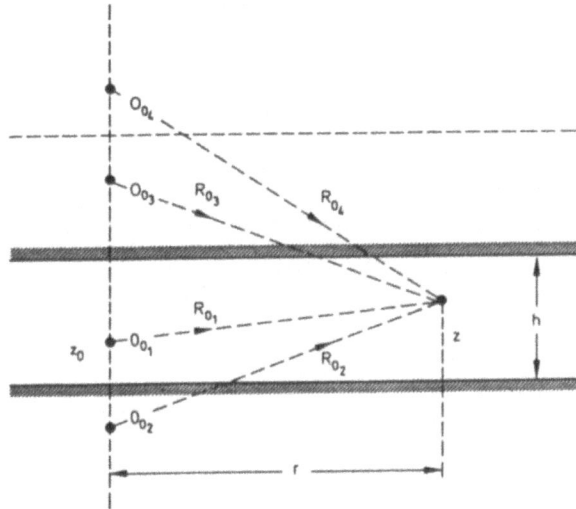

Fig.3.10. Building the received field in the channel out of a superposition of rays reflected at the two boundaries

Formally we have to solve for the harmonic field u the wave equation

$$\nabla^2 u + K^2 u = 0 \quad , \qquad K = \frac{\omega}{c} \quad , \tag{3.3.19}$$

subjected to the boundary conditions that the field must obey at $z = 0$ and $z = h$.

If the field stands for the z components of the Hertz vector (vertical polarization, reflection coefficients $= 1$), the boundary conditions are

$$\frac{\partial u}{\partial z} = 0 \quad \begin{cases} z = 0 \\ z = h \end{cases} \quad , \tag{3.3.20}$$

implying perfectly reflecting interfaces. It is instructive to search for the solutions by trial and error as follows. Suppose that the direct ray R_{0_1} *is* the solution: the function $\exp(jKR_{0_1})/R_{0_1}$ satisfies the wave equation but respects neither boundary condition, so it cannot be the solution after all. If we add to the above component the one coming from the image of the source with respect to the lower boundary (i.e., the ray reflected at the lower interface) the field $\exp(jKR_{0_1})/R_{0_1} + \exp(jKR_{0_2})/R_{0_2}$ satisfies the wave equation and respects the *lower* boundary condition, but violates the *upper* one. We proceed next to adding up also the rays which emanate from the images of the source 0_{0_1}, *and* its image 0_{0_2}, with respect to the upper boundary (which amounts to considering the sum of the direct component, the component reflected once at the lower boundary, the component reflected once at the upper boundary, and the component reflected once at the lower *and* the upper boundary).

The resulting field

$$\frac{e^{jKR_{01}}}{R_{0_1}} + \frac{e^{jKR_{02}}}{R_{0_2}} + \frac{e^{jKR_{03}}}{R_{0_3}} + \frac{e^{jKR_{04}}}{R_{0_4}}$$

satisfies the wave equation, satisfies the *upper* boundary condition, but violates the *lower* boundary condition. What do we learn from this algorithm?

By successively adding pairs of components emanating from the image of the source, its image, and so on, with respect to one boundary and then the other, we *alternately* satisfy the condition at the boundary with respect to which the array of the sources is symmetrical, but we violate the other boundary condition. After adding *one tetrad*, the cycle starts again. Nevertheless, as we proceed adding in fact more and more zigzagging tetrads of rays, the disparity between the two boundary conditions becomes smaller and smaller, and in the limit, when we have added infinite tetrads of rays, the resulting field

$$u = \sum_{1=0}^{\infty} \left[\frac{e^{jKR_{1_1}}}{R_{1_1}} + \frac{e^{jKR_{1_2}}}{R_{1_2}} + \frac{e^{jKR_{1_3}}}{R_{1_3}} + \frac{e^{jKR_{1_4}}}{R_{1_4}} \right] , \tag{3.3.21}$$

where

$$R_{1_1} = \sqrt{r^2 + (21h + z - z_0)^2} ,$$

$$R_{1_2} = \sqrt{r^2 + (21h + z + z_0)^2} ,$$

$$R_{1_3} = \sqrt{r^2 + [2(1 + 1)h - z - z_0]^2} ,$$

and

$$R_{1_4} = \sqrt{r^2 + [2(1 + 1)h + z_0 - z]^2} ,$$

satisfies exactly both boundary conditions. This can be proved easily if we calculate from (3.3.21) above, the value of du/dz at $z = 0$ and $z = h$; it is zero in both cases.

In order now to proceed *analytically* in the calculation of the field u at the receiving point (r, z) from (3.3.21), we make use of the integral expansion (3.3.18) for each term $\exp(jKR_{1i})/R_{1i}$, namely

$$\frac{e^{jKR_{1_i}}}{R_{1_i}} = \frac{jK}{2} \int_{(-\pi/2)+j\infty}^{(\pi/2)-j\infty} e^{jK|z_{1_i}|\cos\theta} H_0^{(1)}(Kr \sin\theta)\sin\theta \, d\theta , \tag{3.3.22}$$

where $R_{1_i} = \sqrt{r^2 + z_{1_i}^2}$, and z_{1_i} for $i = 1,2,3,4$ takes the corresponding values

$$z_{1_1} = 21h + z - z_0 \quad,$$

$$z_{1_2} = 2hl + z + z_0 \quad,$$

$$z_{1_3} = 2(1 + 1)h - z - z_0 \quad,$$

$$z_{1_4} = 2(1 + 1)h - z + z_0 \quad.$$

(3.3.23)

Substituting in (3.3.21), after some trivial algebra we obtain the resulting field u, in integral form 3.2 :

$$u = \int_{-\infty}^{\infty} \frac{\cosh(bz_0)\cosh[(h - z)b]}{b \sin(bh)} H_0^{(1)}(\xi r)\xi \, d\xi \quad, \tag{3.3.24}$$

for $z > z_0$, and

$$u = \int_{-\infty}^{\infty} \frac{\cosh(bz)\cosh[(h - z_0)b]}{b \sin(bh)} H_0^{(1)}(\xi r)\xi \, d\xi \quad, \tag{3.3.25}$$

for $z < z_0$, where $\xi = K \sin\theta$ and $b = \sqrt{\xi^2 - K^2} = jK \cos\theta$. The above integral can be developed using Cauchy's theorem of residues.

The poles of the integrand are the solutions of the equation

$$\sinh(bh) = 0$$

or

$$bh = jl\pi \quad (1 \text{ integer})$$

or

$$\xi_1 = \pm \sqrt{K^2 - \left(\frac{1\pi}{h}\right)^2} = \pm \sqrt{\left(\frac{2\pi}{\lambda}\right)^2 - \left(\frac{1\pi}{h}\right)^2} \quad. \tag{3.3.26}$$

For $2h/\lambda > 1$, the poles are along the real axis of the complex ξ-plane, and for $2h/\lambda < 1$ the poles are along the imaginary axis.

In order to avoid the usual divergences associated with this situation, we assume that the medium in the wave guide is a little bit absorbent (i.e., K has a very small imaginary part); this pushes the poles away from the coordinate's axes, as in Fig.3.11. (We are interested in the poles in the first quadrant since those in the third refer simply to a wave propagating to the *left* of the source.) Our integrand has the form $f(z)/g(z)$; if the roots of $g(z) = 0$ are α_i, then

$$\int_c \frac{f(z)}{g(z)} \, dz = 2\pi j \sum_{i=1}^{\infty} \frac{f(\alpha_i)}{[\partial g(z)/\partial z]_{z=\alpha_i}} \quad. \tag{3.3.27}$$

Applying this to our case, we get

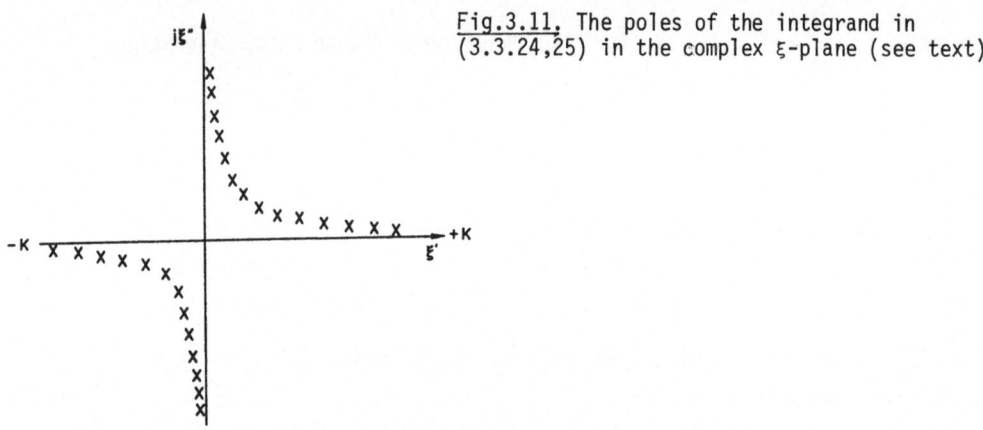

$$u = \frac{2\pi j}{h} \left[\frac{1}{2} H_0^{(1)}(Kr) + \sum_{1=1}^{\infty} \cos\left(\frac{1\pi z}{h}\right)\cos\left(\frac{1\pi z_0}{h}\right)H_0^{(1)}(\xi_1 r) \right] \quad , \qquad (3.3.28)$$

where $\xi_1 = \sqrt{K^2 - (1\pi/h)^2}$.

For $|\xi_1 r| \gg 1$, i.e., for a distance r large with respect to the wavelength, the above result gives the asymptotic expression

$$u \cong \frac{2e^{j\pi/4}}{h} \sqrt{\frac{2\pi}{r}} \left[\frac{1}{2\sqrt{K}} e^{jKr} + \sum_{1=1}^{\infty} \cos\left(\frac{1\pi z}{h}\right)\cos\left(\frac{1\pi z_0}{h}\right) \frac{1}{\sqrt{\xi_1}} e^{j\xi_1 r} \right] \quad . \qquad (3.3.29)$$

This outcome refers to vertical polarization: reflection coefficients of +1. For horizontal polarization —reflection coefficients of -1— the result is different.

Let us interpret this final result. Equation (3.3.29) gives the total received field as a summation of distinct energetic entities, the *normal modes*. Obviously each normal mode is constructed out of one tetrad of rays (3.3.21). The zeroth normal mode cannot be accommodated under the summation symbol like the others. Why? Simply because this particular mode has a "handicap", namely its own "gang of four" contains as a member the ray which goes straight from the source to the receiving point, displaying utter aristrocratic disregard for both boundaries.

For $1 > 2h/\lambda$ the corresponding normal modes have wave numbers ξ_1 which are purely imaginary; hence they constitute evanescent waves, decaying exponentially along the r-direction without propagating. For $1 \leqslant 2h/\lambda$ we come up with 1 discrete propagating normal modes, each travelling with its own phase and group velocity. Thus the wave guide, even in vacuo, produces dispersive phenomena on account of the space quantization (h is finite).

The characteristic velocities for each normal mode are calculated immediately:

$$v_{ph_1} = \frac{\omega}{\xi_1} \frac{c}{\sqrt{1 - (1\lambda/2h)^2}} \qquad (3.3.30)$$

and

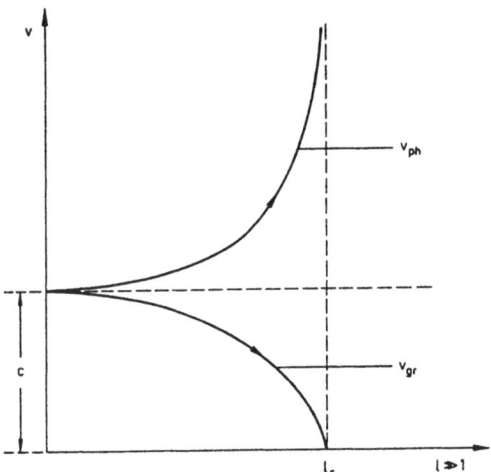

Fig.3.12. Phase velocity and group velocity as a function of the mode number l

$$v_{gr_1} = \frac{\partial \omega}{\partial \xi_1} = c \sqrt{1 - \left(\frac{l\lambda}{2h}\right)^2} = \frac{c^2}{v_{ph_1}} \ . \tag{3.3.31}$$

Their l-dependence is sketched in Fig.3.12.

For horizontal polarization, (3.3.29) becomes

$$u \sim \sum_{l=1}^{\infty} \sin\left(\frac{l\pi z}{h}\right)\sin\left(\frac{l\pi z_0}{h}\right)\frac{1}{\sqrt{\xi_1}} \ e^{j\xi_1 r} \tag{3.3.32}$$

i.e., there is no zeroth normal mode. Equation (3.3.32) can be deduced following the same process as above but now starting from the formula

$$u = \sum_{l=0}^{\infty} \left[\frac{e^{iKR_{11}}}{R_{11}} - \frac{e^{jKR_{12}}}{R_{12}} - \frac{e^{jKR_{13}}}{R_{13}} + \frac{e^{jKR_{14}}}{R_{14}}\right] \tag{3.3.33}$$

on account of the value -1 of the reflection coefficient under horizontal polarization at a perfectly reflecting interface $z = 0$, $z = h$. [Compare (3.3.21)]

3.3.3 A Cavity Resonator

Suppose we now "close" the one-dimensional wave guide by two more pairs of perfectly reflecting interfaces, thereby forming a parallelepiped as in Fig.3.13. What will be the energetic state of affairs within such a cavity in which a point source of given polarization radiates at a frequency ν?

In the first place, within such a cavity no propagation will take place: the *steady state* will involve only standing waves. The question is, in a cavity of volume V, how many normal modes (degrees of freedom) can be sustained within a frequency interval $\Delta\nu$ around an excitation frequency ν? Let us calculate it.

Formally we must solve the wave equation $\nabla^2 u + K^2 u = 0$ under the boundary conditions imposed again by the perfectly reflecting walls of the cavity. Fortunately,

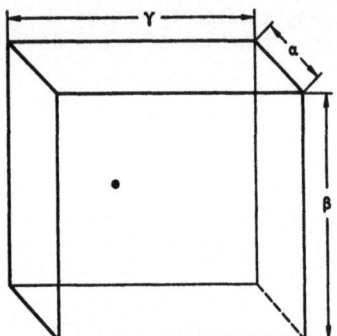

Fig.3.13. A cavity with perfectly reflecting walls

since the medium in the cavity is linear (vacuum), we can write $u = X(x)Y(y)Z(z)$, where $X(x)$, $Y(y)$, $Z(z)$ stand for the corresponding components of the field u. We deduce that

$$X''YZ + XY''Z + XYZ'' + K^2XYZ = 0$$

or

$$\frac{X''}{X} + \frac{Y''}{Y} + \frac{Z''}{Z} + (K_x^2 + K_y^2 + K_z^2) = 0 \quad . \tag{3.3.34}$$

Since each of the above differential terms is a function of only x,y, or z, respectively, the differential equation (3.3.34) splits into three equations as follows:

$$\frac{d^2X}{dx^2} + K_x^2X = 0 \quad , \quad \frac{d^2Y}{dy^2} + K_y^2Y = 0 \quad , \quad \frac{d^2Z}{dz^2} = K_z^2Z = 0 \quad ; \tag{3.3.35}$$

for K_x, K_y, K_z being real wave numbers, we get

$$X = A_1 \cos(K_x x) + A_2 \sin(K_x x)$$

$$Y = B_1 \cos(K_y y) + B_2 \sin(K_y y) \tag{3.3.36}$$

$$Z = \Gamma_1 \cos(K_z z) + \Gamma_2 \sin(K_z z) \quad .$$

For cavity walls which are perfect conductors, we have

$$K_x = \frac{l\pi}{\alpha} \quad , \quad K_y = \frac{m\pi}{\beta} \quad , \quad K_z = \frac{n\pi}{\gamma} \quad , \tag{3.3.37}$$

where α,β,γ are the dimensions of the orthogonal cavity, and l,m,n are integers standing for "modal" numbers. Then for x,y,z components of the intensity say of the electric field at any point in the cavity, we have

$$E_x(x,y,z) = E_1 \cos(K_x x)\sin(K_y y)\sin(K_z z) \quad ,$$

$$E_y(x,y,z) = E_2 \sin(K_y y)\cos(K_y y)\sin(K_z z) \quad , \tag{3.3.38}$$

$$E_z(x,y,z) = E_3 \sin(K_x x)\sin(K_y y)\cos(K_z z) \quad ,$$

where

$$K^2 = K_x^2 + K_y^2 + K_z^2 = \pi^2\left(\frac{1^2}{\alpha^2} + \frac{m^2}{\beta^2} + \frac{n^2}{\gamma^2}\right) . \tag{3.3.39}$$

These expressions for the x,y,z components of the field arise if we take into consideration that since E_x, say, is *vertically* polarized with respect to x, the cosine
term is appropriate as far as the x dependence is concerned.

At the same time, E_x is *horizontally* polarized with respect to both y and z —
hence the sine dependences. The same argument applies for E_y and E_z. The amplitudes
E_1, E_2, E_3 of the field are not independent but are related through Gauss's law
$\nabla \cdot E = 0$, which on the present occasion gives

$$K_x E_1 + K_y E_2 + K_z E_3 = 0 \tag{3.3.40}$$

or, from (3.3.37)

$$\frac{1}{\alpha} E_1 + \frac{m}{\beta} E_2 + \frac{n}{\gamma} E_3 = 0 , \tag{3.3.41}$$

For a cubic cavity of dimension α, the above relation gives

$$1E_1 + mE_2 + nE_3 = 0 . \tag{3.3.42}$$

Concerning now the number of normal modes that can be excited in such a cubic cavity,
let us consider a given mode characterized by one set of integers l,m,n. If $\cos\theta_1$,
$\cos\theta_2$, $\cos\theta_3$ are the direction cosines of this mode and λ the wavelength of the excitation in the cavity, we must have, on account of perfectly reflecting walls,

$$\alpha\cos\theta_1 = 1\frac{\lambda}{2} , \qquad \beta\cos\theta_2 = m\frac{\lambda}{2} , \qquad \gamma\cos\theta_3 = n\frac{\lambda}{2} , \tag{3.3.43}$$

or, since $\cos^2\theta_1 + \cos^2\theta_2 + \cos^2\theta_3 = 1$,

$$\frac{4\alpha^2}{\lambda^2} = 1^2 + m^2 + n^2 . \tag{3.3.44}$$

We can represent each normal mode as a point in a three-dimensional space with
(discrete) coordinates l,m,n (l,m,n = 0,1,2,...). The total number N of normal modes
will approximately equal 1/8 of the spherical volume of a sphere of radius $2\alpha/\lambda$ so

$$N \sim \frac{1}{8} \frac{4\pi}{3} \left(\frac{2\alpha}{\lambda}\right)^3 \tag{3.3.45}$$

or, since frequency $\nu = c/\lambda$ and $\alpha^3 = V$,

$$N \sim \frac{4\pi}{3} \frac{V}{c^3} \nu^3 \tag{3.3.46}$$

and the number of normal modes in a frequency interval $d\nu$ will be

$$dN = \frac{4\pi}{c^3} V\nu^2 d\nu . \tag{3.3.47}$$

This result is for a single source polarization. For *both* polarizations simultaneously excited, the number of normal modes per $d\nu$ frequency interval will be

$$dN = \frac{8\pi}{c^3} V\nu^2 \, d\nu \quad . \tag{3.3.48}$$

How is energy distributed among these modes? When $dN/d\nu$ is very large (and this assumption underlies the way $dN/d\nu$ has been derived) a "reasonable" hypothesis is equipartition of energy — namely that each degree of freedom carries $\sim kT$ joules, where T is the temperature within the cavity and k is Boltzmann's constant. Therefore, the energy per unit volume is

$$u(\nu,T) = \frac{8\pi\nu^2 kT}{c^3} \, d\nu \quad , \tag{3.3.49}$$

and the total energy in the cavity

$$u(T) = \int_0^\infty \frac{8\pi kT}{c^3} \nu^2 d\nu = \infty \quad . \tag{3.3.50}$$

Accordingly, the specific heat of a cavity containing a vacuum, which is $[\partial u(T)/\partial T]_V$, would also be infinite. This implies that it should be impossible to raise the temperature of evacuated cavity! Since the divergence in the above integral is caused by modes of high frequencies, this was fashionably referred to 100 years ago as the "ultraviolet catastrophe".

The ultraviolet catastrophe was ultimately resolved by Max Planck. The intricacies of quantum theory are irrelevant to this lecture course, so we will refer only to the resulting *quantization of energy levels that each discrete mode can acquire*, which idea Planck called "an act of desperation" on his part. In fact, Planck was forced to assume that a mode (an "oscillator") can take up energy only discontinuously in multiples of a unit quantum ε_0. Briefly the syllogism goes as follows: Let N_0 be the number of modes ("oscillators") of the lowest energy and N be the total number of modes possessing different amounts of energy $0, \varepsilon_0, 2\varepsilon_0,\ldots$, etc. Then, according to Boltzmann's distribution law which we derived in Sect.2.3.2, (2.3.49),

$$N = N_0 \sum_{n=0}^\infty \exp\left(-\frac{n\varepsilon_0}{kT}\right) \quad . \tag{3.3.51}$$

Planck further assumed that each unit quantum energy which a mode can take up or exchange is proportional to its frequency ν, that is, $\varepsilon_0 = h\nu$, where h is a universal constant. Thus the total energy E of the normal modes is

$$E = N_0 \sum_{n=0}^\infty n\varepsilon_0 \exp\left(-\frac{n\varepsilon_0}{kT}\right) = N_0 \sum_{n=0}^\infty nh\nu \exp\left(-\frac{nh\nu}{kT}\right) \quad . \tag{3.3.52}$$

The average energy of each mode (in excess of the zero-point energy, see below) is then

$$\bar{\varepsilon} = \frac{E}{N} = \frac{h\nu \sum_{n=0}^\infty n \exp[-n(h\nu/kT)]}{\sum_{n=0}^\infty \exp[-n(h\nu/kT)]} = \frac{h\nu}{\exp(h\nu/kT) - I} \quad , \tag{3.3.53}$$

and the total energy per unit volume and frequency interval $d\nu$ will equal $\bar{\varepsilon}$ multiplied by the number of modes in this interval, i.e.,

$$u(\nu,T) = \frac{8\pi}{c^3} \nu^2 \frac{h\nu}{\exp(h\nu/kT) - 1} d\nu \quad . \tag{3.3.54}$$

So the average number of *photons* per normal mode equals

$$<n> = \frac{1}{\exp(h\nu/kT) - 1} \quad . \tag{3.3.55}$$

Before discussing (3.3.55) in some detail, let us see what happens at zero temperature. The issue is relevant to this course in the sense that in classical communication theory finite temperature is synonymous with a finite amount of thermal noise. The communication engineer believes that by designing devices which work at extremely low temperatures, he can practically "exorcise" all noise. This is not so, however, at low temperatures and high frequencies where quantum noise replaces, so to speak, thermal noise as the basic obstacle to achieving errorless information transfer.

We found that the average energy per normal mode in a cavity at equilibrium is

$$\bar{\varepsilon} = \frac{h\nu}{\exp(h\nu/kT) - 1} \quad . \tag{3.3.53}$$

This can also be written as

$$\bar{\varepsilon} = kT \frac{x}{e^x - 1} \cong kT \quad , \tag{3.3.56}$$

in which $x = h\nu/kT \ll 1$. At sufficiently high temperatures $(e^x \sim 1 + x + x^2/2 + \dots)$ we can also write

$$\bar{\varepsilon} = \frac{h\nu}{\exp(h\nu/kT) - 1} \cong kT \frac{1}{1 + x/2} \sim kT\left(1 - \frac{x}{2}\right) = kT\left(1 - \frac{1}{2}\frac{h\nu}{kT}\right) = kT - \frac{h\nu}{2} \quad . \tag{3.3.57}$$

The last expression tells us that at high temperatures the average energy $\bar{\varepsilon}$ of a normal mode is $kT - h\nu/2$, instead of the equipartition energy value kT. But we do know that at high temperatures the principle of equipartition of energy holds. In order to be kT at high temperatures, $\bar{\varepsilon}$ must have the following form:

$$\bar{\varepsilon} = \frac{h\nu}{2} + \frac{h\nu}{\exp(h\nu/kT) - 1} \quad . \tag{3.3.58}$$

The quantity $h\nu/2$ is the zero-point energy.

We now return to (3.3.55) giving the average number of photons per unit volume per normal mode ("electromagnetic degrees of freedom").

Photons on earth primarily come from the sun. Now the earth's atmosphere is a "membrane" which acts as a cutoff filter to *all* electromagnetic radiation (from $= 10^8$ to 10^{-14} cm) emanating from the sun and which has a bandwidth of $\sim 10^{22}$ or 2^{73} Hz or 73 *octaves*, except for two small "windows": one is the microwave band with a width ~ 10 *octaves* and the other is the optical band with a width of only ~ 1 *octave*. An obvious question is: why did evolution not make use of the microwave window instead of developing receptors at the optical band in all biological species? Today, of course, man is using microwave radiation as an artifact. Again, why did technology proceed in a way which is just *opposite* to biological evolution? That is to say, why did microwave communication devices and techniques *precede* laser technology and integrated optics?

3.4 The Entropy of Electromagnetic Radiation. Information Received by an Electromagnetic Wave Impinging on a Finite Aperture. Ambiguity of Perception

How is an electromagnetic *wave* built up from photons? We just do not know yet. What we do know is that an electromagnetic wave is a *collective property of photons*. Photon emission possesses three parameters: intensity, frequency, and polarization. An electromagnetic wave possesses four parameters: amplitude, frequency, *phase* and polarization. It is *phase* which clearly has all the characteristics of an *emergent* macroscopic property when from individual photons we synthesize an electromagnetic wave.

Inspection of the basic formula (3.3.55) explains why technology had, for a long time, difficulties in creating optical waves while devices for microwaves as well as even lower frequencies were most easily designed and constructed. (Fifty years ago telecommunications were exclusively dependent on *long* waves $\sim 10^4$ Hz and *short* waves $\sim 10^6$ Hz).

Since the *occupation* (average) photon number per degree of freedom <n> and the phase φ are conjugate quantities which obey the uncertainty relation $(\Delta n)(\Delta \varphi) \geqslant 1/2$, an accurate determination of the absolute phase would require $<n> \gg 1$. Now (3.3.55) tells us that at a reasonably low frequency (not *too* low) <n> may become very high. For example [3.3], a power station generating 10000 kW at 50 ± 0.01 Hz puts out $\sim 10^{41}$ photons per degree of freedom. A large magnetron in pulsed operation at a wavelength ~ 10 cm ($\sim 3 \times 10^9$ ± 0.5 × 10^6 Hz) produces $\sim 10^{24}$ photons per degree of freedom; but a powerful (high-pressure) mercury lamp emitting 1 W/cm^2 in the form of green light ($\lambda \sim 5461$ ± 10 Å) gives less than 10^{-3} photons per degree of freedom. Thus a conventional source in the optical region cannot be used for communication purposes due to the inability to display a measurable phase —absolutely indispensable for *synchronization* purposes. It took the invention of laser (coherent) light to accomplish this task: a Q-switched ruby laser can put out up to $\sim 10^{15}$ photons per degree of freedom.

Let us now try to answer the first question, that is, why evolution favored the optical band against the microwave one. Physiological constraints aside (which admittedly play a predominant role) we can perhaps answer the question as follows.

In order to have unambiguous reception (and perception) the "signal-to-noise ratio" must be high. This means that when we observe an object via the radiation that it scatters from an illuminating source (signal), we like to have the number of scattered photons far above the number of photons spontaneously emitted by the object itself via blackbody radiation (noise). Well, from (3.3.55), for optical frequencies, we have $h\nu/kT \gg 1$ and so

$$<n> \sim \exp\left[-\frac{h\nu}{kT}\right] \, . \tag{3.4.1}$$

The number of photons scattered from an object illuminated by the sum ($T_s = 6000$ K) will be

$$<n>_s \cong \exp\left[-\frac{h\nu}{6000\ k}\right] \, . \tag{3.4.2}$$

This will constitute the signal.

The number of photons spontaneously emitted from the object at the same high frequency will be

$$<n>_n \cong \exp\left[-\frac{h\nu}{300\ k}\right] \, , \tag{3.4.3}$$

where $T_n = 300$ K is the temperature of the object. This constitutes the noise.

Then the signal-to-noise ratio is

$$\frac{<n>_s}{<n>_n} \cong \exp\left[\frac{h\nu}{k}\left(\frac{1}{T_n} - \frac{1}{T_s}\right)\right] = \exp\left[\frac{5700\ h\nu}{18 \times 10^5 k}\right] \tag{3.4.4}$$

which for the optical frequency band is of the order of $\sim 10^{28}$ — very high indeed to make reception unambiguous in the optical region. Let us now perform the same calculation in the microwave region. In this case $h\nu/kT \ll 1$, and from (3.3.55) we get

$$<n> \sim \frac{kT}{h\nu} \, . \tag{3.4.5}$$

So

$$<n>_s = \frac{6000\ K}{h\nu} \tag{3.4.6}$$

and

$$<n>_n = \frac{300\ K}{h\nu} \, . \tag{3.4.7}$$

The signal-to-noise ratio is now only

$$\frac{<n>_s}{<n>_n} \sim \frac{6000}{300} = 20 \, , \tag{3.4.8}$$

that is, too poor to allow good reception.

Equation (3.3.55) gives only the *average* number of photons per EM degree of freedom (per normal mode). Inevitably, fluctuations occur around this average number. What is the statistics of these fluctuations? This question is crucial because fluctuations in photon intensity constitute the principal (although not the only) cause for ambiguity in image formation when the information-carrying EM wave impinges on a finite aperture.

We are specifically interested in calculating the second moment (the variance) of the photons' fluctuations around the mean density $<n> = \bar{E}$. Let us recall the simple relation between mean value, mean square value, and mean square fluctuations from the mean value:

$$\overline{\Delta E^2} = \overline{(E - \bar{E})^2} = \overline{E^2 - 2E\bar{E} + \bar{E}^2} = \overline{E^2} - 2\bar{E}^2 + \bar{E}^2 = \overline{E^2} - \bar{E}^2 \quad . \tag{3.4.9}$$

Now any system in contact with a heat bath at temperature T follows the Boltzmann distribution. The probability P_r of the system being in its rth (quantum) state with energy E_r is given by the canonical expression

$$P_r = \frac{\exp(-E_r/kT)}{\Sigma_s \exp(-E_s/kT)} \quad , \tag{3.4.10}$$

where the summation is performed over the available discrete states of the system. Obviously

$$\bar{E} = \Sigma_r E_r P_r \quad . \tag{3.4.11}$$

Substituting the expression for P_r,

$$\bar{E} \, \Sigma_s \exp -\left(\frac{E_s}{kT}\right) = \Sigma_s E_s \exp\left(-\frac{E_s}{kT}\right) \quad . \tag{3.4.12}$$

Differentiating both sides with respect to T and dividing by $\Sigma_s \exp(-E_s/kT)$, we get

$$\frac{1}{kT^2} \bar{E}^2 + \frac{d\bar{E}}{dT} = \frac{1}{kT^2} \overline{E^2} \tag{3.4.13}$$

or finally

$$kT^2 \frac{d\bar{E}}{dT} = \overline{E^2} - \bar{E}^2 = \overline{\Delta E^2} \quad . \tag{3.4.14}$$

This result was first obtained by Einstein. Now let us apply this result to a system characterized by the Planck distribution that we have from (3.3.55):

$$<n> = \bar{E} = \sum_{r=0}^{\infty} E_r P_r = \frac{h\nu}{\exp(h\nu/kT) - 1} \tag{3.4.15}$$

so that, using (3.4.14),

$$\overline{\Delta E^2} = kT^2 \frac{d\bar{E}}{dT} = (h\nu)^2 \frac{\exp(h\nu/kT)}{[\exp(h\nu/kT) - 1]^2} \quad . \tag{3.4.16}$$

The *fractional* fluctuation $\Delta E/\bar{E}$ has a mean square value

$$\frac{\overline{\Delta E^2}}{\bar{E}^2} = \exp \frac{h\nu}{kT} \quad . \tag{3.4.17}$$

Expressed in terms of $\bar{E}$ this becomes

$$\frac{\overline{\Delta E^2}}{\overline{E}^2} = 1 + \frac{h\nu}{\overline{E}} \quad . \tag{3.4.18}$$

At high temperatures, $h\nu/kT \ll 1$, equipartition takes over, so

$$\overline{E} \sim kT \quad , \tag{3.4.19}$$

$$\overline{\Delta E^2} = (kT)^2 \quad , \tag{3.4.20}$$

and

$$\frac{\overline{\Delta E^2}}{\overline{E}^2} = 1 \tag{3.4.21}$$

or $\overline{E^2} = 2\overline{E}^2$, $\tag{3.4.22}$

the fractional fluctuation tends to unity as the mean energy becomes large.

We thus observe that the expression for the mean square value of the fractional fluctuations (3.4.18) can be split into two terms: a "classical term", referring essentially to interference of waves with random phases, leading to (3.4.22), and a "quantum term". This quantum term of fluctuations $(h\nu/kT \gg 1)$,

$$\overline{\Delta E^2_{qu.}} \sim h\nu \overline{E} \quad , \tag{3.4.23}$$

refers to the case where the energy is presented in the form of independent particles (photons) jumping among the available discrete states of the system and thereby giving rise to fluctuations. Consider now N *normal modes* interacting with each other incoherently. Writing E_N for the energy of N normal modes and E_0 for the energy of each one of them, we have $\overline{E}_N = N\overline{E}_0$,

$$\overline{\Delta E^2_N} = kT^2 \frac{d\overline{E}_N}{dT} = NkT^2 \frac{d\overline{E}_0}{dT} = N\overline{\Delta E^2_0} \quad , \tag{3.4.24}$$

and

$$\frac{\overline{\Delta E^2_N}}{\overline{E}^2_N} = \frac{1}{N} \frac{\overline{\Delta E^2_0}}{\overline{E}^2_0} = \frac{1}{N}\left(1 + \frac{h\nu}{\overline{E}_0}\right) \quad . \tag{3.4.25}$$

When N is large, the fractional fluctuations are small. This is consistent with the requirement that the fractional fluctuations in the energy of a macroscopic system in contact with a heat bath should be small —in accordance with experience.

Let us now come to the realistic case where in a communication process we have a *superposition* of a deterministic signal E_s and a component of thermal noise E_n. Consider the interaction between the circular frequency ω_1 of a Fourier component of the signal and the circular frequency ω_2 of a Fourier component of the noise. The following treatment was first put forward by *Gabor* [3.4]. The instantaneous energy density resulting from the interference is going to be proportional to

$$E_s E_s^* + E_n E_n^* + \{E_s E_n^* \exp[j(\omega_1 - \omega_2)t] + E_s^* E_n \exp[-j(\omega_1 - \omega_2)t]\} \quad ,$$

where the asterisks stand for the complex conjugate values. The first two terms are the energy of the signal and the energy of the noise, and the rest arises from interference. Let us write, following Gabor,

$$\varepsilon = \varepsilon_s + \varepsilon_n + \varepsilon_{sn} \qquad (3.4.26)$$

for the energy densities of the fields involved. The mean value of interference energy $\bar{\varepsilon}_{sn}$ is zero, due to the (assumed) total lack of coherence between signal and noise. The mean square interference energy is

$$\overline{\varepsilon_{sn}^2} = \overline{2E_s E_s^* \cdot E_n E_n^*} = 2\varepsilon_s \bar{\varepsilon}_n \qquad (3.4.27)$$

since ε_s is a deterministic signal. The mean square fluctuations of the total intensity are

$$\overline{(\varepsilon - \bar{\varepsilon})^2} = \overline{[(\varepsilon_s + \varepsilon_n + \varepsilon_{sn}) - (\bar{\varepsilon}_s + \bar{\varepsilon}_n)]^2} \quad , \qquad (3.4.28)$$

but since $\varepsilon_s = \bar{\varepsilon}_s$,

$$\overline{(\varepsilon - \bar{\varepsilon})^2} = \overline{(\varepsilon_n + \varepsilon_{sn} - \bar{\varepsilon}_n)^2}$$

$$= \overline{\varepsilon_n^2} + \overline{\varepsilon_{sn}^2} + \bar{\varepsilon}_n^2 + 2\overline{\varepsilon_n \varepsilon_{sn}} - 2\overline{\varepsilon_n}\bar{\varepsilon}_n - 2\overline{\varepsilon_{sn}}\bar{\varepsilon}_n \quad . \qquad (3.4.29)$$

Using the relations $\overline{\varepsilon_s \varepsilon_{sn}} = \overline{\varepsilon_n \varepsilon_{sn}} = 0$ —which again are due to the absence of correlation between the signal and the noise —and (3.4.27), we obtain

$$\overline{(\varepsilon - \bar{\varepsilon})^2} = \overline{\varepsilon_n^2} + 2\varepsilon_s \bar{\varepsilon}_n + \bar{\varepsilon}_n^2 - 2\overline{\varepsilon_n}\bar{\varepsilon}_n$$

$$= \overline{\varepsilon_n^2} - \bar{\varepsilon}_n^2 + 2\varepsilon_s \bar{\varepsilon}_n \quad . \qquad (3.4.30)$$

From (3.4.22) we have, however, for the classical term of the fluctuations

$$\overline{\varepsilon_n^2} = 2\bar{\varepsilon}_n^2$$

so

$$\overline{(\varepsilon - \bar{\varepsilon})^2}_{class.} = 2\varepsilon_s \bar{\varepsilon}_n + \bar{\varepsilon}_n^2 \quad , \qquad (3.4.31)$$

or, since $\varepsilon_s = \bar{\varepsilon} - \bar{\varepsilon}_n$, $(\bar{\varepsilon}_{sn} = 0)$, we finally have

$$\overline{(\varepsilon - \bar{\varepsilon})^2}_{class.} = 2(\bar{\varepsilon} - \bar{\varepsilon}_n)\bar{\varepsilon}_n + \bar{\varepsilon}_n^2 = (2\bar{\varepsilon} - \bar{\varepsilon}_n)\bar{\varepsilon}_n \quad . \qquad (3.4.32)$$

The quantum term of the fluctuations has been obtained already (3.4.23), so

$$\overline{(\varepsilon - \bar{\varepsilon})^2}_{qu.} = h\nu\bar{\varepsilon} \quad . \qquad (3.4.33)$$

Adding (3.4.32 and 33) we obtain the total intensity fluctuations in a system of a deterministic wave surrounded by thermal noise, namely,

$$\overline{(\epsilon - \bar{\epsilon})^2} = h\nu\bar{\epsilon} + 2\bar{\epsilon}\bar{\epsilon}_n - \bar{\epsilon}_n^2 \ . \tag{3.4.34}$$

Expressing the energies $\bar{\epsilon}$, $\bar{\epsilon}_n$ in terms of the number of photons N per normal mode, so that $\epsilon = Nh\nu$ and $\bar{\epsilon}_n = \bar{N}_n h\nu$, we get from (3.4.34):

$$\overline{\delta N^2} = \overline{(N - \bar{N})^2} = \bar{N}(1 + 2\bar{N}_n) - \bar{N}_n^2 \ , \tag{3.4.35}$$

where $\bar{N}_n = 1/[\exp(h\nu/kT) - 1] = <n>$.

Now the problem is: What is the number of distinguishable ways (steps, energy levels) in which we can distribute dN photons in a normal mode with uncertainty $(\overline{\delta N^2})^{\frac{1}{2}}$ between successive levels? The answer is $dN/(\overline{\delta N^2})^{\frac{1}{2}}$. The number of ways (levels) in which up to N_i photons can be distributed to a single normal mode (i) is therefore equal to

$$S_i = \int_{\bar{N}_n}^{N_i} \frac{dN}{(\overline{\delta N^2})^{\frac{1}{2}}} = \int_{\bar{N}_n}^{N_i} \frac{dN}{[\bar{N}(1 + 2\bar{N}_n) - \bar{N}_n^2]^{\frac{1}{2}}} \cong \frac{2N_i^{\frac{1}{2}}}{(1 + 2\bar{N}_n)^{\frac{1}{2}}} \tag{3.4.36}$$

for N very large. The logarithm to base 2 of S_i, $\log_2 S_i$, equals the *entropy of each normal mode (i), i.e., the a priori information carried by each discrete elec- tromagnetic degree of freedom.*

Two things remain to be calculated before we arrive at our final result (i.e., the number of information units carried by a spherical wave surrounded by noise and impinging on a finite receptor area).

We have *first* to calculate the number of degrees of freedom F of a finite aper- ture —which coincides with the degrees of freedom of the incident wave, after re- ception —and *finally* to calculate the ways N photons can be distributed among these F degrees of freedom.

The question is, how many numbers are necessary to specify completely an (opti- cal) image? Consider, following *di Francia* [3.5], a perfect (optical) instrument having a one-dimensional pupil of width a (Fig.3.14).

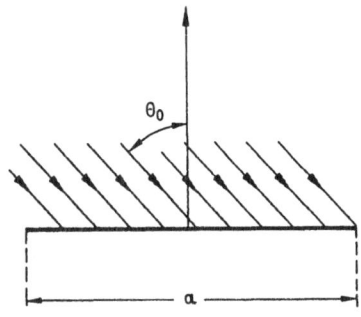

Fig.3.14. A plane wave falling on an aperture of width a at an angle of incidence θ_0

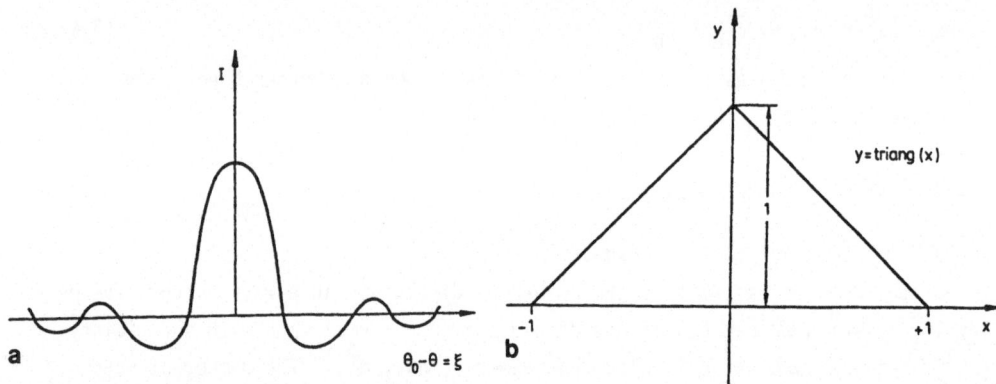

Fig.3.15. (a) Sketch of the function sincx. (b) The function $y = \mathrm{triang}(x)$ (see text)

A spherical wave of wavelength λ impinges at an angle of incidence, i.e., the angle between the line of sight and the normal to the aperture, θ_0. The intensity of illumination as a function of angular distance ξ will be

$$I \sim \mathrm{sinc}^2 x = \left(\frac{\sin \pi x}{\pi x}\right)^2 \,, \tag{3.4.37}$$

where

$$x = \frac{|\theta_0 - \theta|a}{\lambda} = \xi \frac{a}{\lambda} \,.$$

The function sincx is sketched in Fig.3.15a. Therefore

$$I = \mathrm{sinc}^2\left(\xi \frac{a}{\lambda}\right) \,. \tag{3.4.38}$$

The Fourier transform of $\mathrm{sinc}^2(\xi a/\lambda)$ is given by $(\lambda/a)\mathrm{triang}(f\lambda/a)$, drawn in Fig. 3.15b, where f stands for the frequency. It is evident that the spectrum of $\sin^2(\xi a/\lambda)$ has a cutoff frequency $f_c \sim a/\lambda$. Application of Shannon's sampling theorem (see Sect.4.1) leads to the conclusion that the illumination of the aperture is completely determined by giving its values at discrete points (sampling points) spaced $\lambda/2a$ apart. If $\bar{\xi}$ represents the total field angle of the instrument, the total number of sampling points will be $2\bar{\xi}a/\lambda$.

Consider now a two-dimensional pupil, say a rectangle of sides a and b. The image of a point (spherical wave) source will be represented by

$$I = \mathrm{sinc}^2\left(\xi \frac{a}{\lambda}\right) \mathrm{sinc}^2\left(\mu \frac{b}{\lambda}\right) \,, \tag{3.4.39}$$

where μ is the angular distance along the y-direction. By a simple extension of the previous application, we arrive at the number of sampling points needed to specify the image completely; it is now $4\bar{\xi}\bar{\mu}(ab/\lambda^2)$. If Ω stands for the total solid angle of the field and S is the pupil area of the instrument, the number of degrees of freedom of the image will be

$$F = 4\Omega \frac{S}{\lambda^2} \,. \tag{3.4.40}$$

We now come to the last part of our calculation: in how many ways can we distribute N photons among F degrees of freedom, taking into account the ways that N_i photons are distributed in each *single* i^{th} degree of freedom (3.4.36)?

Let us imagine the S_i's (3.4.36) as certain coordinates in an F-dimensional space. The number of distinguishable configurations is now equal to the number of integer lattice points inside a hypersphere of radius $R = 2(1 + 2\bar{N}_n)^{-\frac{1}{2}} N^{\frac{1}{2}}$, in the sector where all S_i's are positive. In other words, the number of distinguishable configurations (for large N) coincides with the volume of this sector of the hypersphere which is (see Sect.4.1 for a detailed treatment of n-dimensional Euclidean space)

$$ P = \frac{(\pi/4)^{F/2}}{\Gamma[(F/2) + 1]} R^F = \frac{1}{\Gamma[(F/2) + 1]} \left(\frac{\pi N}{1 + 2\bar{N}_n}\right)^{F/2} , \tag{3.4.41}$$

where $\Gamma(x)$ is the gamma function.

The information capacity of the wave is $\log_2 P$. Using Stirling's formula,

$$ \ln\Gamma\left(\frac{F}{2} + 1\right) \sim \frac{F}{2}\left[\ln\left(\frac{F}{2}\right) - 1\right] , \quad F \gg 1 , $$

which becomes

$$ \log_2 \Gamma\left(\frac{F}{2} + 1\right) \sim \frac{F}{2}\left[\log_2\left(\frac{F}{2}\right) - \log_2 e\right] , \tag{3.4.42}$$

we get

$$ \log_2 P = \frac{F}{2} \log_2 \frac{2\pi e N}{F(1 + 2\bar{N}_n)} \text{ bits} , \tag{3.4.43}$$

or

$$ \frac{1}{2} \log_2 \frac{2\pi e N}{F(1 + 2\bar{N}_n)} \text{ bits per degree of freedom} . \tag{3.4.44}$$

This is the physical entropy of a spherical electromagnetic wave carrying N photons, in the presence of noise of mean value $\bar{N}_n$, creating an image of F degrees of freedom on a finite aperture. This entropy has nothing to do with Shannon's Theorem[1], as we shall see.

So, we conclude, physical reception is *ambiguous*. There are four distinct reasons for this:

a) The existence of fluctuations of photons within each normal mode (electromagnetic degrees of freedom).

---

[1] *The physical entropy* of the wave has to do with the number of ways energy, up to a certain predetermined limit, can be distributed over the degrees of freedom of the created image.

The selective or informational entropy, on the other hand, has to do with the number of ways in which any given distribution of distinguishable patterns can be realized in a given number of repetitions.

b) The finite aperture of the perceiving instrument. This finiteness has two results: it distorts, like a spatial filter, a proportion of the propagating components of the impinging spherical wave, and second, it attributes a finite number of degrees of freedom to the received image. The ambiguity here is due to the fact that the spherical wave possesses an infinite number of degrees of freedom.

c) The non-detectability of phase: in many conventional communication systems the phase of the wave is not detected, so the information associated with it is irretrievably lost (in "holographic" systems, however, phase is preserved and the information contained therein is decoded).

d) The loss of evanescent waves: "half" of the encoded information in the spherical wave is irretrievably lost due to the impossibility of designing receptors for both propagating *and* evanescent waves.

4. Elements of Information and Coding Theory, with Applications

4.1 Information Transfer and the Concept of Channel Capacity for Discrete and Continuous Memoryless Signals

We are somewhat familiar with the basic notion of information —for both discrete and continuous distributions —from the introduction to entropy in Sect.2.3. Let us continue here first by calculating the information transfer from a source X to a receiver Y in the presence of environmental disturbances in the "channel", i.e., in the medium between X and Y (Fig.4.1).

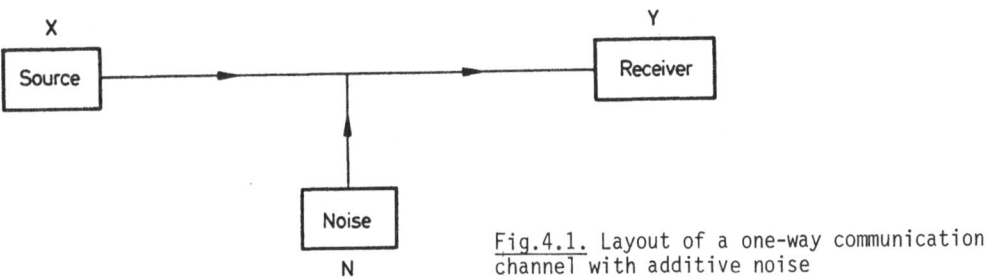

Fig.4.1. Layout of a one-way communication channel with additive noise

We assume that X has $\wedge$ discrete states, i.e., at each moment it can send a mes-sage/pattern x_i with a priori probability (as far as the receiver is concerned) $P(x_i)$, $i \in (1,...,\wedge)$. The a priori entropy of the source equals then

$$S(X) = - \sum_{i=1}^{\wedge} P(x_i)\log_2 P(x_i) \text{ bits} , \tag{4.1.1}$$

and we already know that it is maximum when all $P(x_i)$ are equal to $1/\wedge$, so

$$S_{max}(X) = \log_2\wedge \text{ bits} . \tag{4.1.2}$$

We can also see immediately that $S(X) = 0$, when all $P(x_i)$ but one are zero. We have to prove essentially that $\lim_{\xi \to 0}(\xi\ln\xi) = 0$. To do this we write the above as $\lim_{\xi \to 0}(\ln\xi/\xi^{-1})$ and apply l'Hospital's rule

$$\lim_{\xi \to 0} \frac{1/\xi}{-1/\xi^2} = 0 .$$

So all terms in $S(X)$ for which $P(x_i) = 0$ are zero, and the single term for which $P(x_j) = 1$ is zero as well, as $\log_2 1 = 0$. We understand then that the information carried by a source transmitting one message/symbol is *stereotypically* just zero.

Let us now examine how the information *transfer* takes place from X to Y. System Y possesses also a number of discrete states (say Λ again). For each distinct message/symbol x_i emitted from X and arriving as an input at Y, the receiving system assumes a state y_j. We call this correspondence $x_i \rightarrow y_j$ a "one-to-one mapping".

By observing the state y_j of Y, let the probability of X being in state x_i be $P(x_i/y_j)$. In the absence of "noise" from the channel, $P(x_i/y_j) = 1$ for $i = j$, and $P(x_i/y_j) = 0$ for $i \neq j$. So the $\Lambda \times \Lambda$ matrix of elements $P(x_i/y_j)$ would be diagonal.

However, in the presence of noise, the "channel matrix", as we call it, contains in principle $\Lambda \times \Lambda$ nonzero elements $P(x_i/y_j)$. The information gained by system Y, upon assuming the state y_j, is therefore

$$I(x_i/y_j) = -\log_2 P(x_i/y_j) \text{ bits} \quad . \tag{4.1.3}$$

In order to calculate the *conditional entropy* $S(X/Y)$, i.e., the remaining uncertainty that system Y still experiences after performing all possible observations after the complete repertoire of X, we average (4.1.3) over all possible states of X and Y; first we average over all possible —and unknown —states of X, and we get the conditional entropy of X given a certain state y_j of Y:

$$S(X/y_j) = \langle I(X/y_j) \rangle = - \sum_{i=1}^{\Lambda} P(x_i/y_j) \log_2 P(x_i/y_j) \quad . \tag{4.1.4}$$

Further, we average over all states of Y

$$S(X/Y) = \langle I(X/Y) \rangle = \sum_{j=1}^{\Lambda} P(y_j) S(X/y_j)$$

$$= - \sum_{i=1}^{\Lambda} \sum_{j=1}^{\Lambda} P(y_j) P(x_i/y_j) \log_2 P(x_i/y_j) \quad . \tag{4.1.5}$$

After Bayes's rule,

$$P(y_j) P(x_i/y_j) = P(x_i, y_j) \quad ,$$

so

$$S(X/Y) = - \sum_{i=1}^{\Lambda} \sum_{j=1}^{\Lambda} P(x_i, y_j) \log_2 P(x_i/y_j) \quad , \tag{4.1.6a}$$

where $P(x_i, y_j)$ is the joint probability of the states x_i (in X) and y_j (in Y). The conditional entropy $S(X/Y)$ is thus a measure of uncertainty about source X when Y is accessible to direct observation — in the presence of interfering noise.

The information transfer (from X to Y) equals simply the difference between the initial uncertainty $S(X)$ and the final uncertainty $S(X/Y)$. Here

$$I(X \rightarrow Y) = S(X) - S(X/Y)$$

$$= - \sum_{i=1}^{\wedge} P(x_i)\log_2 P(x_i) + \sum_{i=1}^{\wedge} \sum_{j=1}^{\wedge} P(x_i,y_j)\log_2 P(x_i/y_j)$$

$$= - \sum_{i=1}^{\wedge} \left[\sum_{j=1}^{\wedge} P(x_i,y_j) \right] \log_2 P(x_i) + \sum_{i=1}^{\wedge} \sum_{j=1}^{\wedge} P(x_i,y_j)\log_2 P(x_i/y_j)$$

$$= - \sum_{i=1}^{\wedge} \sum_{j=1}^{\wedge} P(x_i,y_j)\log_2 \left(\frac{P(x_i)}{P(x_i/y_j)} \right) \quad \text{bits} \quad . \tag{4.1.7a}$$

It is quite easy to see that $I(X \rightarrow Y) \geqslant 0$.

In cases where there is no noise in the channel, $P(x_i/y_j) = 1$ for $i = j$, and $P(x_i/y_j) = 0$ for $i \neq j$, so (4.1.6) becomes zero since the terms with $i = j$ will go to zero, as $\log_2 1 = 0$, and the terms with $i \neq j$ will go to zero as well, on the basis of the previously given proof that $\lim(\xi \ln \xi) = 0$, where now $\xi = P(x_i/y_j)$. Then

$$I(X \rightarrow Y) = S(X) = I_{max} \quad . \tag{4.1.8}$$

In the opposite case, where the noise is so disturbing that x_i and y_j become completely uncorrelated,

$$P(x_i,y_j) = P(x_i)P(y_j) \quad , \tag{4.1.9}$$

and so in (4.1.6a)

$$P(x_i) = P(x_i/y_j) \tag{4.1.10}$$

[since Bayes's rule for all i and j in general gives $P(x_i,y_j) = P(y_j)P(x_i/y_j)$]. Equation (4.1.7a) then gives zero ($\log_2 1 = 0$), and this means that $S(X/Y) = S(X)$ and $I(X \rightarrow Y) = 0$, as expected.

In the case of a source, a receiver, and a channel characterized by continuous probability density functions, (4.1.6a,7a) become

$$S(X/Y) = - \iint P(x,y)\log_2 P(x/y)dx \, dy \quad , \tag{4.1.6b}$$

and

$$I(X \rightarrow Y) = S(X) - S(X/Y) = - \iint P(x,y)\log_2 \frac{P(x)}{P(x/y)} dx \, dy \quad . \tag{4.1.7b}$$

Let us now occupy ourselves with the important problem of transforming a continuous wave form that possesses an infinite number of degrees of freedom (and so, if it represents the information pattern, it would take infinite time for its reception) into a discrete time sequence: in other words let us deal with the so-called sampling theorem.

Consider a continuous, time-bounded piece of waveform X(t), as in Fig.4.2a, smooth enough so that its Fourier transform — its spectrum — has a finite bandwidth with a cutoff frequency W (Fig.4.2b). How should we sample X(t) along the time axis as accurately as possible in order to come up with a discrete time series of a finite number of degrees of freedom? If this analogue to digital transformation

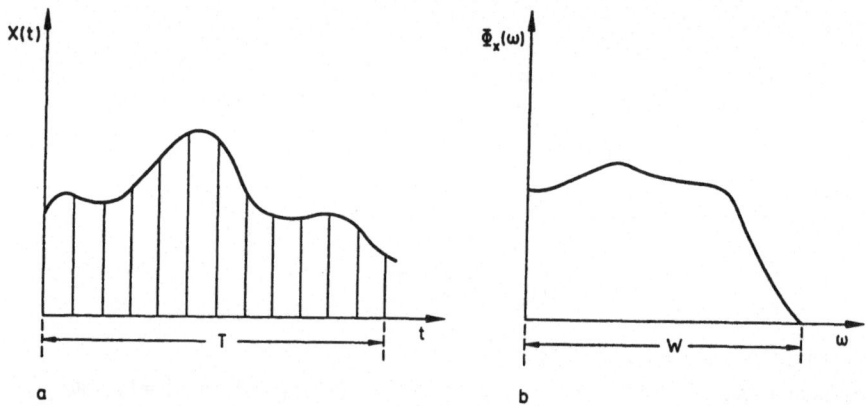

<u>Fig.4.2a,b.</u> A function (**a**) and its Fourier transform (**b**) used to illustrate the sampling theorem (see text)

can be accomplished, then we can replace the one-dimensional waveform of an infinite number of degrees of freedom by a hypervector (in a multidimensional space) whose components are the sample coordinates of $X(t)$.

The proof of the theorem (which is not reproduced here but the interested reader can find in any good textbook on information theory) gives the number of samples equal to 2WT (which implies a sampling at a rate of one coordinate every $1/2W$ seconds). So from now on we can speak of hypervectors in n(=2WT)-dimensional state spaces instead of continuous wave forms in one-dimensional (time) space. Some very "tricky" properties are presented by n-dimensional Euclidean spaces, defying in many respects intuition and common sense. Therefore, as a preparation to what will follow, i.e., to the treatment of information transfer and optimum reception of continuous waveforms, we think that a small detour to describe some salient characteristics in *n-dimensional Euclidean spaces* is in order. We follow *Hamming* [4.1].

The term n-dimensional space means merely that we have n independent variables: $x_1, x_2, \ldots, x_n$. Euclidean space means that we use the Pythagorean distance $x_1^2 + x_2^2 + \ldots + x_n^2 = r^2$, and we can define a sphere of radius r by this expression. We intend specifically to calculate the volume of an n-dimensional sphere. This volume depends on the radius r as r^n. We write, for the time being, $V_n(r) = C_n r^n$, where C_n is some constant—to be calculated—which depends on n. (For example, $C_1 = 2$, $C_2 = \pi$, and $C_3 = 4\pi/3$.)

To find the value of C_n we have to introduce—sorry, it cannot be done otherwise!—the famous *gamma function* $\Gamma(n)$. Essentially we deal with the study of a certain definite integral as a function of a parameter n. Consider the integral

$$\Gamma(n) = \int_0^\infty e^{-x} x^{n-1} \, dx \quad .$$

(4.1.11)

For n > 1 we can integrate by parts to get

$$\Gamma(n) = x^{n-1} \frac{e^{-x}}{-1} \Big|_0^\infty + (n - 1) \int_0^\infty e^{-x}x^{n-2} \, dx$$

or

$$\Gamma(n) = (n - 1)\Gamma(n - 1) \quad ; \tag{4.1.12}$$

for n = 1 we have

$$\Gamma(1) = \int_0^\infty e^{-x} \, dx = -e^{-x}\Big|_0^\infty = 1 \quad , \tag{4.1.13}$$

and it follows that for an integer n,

$$\Gamma(2) = 1 \ , \quad \Gamma(3) = 2! \ , \quad \Gamma(4) = 3! \ ,..., \ \Gamma(n) = (n - 1)! \ , \quad (n = 1,2,...) \ . \tag{4.1.14}$$

Let us note in passing that the gamma function provides the natural generalization of the factorial function, since the above integral also exists for *non*-integer n.

We consider next the gamma function for argument 1/2. We have

$$\Gamma\left(\frac{1}{2}\right) = \int_0^\infty e^{-x}x^{-1/2} \, dx \quad . \tag{4.1.15}$$

Putting $x = t^2$ gives

$$\Gamma\left(\frac{1}{2}\right) = 2 \int_0^\infty e^{-t^2} \, dt = \int_{-\infty}^\infty e^{-t^2} \, dt \tag{4.1.16}$$

(this is just the so-called error integral).

Consider now the product

$$\Gamma\left(\frac{1}{2}\right)\Gamma\left(\frac{1}{2}\right) = \int_{-\infty}^\infty e^{-x^2} \, dx \int_{-\infty}^\infty e^{-y^2} \, dy \quad . \tag{4.1.17}$$

We transform to polar coordinates:

$$\left[\Gamma\left(\frac{1}{2}\right)\right]^2 = \int_0^{2\pi} \int_0^\infty e^{-r^2}r \, dr \, d\theta = 2\pi \frac{e^{-r^2}}{-2}\Big|_0^\infty = \pi \quad . \tag{4.1.18}$$

Therefore

$$\Gamma\left(\frac{1}{2}\right) = \sqrt{\pi} \quad . \tag{4.1.19}$$

To find now the values of our constant C_n we can use a similar trick, namely that of multiplying the gamma integral by itself and then changing to polar coordinates.

Let us consider the product of n of such integrals:

$$\left[\Gamma\left(\frac{1}{2}\right)\right]^n = \pi^{n/2} = \int_{-\infty}^{\infty}\int_{-\infty}^{\infty} \cdots \int_{-\infty}^{\infty} e^{-r^2}\, dx_1\, dx_2 \ldots dx_n$$

$$= \int_0^{\infty} e^{-r^2}\frac{dV_n(r)}{dr}\, dr \quad , \tag{4.1.20}$$

where $dV_n = dx_1\, dx_2 \ldots dx_n$ is the volume element.

Now compare this with the result for $n = 2$. For $n = 2$ we had (4.1.18)

$$\left[\Gamma\left(\frac{1}{2}\right)\right]^2 = 2\pi\int_0^{\infty} e^{-r^2}\, r\, dr = \int_0^{\infty} e^{-r^2}\frac{d(\pi r^2)}{dr}\, dr \quad . \tag{4.1.21}$$

It is now evident that we can write

$$\left[\Gamma\left(\frac{1}{2}\right)\right]^n = \pi^{n/2} = C_n\int_0^{\infty} e^{-r^2} n r^{n-1}\, dr \quad . \tag{4.1.22}$$

Setting $r^2 = t$, $dr = (t^{-1/2}/2)dt$, we get

$$\pi^{n/2} = \frac{nC_n}{2}\int_0^{\infty} e^{-t}\frac{t^{(n-1)/2}}{t^{1/2}}\, dt = \frac{nC_n}{2}\int_0^{\infty} e^{-t} t^{(n/2)-1}\, dt$$

$$= \frac{nC_n}{2}\Gamma\left(\frac{n}{2}\right) = C_n\,\Gamma\left(\frac{n}{2}+1\right) \quad . \tag{4.1.23}$$

Therefore,

$$C_n = \frac{\pi^{n/2}}{\Gamma\left(\frac{n}{2}+1\right)} \tag{4.1.24}$$

and

$$C_n = \frac{2\pi}{n} C_{n-2} \quad . \tag{4.1.25}$$

So we get

$$V_n(r) = C_n r^n = \frac{\pi^{n/2} r^n}{\Gamma\left(\frac{n}{2}+1\right)} \quad , \tag{4.1.26}$$

a formula we have already used in our calculations for the entropy of an n-dimensional optical image (Sect.3.4).

An interesting result refers to the way the volume of the unit sphere ($r = 1$), i.e., the coefficient C_n, changes with the number of dimensions of the Euclidean space (Table 4.1).

To our surprise we discover that the volume of the unit-radius sphere increases with n and reaches a maximum at $n = 5$, but falls off rapidly toward zero as n approaches infinity! For $n = 2K$ the volume of an n-dimensional sphere of radius r is

$$\frac{\pi^K}{K!} r^{2K} = \frac{(\pi r^2)^K}{K!} \quad . \tag{4.1.27}$$

n	C_n
1	2
2	3.4
3	4.188
4	4.93
5	5.26
6	5.16
7	4.72
8	4.05
2K	$\dfrac{\pi^K}{K!} \to 0$ as $K \to \infty$

Once $K > \pi r^2$, then increasing K (or n) will *decrease* the volume. Indeed, given any radius r, no matter how large, the number of dimensions of the space can be increased until the volume of the sphere is arbitrarily small. [For n = odd, $V_n(r)$ from (4.1.26) changes smoothly with increasing n.]

Let us now consider the fraction of the volume of an n-dimensional sphere that is within a layer (or shell) of any arbitrarily small thickness at the surface. We get

$$\frac{\text{shell}}{\text{volume}} = \frac{C_n r^n - C_n(r - \delta)^n}{C_n r^n} = 1 - \left(1 - \frac{\delta}{r}\right)^n , \qquad (4.1.28)$$

which tends to 1 as n gets larger for given δ. Thus we discover another surprising characteristic of an n-dimensional sphere, namely that almost all the volume of such a sphere is concentrated, so to speak, arbitrarily close to the surface: there is simply not much volume inside a high-dimensional sphere; almost all of it is *on the surface*!

Let us end this digression with one more staggering "paradox" in the world of n-dimensional spheres (see Fig.4.3). Suppose we are given a 4 × 4 square centered about the origin (0,0) and consider four unit-radius circles, one in each of the four corners. Next consider the radius of the circle about the origin that is tangent to the points of the four circles nearest the origin. This circle has a radius

$$r_2 = \sqrt{(1 - 0)^2 + (1 - 0)^2} - 1 = \sqrt{2} - 1 = 0.414... \quad . \qquad (4.1.29)$$

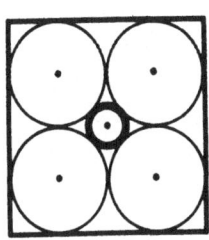

Fig.4.3. The "paradox" of the central sphere in n-dimensional Euclidean space (see text)

Next consider the similar problem in three dimensions, i.e., a $4 \times 4 \times 4$ cube with *eight* unit spheres in its corners. The inner central sphere about the origin has a radius

$$r_3 = \sqrt{(1 - 0)^2 + (1 - 0)^2 + (1 - 0)^2} - 1 = \sqrt{3} - 1 = 0.732\ldots \quad . \tag{4.1.30}$$

Finally, let us go to n dimensions. We have a "hyper"-cube $4 \times 4 \times \ldots 4$ with 2^n unit spheres in the corners, each one touching all its n neighboring spheres. The radius of the inner sphere is going to be $r_n = \sqrt{n} - 1$, with say $n = 10$; this gives

$$r_{10} = \sqrt{10} - 1 = 3.16 - 1 = 2.16 > 2 \quad , \tag{4.1.31}$$

which means that the inner sphere extends itself outside the cube!

To accentuate the apparent paradox further, consider the volume of this n-dimensional central sphere relative to the whole cube volume as a function of n. Let us take (to simplify the calculations) n even, i.e., $n = 2K$. We have

$$
\frac{\text{volume of the central sphere}}{\text{volume of the cube}} = C_n \frac{(\sqrt{n} - 1)^n}{4^n} = C_{2K} \frac{(\sqrt{2K} - 1)^{2K}}{4^{2K}}
$$

$$
= \frac{\pi^K (\sqrt{2})^{2K}}{K!} \frac{(\sqrt{K})^{2K}}{4^{2K}} \left[\left(1 - \frac{1}{\sqrt{2K}}\right)^{\sqrt{2K}} \right]^{\sqrt{2K}}
$$

$$
= \frac{\pi^K 2^K K^K}{4^{2K} K!} \left[\left(1 - \frac{1}{\sqrt{2K}}\right)^{\sqrt{2K}} \right]^{\sqrt{2K}} \quad . \tag{4.1.32}
$$

Using Stirling's approximation for $K!$, namely $K! \cong K^K e^{-K} \sqrt{2\pi K}$ and the result $\lim_{X \to \infty} [1 - (1/x)]^X = 1/e$, we get

$$
\frac{\text{volume of the central sphere}}{\text{volume of the cube}} \cong \frac{\pi^K 2^K K^K}{4^{2K} K^K e^{-K} \sqrt{2\pi K}} e^{-\sqrt{2K}}
$$

$$
= \left(\frac{\pi e}{8}\right)^K \frac{e^{-\sqrt{2K}}}{\sqrt{2\pi K}} \quad , \tag{4.1.33}
$$

and since

$$\frac{\pi e}{8} \sim 1.0675 \quad ,$$

we observe that as K increases, the factor $(\pi e/8)^K$ will go to infinity *faster* than the other two factors $\exp(-\sqrt{2K})$ and $1/\sqrt{2\pi K}$ go to zero. We therefore conclude that as $K \to \infty$, the volume of the central (inner) sphere becomes arbitrarily larger than the volume of the cube which contains all the $2^{2K} = 2^n$ unit spheres in its corners. (The same conclusion is reached when $n = 2K - 1$; the calculations are lengthier.)

Finally, imagine what happens to the hypervector from the origin to the point $(1,1,1,\ldots,1)$ as n increases. Since the direction cosines of the hypervector are all equal to $\cos\theta = 1/\sqrt{n}$, we observe that as $n \to \infty$ the hypervector becomes "almost" perpendicular to *all* coordinate axes! This is a result of great significance in designing efficient codes to combat noise in the channel (Sect.4.2).

At this point we close the "parenthesis" on the intricacies of the multidimensional Euclidean space and go back to the treatment of information transfer and optimum reception of continuous waveforms —which by now are *all substituted by hypervectors in n(=2WT)-dimensional Euclidean space.*

Specifically, our emitted pattern X(t) stands now as a hypervector $\mathbf{X}(t)$ with coordinates $x_1, x_2, \ldots, x_{2WT}$ whose amplitudes are exactly the amplitudes of X(t) at the (sampling) moments 0, 1/2W, 2/2W,..., 2WT/2W = T s. The power of the (deterministic) transmitted signal (wave form) will be equal to

$$<X^2> = \sum_{i=1}^{2WT} <x_i^2> = 2WT<x_i^2> = 2WT\sigma_0^2 , \qquad (4.1.34)$$

where σ_0^2 is the variance (mean signal power) of $\mathbf{X}(t)$. Let us come now to noise: the noise in the channel can also be sampled in the same way as the deterministic signal and gives rise to a hypervector $\mathbf{N}(t)$ in the same 2WT-dimensional space. The obvious difference is that now the samples of noise $n_1, n_2, \ldots, n_{2WT}$ are not deterministic coordinates but (each one of them) one-dimensional probability density functions (p.d.fs). The noise hypervector will play then the role of error-vector in the state space.

The received signal Y(t) also will be represented by a hypervector $\mathbf{Y}(t)$. Provided that the noise has relatively small power so that it can be considered as *additive*, i.e., as affecting only the amplitude and not the phase of the transmitted signal $\mathbf{X}(t)$, we can write at the receiving end, after the high-frequency EM carrier has been filtered out,

$$\mathbf{Y}(t) = \mathbf{X}(t) + \mathbf{N}(T) , \qquad (4.1.35)$$

or, in terms of power

$$<Y^2> = X^2 + <N^2> = X^2 + \sum_{i=1}^{2WT} <n_i^2> = X^2 + 2WT<n_i^2> = X^2 + 2WT\sigma^2 . \qquad (4.1.36)$$

In the above expression it has been assumed that the signal and the noise are completely uncorrelated so that the cross term $2<X \cdot N>$ can be dropped. This assumption is valid for thermal noise. Also we assumed that all the p.d.f. of the noise samples $p(n_i)$ have zero mean and the same variance σ^2. We conclude that the result of the noise's action on the transmitted signal is the surrounding, in 2WT state space, of the edge of the hypervector $\mathbf{X}$ with a sphere, or rather a cloud of "uncertainty", with radius r_N having some (still not known) p.d.f. $p(r_N)$ and a root-mean-square value $r = \sigma\sqrt{2WT}$.

Let us now define a new parameter (and subsequently calculate it in a number of representative cases): the *channel capacity* of a given medium/environment, mediating the transactions between the source of information and the receiver.

We call the *channel capacity* of a medium the maximum number of information units (bits) that the medium can allow to pass through it per unit time, without interference error caused by the thermodynamic state of the channel (that is, by the "turbulence" in the channel). In other words, channel capacity is the maximum rate of errorless signaling through the medium concerned, measured in bits/s.

For example, suppose a medium allows the transmission of equal-length rectangular pulses of ξ different amplitudes and unit width at a rate up to ν pulses per second —without any observable distortion. Then the channel capacity of such a medium equals

$$C = \nu \log_2 \xi (= \log_2 \xi^\nu) \quad \text{bits/s} \quad , \tag{4.1.37}$$

as this the *maximum* information content of the above signal.

In the general case the channel capacity is defined as

$$C = \frac{1}{T} \max I(X \rightarrow Y) \quad , \tag{4.1.38}$$

provided that the rate of information transmission is constant. Substituting for the expression $I(X \rightarrow Y)$ from (4.1.7a), we get

$$C = \frac{1}{T} \max\{S(X) - S(X/Y)\} = \frac{1}{T} \max\{S(Y) - S(Y/X)\} \quad , \tag{4.1.39}$$

where the maximization operation will take place with reference to all possible mappings (coding) of the signal $X(t)$. Since the noise N is considered here as statistically independent of the signal X, we can write

$$S(Y) = S(X) + S(N) \quad . \tag{4.1.40}$$

In order to have errorless information transfer we should minimize the conditional entropy of the signal, i.e., put $S(X/Y) = 0$, which implies

$$S(Y/X) = S(Y,X) - S(X) = S(Y) + S(X/Y) - S(X)$$
$$= S(Y) - S(X) = S(X) + S(N) - S(X) = S(N) \quad . \tag{4.1.41}$$

This simply means that the existence of the channel noise N is the only reason the transmitter does not know exactly what the receiver will "see". Therefore we have

$$C = \frac{1}{T} \max [S(Y) - S(N)] \quad \text{bits/s} \quad . \tag{4.1.42}$$

We will now maximize $S(Y) - S(N)$, subject to the worst possible case of additive noise, that is, for noise making $S(N)$ maximum; $S(N)$ stands for the entropy of a 2WT-dimensional random process.

We can write $S(N) = S(n_1, n_2, \ldots, n_{2WT})$; in general,

$$S(n_1, n_2, \ldots, n_{2WT}) \leq \sum_{i=1}^{2WT} S(n_i) \quad , \tag{4.1.43}$$

the equality happening only when the random variables $n_1, n_2, \ldots, n_{2WT}$ are statistically independent. The worst case is therefore realized when

$$S(N) = \sum_{i=1}^{2WT} S(n_i) \quad . \tag{4.1.44}$$

We have now to search for the p.d.f. which under the given power σ^2 gives the maximum $S(n_i)$. But this problem has already been solved in Sect.2.3.2. To have for all $i \in (1,\ldots,2WT)$, $S(n_i)$ a maximum for given power σ^2, n_i must be Gaussian distributed. If so, then

$$S(n_i) = \log_2(\sigma\sqrt{2\pi e}) \quad , \tag{4.1.45}$$

where σ is the common variance for all one-dimensional p.d.f.s $p(n_i)$, $i \in (1,\ldots,2WT)$. Finally

$$S(\mathbf{N})_{max} = \sum_{i=1}^{2WT} \log_2(\sigma\sqrt{2\pi e}) = WT \log_2(2\pi e\sigma^2) \quad . \tag{4.1.46}$$

We conclude then that the worst form of additive multidimensional noise is the one where all the one-dimensional components n_i are Gaussian and *stochastically independent* of each other. This stochastic independence in turn implies that the autocorrelation function $B(\tau)$ of $N(t)$ should go to zero only at the discrete instances

$$\tau = \frac{1}{2W} \quad , \quad \frac{2}{2W} \quad , \quad \frac{3}{2W} \quad ,\ldots, T \quad .$$

Such an autocorrelation function has the form

$$B(\tau) = \sigma^2 \frac{\sin(W\tau)}{W\tau} \quad . \tag{4.1.47}$$

Moreover, if we take the Fourier transform of $B(\tau)$ in order to arrive at the noise spectrum, we find that the spectrum of $N(t)$ must be flat for all frequencies W. So, we conclude, the worst possible additive noise has a multidimensional p.d.f. which is *Gaussian* and a corresponding frequency spectrum which is *flat* ("white" noise), within the frequency band W.

In order now to carry out the maximization process in (4.1.42), it is obvious that the hypervector $\mathbf{Y} = \mathbf{X} + \mathbf{N}$ must also possess a multidimensional Gaussian distribution and a flat spectrum. In turn, this means that the information-carrying hypervector $\mathbf{X}$ must be *coded* (Sect.4.2), i.e., mapped in such a way as to be transformed into a canonical p.d.f. with a white spectrum as well [at the receiving end the inverse operation —decoding —must be carried out in order to detect the original signal $X(t)$]. Therefore, we must have

$$S(\mathbf{Y})_{max} = WT \log_2(2\pi e\langle y^2 \rangle)$$

$$= WT \log_2[2\pi e(\langle x^2 \rangle + \sigma^2)] = WT \log_2[2\pi e\sigma^2(1 + \Gamma)] \quad , \tag{4.1.48}$$

where

$$\Gamma = \frac{\langle x^2 \rangle}{\sigma^2} = \text{signal-to-noise ratio} \quad . \tag{4.1.49}$$

Substituting (4.1.46,48) in (4.1.42) we finally get the expression for the channel capacity

$$C = W \log_2(1 + \Gamma) \quad \text{bits/s} \quad . \tag{4.1.50}$$

For $\Gamma \gg 1$ (strong signal and/or weak noise)

$$C \sim W \log_2 \Gamma ,$$

and for $\Gamma \ll 1$ (weak signal, strong noise)

$$1 + \Gamma \sim e^{\Gamma} , \qquad C \sim W\Gamma \log_2 e , \qquad \text{or} \qquad C \sim 1.443 \, W\Gamma .$$

What is remarkable in (4.1.50) —known as Shannon's formula— is that the channel capacity *remains finite* even in cases where the signal is many orders of magnitude *below* the ambient noise, and that the value of C can be improved by increasing the bandwidth W of the information-carrying signal.

Before proceeding further, let us give a specific example involving the use of (4.1.50). Suppose that the signal consists of a string of discrete rectangular pulses with equiprobable amplitudes +1V or -1V (Fig.4.4). The bandwidth of the channel W equals 100 Hz, and the transmission takes place in the presence of Gaussian white noise of spectral density $\sim 0.15 \times 10^{-3}$ W/Hz. The total noise power is $\sigma^2 = 0.3$ W (on a bandwidth ±1000 Hz) and the one-dimensional p.d.f. of the noise is

$$P(n) = \frac{1}{\sqrt{2\pi} \times 0.3} \exp\!\left(\frac{-n^2}{2 \times 0.3}\right) . \tag{4.1.51}$$

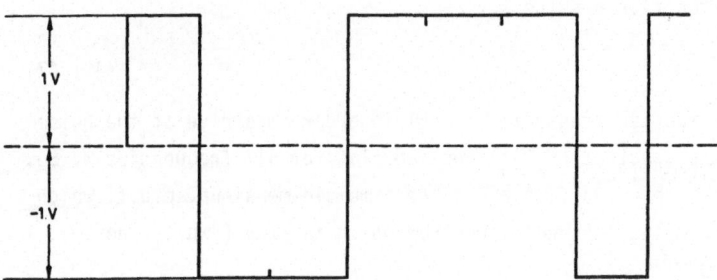

Fig.4.4. A random sequence of +1, -1 pulses

The received string of pulses —more or less contaminated with noise (whose action involves the change of a positive pulse into a negative one or vice versa) —is interpreted by considering the string at any point (see Fig.4.4) between two successive pulses.

We consider that the noise has distorted a positive pulse when the noise is more negative than -1V, and that the distortion involves a negative pulse if the noise exceeds +1V.

Since we consider the probabilities of affecting a positive or a negative pulse equal, the overall probability of receiving an erroneous pulse is

$$P = \frac{1}{2} \int_{-\infty}^{-1} P(n)dn + \frac{1}{2} \int_{1}^{\infty} P(n)dn = \int_{-\infty}^{-1} P(n)dn$$

$$= \frac{1}{\sqrt{2\pi} \times 0.3} \int_{-\infty}^{-1} \exp\!\left(\frac{-n^2}{2 \times 0.3}\right)dn$$

$$= \frac{1}{\sqrt{2\pi}} \int_{-\infty}^{-1/\sqrt{0.3}} e^{-y^2/2} \, dy \sim 0.0340 \quad . \tag{4.1.52}$$

This means that 3.4% of the received pulses in a string are erroneous.

The information transfer $I(X \to Y)$ per pulse from transmitter to receiver will then be calculated from the expression

$$I(X \to Y) = S(X) - S(X/Y) = - \left[\frac{1}{2} \log_2 \frac{1}{2} + \frac{1}{2} \log_2 \frac{1}{2} \right]$$

$$+ (0.034 \log_2 0.034 + 0.966 \log_2 0.966)$$

$$= 0.786 \text{ bits/pulse} \quad . \tag{4.1.53}$$

The rate of pulse transmission is, of course, 2000 pulses per second, so the rate of information transmission of the examined system is just

$$\frac{I(X \to Y)}{T} = 2000 \times 0.786 = 1572 \text{ bits/s} \quad . \tag{4.1.54}$$

Let us see how this rate fares versus the "theoretical maximum" given by Shannon's formula. The signal-to-noise ratio is

$$\Gamma = \frac{1}{0.3}$$

and

$$C = W \log_2(1 + \Gamma) = 1000 \log_2(1 + 3.333) = 2115.5 \text{ bits/s} \quad . \tag{4.1.55}$$

So the system transmits at a suboptimal rate equal to 74.5% of the theoretical maximum, and furthermore it errs by 3.4%. (We are going to see that with the proper coding, the actual rate can asymptotically approach Shannon's limit.)

4.2 Some Ideas from Coding Theory Instrumental in Minimizing Reception Error

Most conventional one-way communication links conform to the "block diagram" given in Fig.4.5. Let us briefly explain the functional significance of each "box", going from left to right.

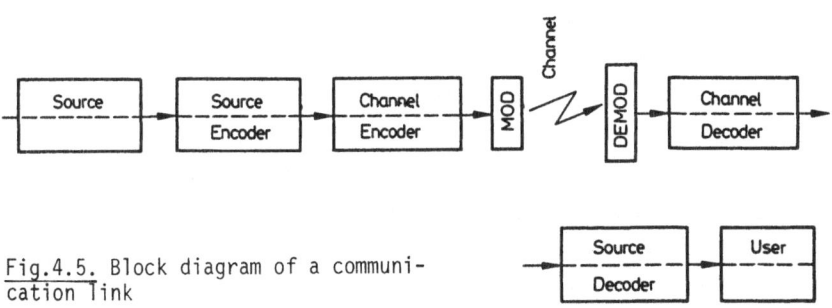

Fig.4.5. Block diagram of a communication link

The source *generates* information (the dynamics of which will be examined in Chap.6) that the sender should communicate to the user. It possesses a continuous or discrete collection of "patterns" whose total number and a priori probability distribution are known to the user in advance. The channel is a physical medium that spatially intervenes between the two communicating partners/systems. Noise in the channel can be additive or multiplicative, memoryless or Markovian. (The last property means that the errors inflicted by the noise on the transmitted message may occur either independently of each other or in "groups" or in "bursts". This implies some degree of stochastic interdependence among them.) In any case, the *only* parameter of a channel of strict operational significance is its capacity.

The source encoder is a device that transforms the source output into a form suitable for actual transmission over the channel. Likewise the source decoder converts the channel output into a form that can be *interpreted* by the user. More specifically, the function of the source encoder is to *map* the output $x(t)$ produced by the source over some time interval, into one of a (finite) set of preselected messages (strings of pulses or "symbols"). That is to say, the space of possible source outputs over the time interval in question is partitioned into a set of equivalent classes. The source encoder indicates to the *channel encoder* which of these classes contains the particular source output observed. The operation is repeated in consecutive time intervals. The *channel encoder*, upon being informed that the source output belongs to the mth equivalent class, transmits the wave form $\bar{x}_m(t)$ over the channel. The *channel decoder* examines the received noise-contaminated wave form $y(t)$ over an appropriate time interval, and makes a decision regarding which message was sent.

The channel decoder then forwards its estimate m' of the message number to the source decoder, which in turn presents $\bar{y}(t)$ to the user as the system's estimate of $x(t)$ over the m' time interval in question. (For a detailed example of the above process see Sect.4.7 and also Appendix A.)

Finally the MOD operation involves the modulation of one out of the four possible parameters of a high-frequency EM carrier (amplitude, frequency, phase, polarization) by the instantaneous amplitude of the wave form (analogue or digital) emanating from the output of the channel encoder. The DEMOD involves the inverse operation, the rejection of the carrier EM wave and the forwarding of the demodulated message into the input of the channel decoder.

Perhaps by now the reader understands that the most involved operation takes place in the *channel decoder*. There is essentially a "memory box" in the channel decoder containing an exact replica of all members of the repertoire of the channel encoder. Upon receiving the particular contaminated wave form, the channel decoder cross-correlates it with *all* members of its memory box, and selects as the most probable wave form the wave-form member of its memory box which forms with the incoming wave form the greatest cross-correlation. Let us finally note that the source encoder and the channel decoder perform *many-to-one mapping* operations, while the channel encoder and the source decoder merely implement *one-to-one mappings* that transform the decisions rendered by the source encoder and the channel decoder into inputs that are acceptable to the channel and the user respectively. Thus the source encoder and the channel decoder are *hierarchically higher* than the channel encoder and the source decoder.

Let us approach the problem of coding and decoding in steps, by asking first the following question: how can one understand, dynamically implemented, something that Shannon's equation (4.1.50) formally predicts, namely that *unambiguous* information transfer *can* take place even if the signal is *as small as we want* vis-à-vis the channel's noise? In the first place, what type of wave form do we identify as "signal" and what as "noise"?

We call "signal" a wave form whose phase φ_S remains coherent for all times, Sect.2.2.4, with respect to the *phase* φ_R *of some reference oscillator* of the specific receiver, i.e.,

$$|\varphi_S(t) - \varphi_R(t)| < \Phi \ll 2\pi \quad ; \tag{4.2.1}$$

thus the characterization is strictly contextual.

In a similar way, any wave form whose phase is incoherent with respect to the phase of the reference oscillator of the receiver is termed "noise"; in such cases the phase difference

$$|\varphi_S(t) - \varphi_R(t)| \sim 2\pi \tag{4.2.2}$$

is unbounded.

Suppose we receive a series of pulses highly contaminated by noise, under conditions $\Gamma \ll 1$. The only parameter we know is the period T. How can we detect (and amplify) the weak signal (if any) in a "sea" of overwhelming noise? Suppose the noise is additive and has the following two first moments $\langle n(t) \rangle = 0$, $\langle n^2(t) \rangle = \sigma^2$. The received wave form will be $Y(t) = X(t) + n(t)$. We may use a processing system, essentially adopted from biological organisms, known as the storage and integration procedure. It has to do with storing the Kth pulse for T seconds and then superimposing it on the incoming $(K+1)^{th}$ pulse — in real time. If we use such a "reception algorithm" m times in succession, the total signal at the output of the storage unit will look like

$$Z(t) = \sum_{K=1}^{m} Y(t + KT) = \sum_{K=1}^{m} X(t + KT) + \sum_{K=1}^{m} n(t + KT) \quad . \tag{4.2.3}$$

But the signal is periodic $X(t + KT) = X(t)$, so

$$Z(t) = mX(t) + N(t) \quad , \tag{4.2.4}$$

where

$$N(t) = \sum_{K=1}^{m} n(t + KT) = \sum_{K=1}^{m} n_K \quad . \tag{4.2.5}$$

Here n_K are samples of the noise taken every T seconds. If the correlation time of noise is far smaller than T, then the variance $D(N)$ of N will equal the sum of variances of the individual noise segments

$$D(N) = mD(n(t)) = m\sigma^2 \ . \tag{4.2.6}$$

So, provided also that signal and noise are uncorrelated, we will have at the output of the storage and integration unit

$$<Z^2> = m^2X^2(t) + m\sigma^2 \ , \tag{4.2.7}$$

which means that the signal-to-noise ratio at the output equals

$$\Gamma_{out} = \frac{mX^2}{\sigma^2} = m\Gamma_{input} \ , \tag{4.2.8}$$

i.e., it is *m times higher*. So we may start with a figure of $\Gamma_{input} \sim 10^{-6}$, and if $m = 10^9$, we come up with an output signal-to-noise ratio of the order of 30 dB.

The second question we pose with reference to the problem of efficient decoding is as follows. We know that the received signal **Y** constitutes a hypervector in a 2WT-dimensional Euclidian space. In order to recover the transmitted signal **X**, we have essentially to penetrate a spherical "cloud" of noise surrounding **X** (Fig.4.6). How can we detect **X** from **Y** unambiguously? We sketch the general process as follows.

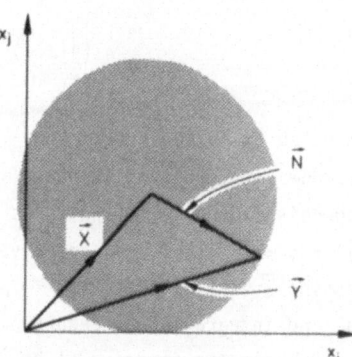

Fig.4.6. The noise (fuzzy) sphere centered at the unknown signal in 2WT-dimensional state space

We must first of all specify as closely as possible the statistics of the "noise cloud". This means we have to discover the probability density function $P(r)$ of the radius r of the noise hypersphere in the 2WT-dimensional space. We have for this radius

$$r^2 = \sum_{i=1}^{2WT} n_i^2 \ ,$$

or

$$r = \sqrt{\sum_{i=1}^{2WT} n_i^2} \tag{4.2.9}$$

where n_i are the individual one-dimensional samples (projections) of the noise hypervector N. The 2WT-dimensional distribution $P(n_1,n_2,\ldots,n_{2WT})$ is simply the product of the individual distributions $P(n_i)$, $i \in (1,\ldots,2WT)$ since the samples n_i are statistically independent. Let us call $2WT = K$; then,

$$P(n_1,n_2,\ldots,n_K) = \frac{1}{(2\pi\sigma^2)^{K/2}} \exp\left(-\sum_{i=1}^{K} \frac{n_i^2}{2\sigma^2}\right) . \tag{4.2.10}$$

Next we transform to spherical coordinates in K-dimensional space with an element of volume in the two coordinate systems related as follows (see also *Beckman* [4.2]):

$$dn_1 dn_2 \ldots dn_K = \frac{\partial(n_1,n_2,\ldots,n_K)}{\partial(r,\theta_1,\theta_2,\ldots,\theta_{K-1})} dr\, d\theta_1\, d\theta_2 \ldots d\theta_{K-1}$$

$$= r^{K-1} F(\theta_1,\theta_2,\ldots,\theta_{K-1})\, dr\, d\theta_1\, d\theta_2 \ldots d\theta_{K-1} , \tag{4.2.11}$$

where $\partial(n_1,n_2,\ldots,n_K)/\partial(r,\theta_1,\theta_2,\ldots,\theta_{K-1})$ is the Jacobian, and $\theta_1,\ldots,\theta_{K-1}$ are the corresponding polar angles.

The p.d.f. of the noise radius r may be calculated from

$$P(r) = \frac{1}{(2\pi\sigma^2)^{K/2}} \int_0^\pi \int_0^\pi \cdots \int_0^\pi e^{-r^2/2\sigma^2} r^{K-1} F\, d\theta_1\, d\theta_2 \ldots d\theta_{K-1}$$

$$= C_K r^{K-1} e^{-r^2/2\sigma^2} , \tag{4.2.12}$$

where

$$C_K = \frac{1}{(2\pi\sigma^2)^{K/2}} \int_0^\pi \int_0^\pi \cdots \int_0^\pi F\, d\theta_1\, d\theta_2 \ldots d\theta_{K-1} \tag{4.2.13}$$

is a constant whose value is found from the normalization condition

$$\int_0^\infty P(r)dr = 1 = C_K \int_0^\infty r^{K-1} e^{-r^2/2\sigma^2} dr = C_K 2^{[(K/2)-1]} \sigma^K \left(\frac{K}{2} - 1\right)!$$

or

$$C_K = \frac{1}{2^{[(K/2)-1]} \sigma^K \left(\frac{K}{2} - 1\right)!} . \tag{4.2.14}$$

Therefore

$$P(r) = \frac{1}{2^{[(K/2)-1]} \sigma^K \left(\frac{K}{2} - 1\right)!} r^{K-1} e^{-r^2/2\sigma^2} , \tag{4.2.15}$$

i.e., it follows what is known as the "chi"-distribution.

Normalizing r to the root-mean-square (rms) σ, i.e., introducing the dimensionless parameter $\rho = r/r_{(rms)}$, where $r_{(rms)} = \sigma\sqrt{K}$, we get for our p.d.f.

$$P(\rho) = C_K \sigma^K K^{K/2} \rho^{K-1} e^{-\rho^2 K/2}$$

$$= \xi_K \exp\left[-\rho^2 \frac{K}{2} + (K - 1)\ln\rho\right] \quad , \tag{4.2.16}$$

where $\xi_K = C_k \sigma^K K^{K/2}$.

Let us examine closely this (final) expression for the p.d.f. of the (normalized) radius of the hypersphere of noise. The p.d.f. has a single maximum for

$$\frac{\partial P(\rho)}{\partial \rho} = 0 \quad , \tag{4.2.17}$$

or

$$\frac{d}{d\rho}\left[-\rho^2 \frac{K}{2} + (K - 1)\ln\rho\right] = 0 \quad , \tag{4.2.18}$$

or

$$\rho = \rho_0 = \sqrt{\frac{K - 1}{K}} = \sqrt{1 - \frac{1}{2WT}} \quad . \tag{4.2.19}$$

As W (or T) increases, $\rho_0 \to 1$. This means that the most probable value for the radius r of the noise hypersphere is always *smaller* than $\sigma\sqrt{2WT}$, but it tends asymptotically to this rms value as either the bandwidth or the time duration of the signal increases. Let us next examine how *sharp* this maximum is, i.e., how "well defined", distinct, or solid the surface of the noise hypersphere can be. This amounts to finding the absolute value of the second derivative of the exponent in (4.2.16) at the point ρ_0, which leads to the value 2K = 4WT. As $K \to \infty$ we see therefore that the sharpness of the maximum ($\rho_0 \to 1$) becomes *infinite*, which means that at this limit all the volume of the noise hypersphere is concentrated on the surface of a "hollow" spherical shell of radius $\sigma\sqrt{2WT}$ —something we already have proved earlier, in Sect.4.1. In such a case, however, the detection of the unknown signal X(t) from the measurement of Y(t) becomes unambiguous, as illustrated in Fig.4.7, since the noise hypersphere behaves no more as a cloud, but as a "hairless" deterministic spherical shell (a sharply defined boundary) of radius $\sigma\sqrt{2WT}$. Thus for large K = 2WT values we are justified in abandoning the probabilistic treatment of distortion and regarding the received signal as *corrigible*, if the noise spheres with deterministic radius $\sigma\sqrt{2WT}$ surrounding the signal points are so distant from each

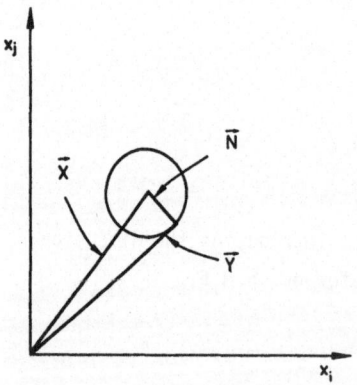

Fig.4.7. As 2WT→∞, the noise sphere loses "hair" and becomes "hard"

other that a distorted received signal can then be unambiguously intepreted as having orginated from the only signal located at a distance $\sigma\sqrt{2WT}$ from the received
signal. So increasing the bandwidth (using frequency modulation for example) has
three effects:

a) It sharpens the boundaries of the noise spheres.
b) It fixes their radii to the value $\sigma\sqrt{2WT}$.
c) It makes the difference signals almost *orthogonal* in state space, thereby minimizing their cross-correlations and facilitating the decision-making process at the
   output of the channel decoder.

Increasing the "other" parameter in Shannon's formula, namely the signal-to-noise
ratio, has the effect of inflating the signal with respect to the noise and separating the noise spheres. Indeed all signals with power X^2 will be mapped onto the surface of a sphere of radius $X\sqrt{2WT}$, *each* surrounded by a noise sphere of radius $\sigma\sqrt{2WT}$.
If σ remains constant but X increases, then the expanding signal sphere will spread
the noise spheres further apart from each other. On the other hand, if Γ decreases,
the radius of the signal sphere will shrink; for $\Gamma < 1$ the noise spheres will certainly overlap, but fortunately this does not cause ambiguity in reliable reception
as we already have proved —provided that W increases. Even if the (sharp) noise
spherical shells do overlap, the distorted message can still be correctly deciphered,
provided we interpret it as shown above, that is, as having originated not (necessarily) from the nearest signal point but from the only one at distance $\sigma\sqrt{2WT}$ from
the observed point[1]. In conclusion then, we can say that the whole process of
transmission and reception of a set of wave forms —hypervectors—involves a sequence
of an "expansion" and a "contraction" of the dimensionality of the state space,
i.e., a *proliferation* and a subsequent *compression* of the degrees of freedom of the
transmitted messages.

Indeed, at the transmitting side, the effort towards efficient coding requires the
orthogonality of the "word"/members of the sender's repertoire. This is accomplished
by expanding the bandwith W or increasing the transmission time T, thereby increas-

---

1 Incidentally, it can be proved that with such a decoding scheme (and for a memoryless channel) the number of decoding steps does *not* increase exponentially (2^n)
with the length n of the code word —as in the case of the minimum-distance decoding —but as $\sim n^2$. Furthermore, it can be shown that using *hierarchical* coding-decoding principles (see, for example, *Forney* [4.3]) one may, in theory, reduce the
computational complexity even further —up to $\sim n$.
   In such a hierarchical coding-decoding scheme one is "nesting" codes within codes.
Specifically, both the coding and decoding processes go stepwise or rather in ascending hierarchy: First, at the lowest level, simple digits ("alphabet") are grouped
into blocks ("words"). Then, at the second level, each of those blocks ("words") is
treated as a new type of digit, and many of them are grouped again into blocks
("sentences"), and so on. In the decoding side the inverse procedure takes place.
The principle of "concatenated" codes is just an attempt to imitate the hierarchy
of natural language.

ing the number of dimensions (2WT) of the state space, where the individual "words" (digitized wave forms via the sampling theorem) figure as hypervectors.

At the receiving end, a "contraction" is performed by the channel decoder amounting to a set of *convolutions* between the incoming (noise-contaminated) signal and each member/word of the transmitter's repertoire. Since the individual noise-free "words" are mutually orthogonal, their pair cross-correlations will be zero. Thus the above operations allow the receiver to detect and correct multiple (albeit a finite number of) errors, which occur due to the channel noise. In the following section we are going to examine exactly how this coding-decoding process is implemented —within the bounds of the Shannon pattern.

4.3 Some Efficient Coding Algorithms for Source-Channel Matching and Single-Error Detection and Correction

The aim of coding is twofold: (a) first, to match the rate of information transmission emanating from the source to the channel capacity, and (b) second, to transform the signal to be transmitted into a string of 0's and 1's in such a way that the ambiguity of selecting the right string or wave form in the channel decoder will be minimized.

4.3.1 Coding for Source-Channel Matching

We are initially concerned with zero-memory sources, i.e., sources with stochastically independent symbols. In a "block code" each state of the source, i.e., each message, is always transformed into the same fixed sequence of code "letters" or symbols called a "code word". We shall be interested in achieving a *maximum* rate of transmission of information; obviously this will be achieved when the code carries maximum information per letter (or per symbol). (With r letters we can form $n = r^\Lambda$ code words of average length Λ.) If we regard the code letters rather than the code words as the states assumed by the code, it follows that for a code using an r-letter alphabet this maximum can be achieved when all the r letters are independent and equiprobable, yielding $S_{max} = \log_2 r$ bits/letter. Let the code word x_i consist of a sequence of λ_i letters. Then the average length of the code will be

$$\Lambda = \sum_{i=1}^{n} \lambda_i P(x_i) \quad \text{symbols/word} \quad , \tag{4.3.1}$$

and the code will carry an average amount of information of

$$\frac{S(X)}{\Lambda} \leq \log_2 r \quad \text{bits/letter} \quad , \tag{4.3.2}$$

where

$$S(X) = - \sum_{i=1}^{n} P(x_i) \log_2 P(x_i) \quad , \tag{4.3.3}$$

$P(x_i)$ being the a priori probability of the message/word x_i. The "code efficiency" is defined as

$$\eta = \frac{S(X)}{\Lambda \log_2 r} \leqslant 1 \quad . \tag{4.3.4}$$

It follows that a 100% efficiency must refer to a code with independent and equiprobable letters. In reality, the code efficiency η *must* be smaller than 100% for two reasons. First, in any language the occurrence of letters is *not* equiprobable (we have variety) —which makes $S(X) < \log_2 r^{\Lambda}$. Second, the letters *are* interdependent (reliability). Thus the information carried by a letter is lowered by the fact that the letter was already expected from the sequence preceding it. This process increases the letter's a priori probability and therefore diminishes its information.

Now let the duration of each state/word of the code be L seconds. The source then puts out messages or code words at a rate $\sim F = 1/L$ messages/s of $f = 1/1$ letters per second, if $f = F\Lambda$, and 1 is the duration of each letter. For a noiseless channel, let the maximum frequency of letters per second that the channel can handle *without error* be f_{max} letters/s.

We would like to know the maximum rate at which the messages x_i of the source can be transmitted through this channel without error, and how to encode the source into the optimum code so as to achieve this maximum —or approach it arbitrarily closely. Since the average information per message put out by the source is $S(X)$, the rate at which the source puts out information is

$$R = FS(X) \quad \text{bits/s} \quad , \tag{4.3.5}$$

so

$$R = \frac{f}{\Lambda} S(X) \quad \text{bits/s} \quad . \tag{4.3.6}$$

The maximum flow of information that the channel can handle without error is

$$C = f_{max} \left(\frac{S(X)}{\Lambda} \right)_{max} = f_{max} \log_2 r \quad \text{bits/s} \quad , \tag{4.3.7}$$

since $S(X)$ cannot exceed $\Lambda \log_2 r$. It therefore follows that messages cannot be transmitted without error at a rate exceeding the channel capacity, i.e., if $F > C/S(X)$ messages/s. Now we observe that the inequality $S(X)/\Lambda \leqslant \log_2 r$ is preserved if we change from binary logarithms to logarithms with base r, i.e., $S_r(X)/\Lambda \leqslant 1$ or

$$- \sum_{i=1}^{n} P(x_i) \log_r P(x_i) \leqslant \sum_{i=1}^{n} \lambda_i P(x_i) \quad . \tag{4.3.8}$$

Thus the channel capacity is achieved for

$$\lambda_i = -\log_r P(x_i) \tag{4.3.9}$$

141

or for

$$P(x_i) = r^{-\lambda_i} \quad , \tag{4.3.10}$$

i.e., for a source whose set of a priori probabilities is coded to follow the above law of making long words more and more improbable. However, even if the state probability distribution of the source does not follow the exponential law (4.3.10), we can still get very close to a 100% efficient code as follows: we rewrite (4.3.9) as

$$-\log_r P(x_i) \leqslant \lambda_i \leqslant -\log_r P(x_i) + 1 \quad , \tag{4.3.11}$$

where the second inequality follows from the fact that we have selected λ_i as the nearest integer to $-\log_2 P(x_i)$.

Multiplying the above relation by $P(x_i)$ and summing over all i, we obtain

$$\frac{S(X)}{\log_2(r)} \leqslant \Lambda \leqslant \frac{S(X)}{\log_2(r)} + 1 \quad . \tag{4.3.12}$$

Now we consider entire sequences of the source, i.e., sequences of messages/states taken m at a time; there will be a total of n^m such sequences. Let us regard each such sequence as a *new* message/state, belonging to a higher hierarchical level, and consider this new source with n^m state. (This is called the mth extension, $X^{(m)}$, of the original source X.) Since we have assumed that the x_i's are independent, we have

$$S(X^m) = mS(X) \quad , \tag{4.3.13}$$

and the mean length of the code words encoding each sequence will be $m\Lambda$ letters. Substituting into (4.3.12), we get

$$\frac{mS(X)}{\log_2 r} \leqslant m\Lambda \leqslant \frac{mS(X)}{\log_2 r} + 1$$

or

$$\eta \leqslant 1 \leqslant \eta + \frac{1}{m\Lambda} \quad . \tag{4.3.14}$$

Since we can make m as large as we please by encoding sufficiently long sequences of messages x_i, we can make the efficiency η of the new code differ from unity as little as possible —by paying, of course, the price of increasing complexity in coding.

Before moving further, let us briefly examine how, in practice, we can implement the optimum distribution (4.3.10) so as to maximize the information transmission rate, that is, make the rate of information transmission asymptotically approach the channel capacity. Suppose that a source has a repertoire of say eight independent messages, A, B, C, D, E, F, G, and H with the a priori probabilities of occurrence given in Table 4.2.

Table 4.2. The a priori probabilities of occurrence of the eight possible independent messages of the source described in the text

x_i	A	B	C	D	E	F	G	H
$P(x_i)$	0.1	0.18	0.4	0.05	0.06	0.1	0.07	0.04

Moreover, suppose that messages occur at a rate of one per second. The entropy of
the source, calculated by using the expression $-\Sigma p_i \log_2 p_i$, gives away ~2.55 bits/s.
If the set of messages is transmitted over a channel in which the signal can assume
any one of eight voltage levels, the source can be coded by letting each of these
levels represent a message. The maximum information content of an eight-level signal
is $\log_2 8 = 3$ bits, and a channel in which independent levels occur every second has
a capacity of $C = 3$ bits/s. The coding efficiency of the present scheme is therefore
$(2.55 \times 100)/3 = 85\%$. To improve this number, according to the preceding theoretical
analysis, we have to rearrange the eight messages in such a way as to make the *less
probable* ones *more lengthy*.

   Thus the messages are arranged in descending order of probability. This series
is divided to give two groups of as nearly equal probability as possible. Then these
groups are similarly subdivided, and the process repeats itself until each message
is isolated. At each division, the messages in the first group are assigned the
symbol 0 and those in the second group the symbol 1. The coded binary form of each
message is then given by the appropriate sequence of these binary digits. The procedure is illustrated in Table 4.3.

Table 4.3. Coding the eight independent messages of the source in Table 4.2

Message	C	B	A	F	G	E	D	H
Probability	0.4	0.18	0.1	0.1	0.07	0.06	0.05	0.04
	0	1	0	1	0	1	0	1
					0		1	
			0			1		
		0				1		
Code	00	01	100	101	1100	1101	1110	1111

   The capacity required by a binary channel to transmit these signals at a rate of
one per second is obtained by finding the average rate of occurrence of the binary
digits. In a long time T, two-digit singals, for example, will occur (0.4 +0.18)T
times, and the total number of digits is given by

$$T[2(0.4 + 0.18) + 3(0.1 + 0.1) + 4(0.07 + 0.06 + 0.05 + 0.04)]$$

$$= 2.64 \ T \ \text{digits} \ .$$

The capacity of a channel transmitting 2.64 digits per second is $C = 2.64$ bits/s, giving a coding efficiency $(2.55 \times 100))/2.64 = 96.6\%$. By introducing slight modifications to the above algorithm, we can asymptotically reach a 100% efficiency (see, for example, *Huffman* [4.4]).

4.3.2 Coding for Error Detection and Correction

We are now interested in the case of coding aimed at recognizing and correcting (single, or multiple but uncorrelated) errors due to the channel noise. Consider a string of n 0's and 1's as a point in n-dimensional space. Each digit gives the value of the corresponding coordinate in the n-dimensional space (where we are assuming that the encoded message is exactly n bits long). Thus we have a "cube" in n-dimensional space; each vertex is a string of n 0's and 1's. The space consists *only* of the 2^n vertices; this is sometimes called a "vector space". Each vertex is a possible received message, but only selected vertices are to be original messages. A *single* error in a message moves the message point along one edge of the hypercube to an immediately adjacent point. If we require that every possible originating message be at least a distance of two sides away from any other message point, then any single error will move the message only one side away and leave the received message as an illegitimate message. If the minimum distance between message points is three sides of the cube, then any single error will leave the received message *closer* to the original message than to any other message. Thus we can have *single* error correction.

Effectively we have to introduce a distance function which is the minimum number of sides of the cube that we have to traverse to get from one point to another. This gives the number of bits in which the two words differ. Thus the distance can be considered as the *logical* (in the Boolean sense) difference or sum of the two points. It is called the *Hamming distance* [4.1].

We can express the minimum distance between vertices of a set of message points in terms of the errors that it is possible to correct. The minimum distance must be at least one for *uniqueness* of the code. A minimum distance of two gives single-error *detectability*. A minimum distance of three gives single-error *correctability*; any single error leaves the point closer to where it was than to any other possible message. A minimum distance of four will give both—single-error correction *plus* double-error detection. A minimum distance of five would allow double-error correction.

Conversely, if a required degree of detection or correction is to be achieved, then the corresponding minimum distance between message points should be observed. Taking the case of single-error correction with the minimum distance of three, we can surround each message point with a unit sphere and not have the spheres overlap. The volume of a "sphere" of radius one is the center plus the n points with just *one* coordinate changed, a volume of $1 + n$. The total volume of the n-dimensional space is 2^n, the number of possible points. Since the "spheres" do not overlap, the maximum number of *message positions* K must satisfy the inequality

$$\frac{\text{total volume}}{\text{volume of a "sphere"}} \geq \text{maximum number of "spheres"} \quad ,$$

or

$$\frac{2^n}{n + 1} \geq 2^K \quad . \tag{4.3.15}$$

We can investigate the corresponding restrictions for higher error correction. For example, for double-error correction we must have a minimum distance of five. We can put non-overlapping "spheres" of radius two about each message position. The volume of the sphere of radius two is the center position plus the n positions a distance one away, plus those with two coordinates out of the n changed —which equals $n(n - 1)/2$.

Dividing the total volume of the space 2^n by the volume of these spheres gives an upper bound on the number K of the *possible code message positions* in the space

$$\frac{2^n}{1 + n + n(n - 1)/2} \geq 2^K \quad . \tag{4.3.16}$$

Similar inequalities can be written for larger spheres, by way of investigating greater than two-error correctability.

Now let us see how such a code can be implemented (Fig.4.8). Consider the following example:

We are given a source possessing a repertoire of eight words, each of three-bit length, made out of an alphabet of two symbols 0 and 1 ($2^3 = 8$). How can we "inflate"

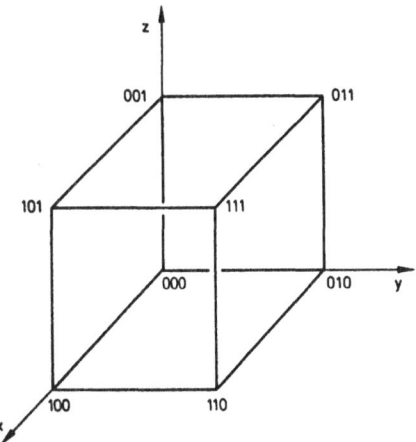

Fig.4.8. A representation of the set of three-digit, two-symbol words

the state space so as to allow single- or multiple-error correction at the receiving site? The idea is, of course, to code in such a way as to make our eight digital words more "orthogonal". This orthogonality is expressed by the fact that the pair cross-correlation between any two such coded words

$$\rho(\tau) = \frac{1}{\xi} \sum_{\substack{i=1 \\ j=1}}^{\xi} x_i x_{j+\tau} \qquad (4.3.17)$$

— where x_i, x_j are strings of n 0's and 1's of length ξ —tends to a delta function as ξ tends to infinity. Let us examine the first step of coding, namely the process of transforming the set of the above three-bit long words to a set of such length that single-error correction is possible.

For such a purpose we use a device called a "shift register" (Fig.4.9). A shift register consists of cascaded binary flip-flops where the binary content of all stages shifts to the neighboring "left" stage when a clock pulse occurs. If a feedback loop is formed (as in Fig.4.9) using modulo-two addition of certain stage outputs to determine the next input digit, one can generate maximal-length sequences. A sequence is maximal length if for $\wedge$ stages (here $\wedge = 3$), the output sequence goes through each possible combination and arrangement of $\wedge$ digits, except the all-zero combination, before it repeats. Thus a maximal sequence has length $\xi = 2^{\wedge} - 1$, where ξ is the bit length of maximal sequence and $\wedge$ is the number of register stages in the generator. Maximal-length sequences are also called "pseudo-noise" sequences because their occurrence statistics resemble a sequence of truly random binary events (coin tosses). Thus there is always just one more "one" than "zero" in a maximal-length sequence. Also the periodic autocorrelation is optimum: it has a value 1 at $\tau = 0$ and a value $-1/\xi$ for all other τ values, and, of course, it is periodic with respect to ξ.

In the present example things work as follows: we put each three-bit word of our repertoire of eight words in the register and allow the logical feedback loop to go on until the word we put in the register reappears. We get the mapping shown in Table 4.4. The pair cross-correlation function can be given as the ratio (A - B)/7, where A is the number of digits in which any two words agree and B is the number of digits in which they differ. Here A = 3 and B = 4, so the pair-correlation function is -1/7.

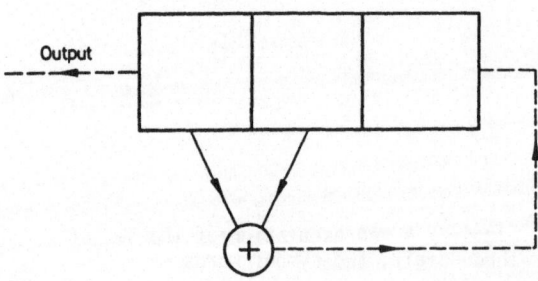

Fig.4.9. The shift register with a feedback loop (see text)

Table 4.4. Transformation of 3-bit words to a set which allows single-error correction

Initial set (Λ)	Final set (ξ)
111	1110010
110	1100101
100	1001011
001	0010111
010	0101110
101	1011100
011	0111001
000	0000000

In general, after many "meta-coding" operations, the pair cross-correlation function assumes the expression

$$\rho = -\frac{1}{\xi} = -\frac{1}{2^\Lambda - 1} \quad , \tag{4.3.18}$$

and as $\Lambda \to \infty$, $\rho \to 0$, that is, orthogonality is achieved.

The resulting eight words of seven-bit length allow obviously for single-error correction. Of course, there are $128 = (2^7)$ words of length seven bits. Out of them, however, only the eight above are "legitimate".

4.4 Information Sources with Memory. Markov Chains

We have seen in the preceding section that the maximum rate of information transmission is accomplished by removing all redundancy from the strings/words to be transmitted, that is, by making all transmitted symbols equiprobable and statistically independent. Nevertheless, we have seen also that in order to make our messages immune to the channel noise, we have to map or *code* them in state space —to achieve error detection and correction. In short, we found that we must *reintroduce* redundancy through the back door, so to speak, in order to allow *reliable* communication.

Clearly the quests for speed and reliability are in conflict. How can we compromise in order to create a language of communication which will possess enough *reliability*, on one hand, and reasonable *variety*, on the other, to avoid stereotypical repetitiousness and therefore stagnation?

Let us define more carefully the above parameters of variety and reliability, and consequently examine under what conditions their interplay can lead to some optimum [4.5].

The *variety factor* (or coefficient) D_1 of a set of N messages is the amount by which the entropy of the system deviates from its maximum value, or the amount by which the set of a priori probabilities P_i deviate from equiprobability:

$$D_1 = S_{max} - S_1 = \log_2 N + \sum_{i=1}^{N} P_i \log_2 P_i \quad . \tag{4.4.1}$$

When all P_i's are equal, D_1 is zero, and D_1 becomes *maximum*, $(\log_2 N)$, when $S_1 = 0$, i.e., when the information source possesses just one state with a priori probability one. The *reliability factor* (or coefficient) D_2 of a set of N messages is the amount of deviation of the symbols of the information source from full independence. To be more specific, we introduce the *joint entropy* of a pair of symbols/states x_i, x_j:

$$S_2 = - \sum_{i=1}^{N} \sum_{j=1}^{N} P(x_i, x_j) \log_2 P(x_i, x_j) \tag{4.4.2}$$

where $P(x_i, x_j)$ is the joint p.d.f. of the pair of symbols (x_i, x_j). We define the reliability factor as the difference

$$D_2 = S_2^{ind} - S_2^{dep} \tag{4.4.3}$$

of the values of joint entropy when (x_i, x_j) are stochastically independent (S_2^{ind}) and when (x_i, x_j) are stochastically dependent (S_2^{dep}).

Let us calculate first S_2^{ind}; in this case, from Bayes's theorem in elementary probability theory, we have $P(x_i, x_j) = P(x_i) P(x_j)$, so

$$S_2^{ind} = - \sum_{i=1}^{N} \sum_{j=1}^{N} P(x_i) P(x_j) \log_2 [P(x_i) P(x_j)]$$

$$= - \sum_{i=1}^{N} \sum_{j=1}^{N} P(x_i) P(x_j) \log_2 P(x_i) - \sum_{i=1}^{N} \sum_{j=1}^{N} P(x_i) P(x_j) \log_2 P(x_j)$$

$$= - \sum_{i=1}^{N} P(x_i) \log_2 P(x_i) \left[\sum_{j=1}^{N} P(x_j) \right] - \sum_{j=1}^{N} P(x_j) \log_2 P(x_j) \left[\sum_{i=1}^{N} P(x_i) \right]$$

$$= +2S_1 \quad . \tag{4.4.4}$$

Let us now calculate S_2^{dep}. From Bayes's rule we have now

$$P(x_i, x_j) = P(x_i) P(x_j / x_i) = P(x_j) P(x_i / x_j) \quad .$$

So

$$S_2^{dep} = - \sum_{i=1}^{N} \sum_{j=1}^{N} P(x_i, x_j) \log_2 [P(x_i) P(x_j / x_i)]$$

$$= - \sum_{i=1}^{N} \sum_{j=1}^{N} P(x_i) P(x_j / x_i) \log_2 P(x_i) - \sum_{i=1}^{N} \sum_{j=1}^{N} P(x_i, x_j) \log_2 P(x_j / x_i)$$

$$= \left[- \sum_{i=1}^{N} P(x_i) \log_2 P(x_i) \right] \left[\sum_{j=1}^{N} P(x_j / x_i) \right] + S_M \tag{4.4.5}$$

where

$$S_M = - \sum_{i=1}^{N} \sum_{j=1}^{N} P(x_i, x_j) \log_2 P(x_j/x_i) \quad . \tag{4.4.6}$$

Since

$$\sum_{j=1}^{N} P(x_j/x_i) = 1 \quad ,$$

we finally get

$$S_2^{dep} = S_1 + S_M \tag{4.4.7}$$

and

$$D_2 = S_2^{ind} - S_2^{dep} = 2S_1 - S_1 - S_M = S_1 - S_M \quad . \tag{4.4.8}$$

To make the interplay between D_1 and D_2 more apparent, we construct the symbolic diagram in Fig.4.10. A *language* without rules possesses no reliability and also has zero variety. For such a "language" the entropy is

$$S = S_{max} = \log_2 N \quad .$$

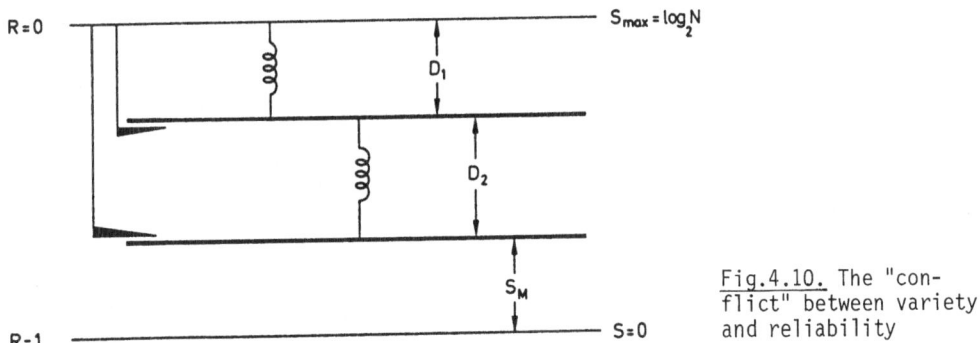

Fig.4.10. The "conflict" between variety and reliability

We start building *order* into such a system by introducing first some "grammatical rules", i.e., an amount of variety D_1. However D_1 is not to be allowed to *grow* beyond a certain limit because of the ensuing stereotype. We next introduce an amount of reliability D_2 by constructing "syntactic rules" which establish a statistical interdependence among the various symbols/states. However, D_2 is not to be allowed to *drop* below a certain limit because of the resulting inability to detect and correct errors, due to the lack of "context". The remaining entropy

$$S_M = \log_2 N - D_1 - D_2 \tag{4.4.9}$$

is the entropy of a language possessing D_1 variety and D_2 reliability.

The redundancy of the language is defined as

$$R = 1 - \frac{S_M}{S_{max}} = \frac{D_1 + D_2}{\log_2 N} \quad , \tag{4.4.10}$$

and changes from 0 (absolute disorder) to 1 (entropy zero —due to *strict* symbol stereotype or *strict* interdependence).

A source with statistically interdependent messages possesses *memory*: we will consider cases where the probability of a symbol depends only on the immediately preceding one. Such an information source will put out a sequence of messages that forms a simple *Markov chain*. A Markov chain of n states is unambiguously defined from the transition $n \times n$ matrix $\mathbf{P}$, with elements P_{ij} standing for the transition probabilities from state i to state j. The chain may possess *steady states*; the necessary condition is that there exists an integer λ such that the matrix $\mathbf{P}^{\lambda}$ has at least one column of nonzero elements. Moreover, the chain is *ergodic* if from any state we can get to any other state. The question now is, what is the entropy of a Markov chain? (That is, what is the entropy of an information source with memory?)

An immediate outcome of the statistical interdependence of the individual symbols/states of the information source is that all the symbols the source can conceivably produce within a given length of time do not carry the same amount of information. This allows us to take advantage of the statistical structure of a particular source to achieve a higher transmission rate *for a given probability of error*.

We define in general the source's uncertainty as the uncertainty of a particular symbol produced by the source, given that we have observed all previous symbols; that is, given an information source $x_1, x_2, \ldots,$ the uncertainty or the entropy of the source denoted by $S\{x_n\}, n = 1, 2, \ldots,$ or $S(X)$ is defined as

$$\lim_{n \to \infty} S(x_n / x_1, x_2, \ldots, x_{n-1}) \tag{4.4.11}$$

or

$$S(X) = \lim_{n \to \infty} \frac{S(x_1, x_2, \ldots, x_n)}{n}$$

$$= \frac{S(x_1) + S(x_2/x_1) + \ldots + S(x_n/x_1, \ldots, x_{n-1})}{n} . \tag{4.4.12}$$

For each state x_K of the Markov chain, let $x_{K(1)}, x_{K(2)}, \ldots, x_{K(n_K)}$ be the n_K states that can be reached in one step from x_K, i.e., the states x_i for which $P_{Ki} \neq 0$. We define the uncertainty of the state x_K as

$$S_K = - \sum_{i=1}^{n_K} P_{Ki} \log_2 P_{Ki} . \tag{4.4.13}$$

The uncertainty of the Markov chain will be calculated as the average value of S_K, namely

$$S(X) = \sum_{K=1}^{n} P_K S_K , \tag{4.4.14}$$

where n is the total number of states of the chain and P_K is the probability of finding the chain in the (steady) state K. The n values of P_K are calculated from the first n - 1 equations of the linear system

$$P_i = \sum_{j=1}^{n} P_j P_{ji} \tag{4.4.15}$$

and the condition

$$\sum_{i=1}^{n} P_i = 1 \quad . \tag{4.4.16}$$

So we see that in a Markov chain the transitional probabilities determine the probabilities of the steady states —what we have called so far "a priori" probabilities.

4.5 Specific Examples of Some Useful Channels and Calculations of Their Capacities

4.5.1 Capacity of a Homogeneously Turbulent Channel

Before we come to the evaluation of the channel capacity of a number of typical and useful communication links, it is useful to examine how the channel capacity in the presence of additive white Gaussian noise (Shannon's formula) is modified by *slow* random variations of the parameters. Specifically, in the expression giving the capacity (4.1.50)

$$C = W \log_2\left(1 + \frac{S}{N}\right) \quad ,$$

we will introduce slow variations in the power of the signal S and/or the power of the noise N. These variations are due to the slow evolution of the refractive index of the medium within which information transfer takes place and are provoked in particular by temperature variations. These fluctuations are slow with respect to the period of the carrier EM wave.

Following *Siforov* [4.6], let us put $\alpha = S/S_m$, where S_m is the average power of signal taken over a very large time interval, and $\beta = S_m/N$. Then the channel capacity becomes

$$C = W \log_2(1 + \alpha\beta) \quad . \tag{4.5.1}$$

In the absence of slow variations the channel capacity is

$$C_0 = W \log_2(1 + \beta) \quad ; \tag{4.5.2}$$

their ratio is

$$x = \frac{C}{C_0} = \frac{\ln(1 + \alpha\beta)}{\ln(1 + \beta)} = f(\alpha) \quad . \tag{4.5.3}$$

Let $p(x)$ denote the probability density function of x which, as seen from above, is a function of the random variable α.

The average value of x equals the ratio of the average rate C_m of transmission of information through a given channel having randomly varying parameters, to the capacity C_0, which refers to a channel having invariable parameters and the same average power of the signal. It is expressed as

$$\eta = \frac{C_m}{C_0} = \int_0^\infty xp(x)dx = F(\beta) \quad . \tag{4.5.4}$$

This treatment applies to channels in which the transmitted signal is split up among a large number of paths. The probability distribution of the signal amplitude at the receiving end can be evaluated from the probability density function of the amplitude of a great number of phasors of equal amplitude and randomly fluctuating phases in the interval $-\pi,\ldots,\pi$. When the p.d.f. for each phase is flat and equal to $1/2\pi$ within the above interval, the probability distribution of the signal amplitude at the output of the channel follows the so-called Rayleigh distribution [4.2], namely

$$P(\varepsilon) = \varepsilon \exp(-\varepsilon^2/2) \quad , \tag{4.5.5}$$

where $\varepsilon = E_r/E_0$, E_r is the random amplitude of the signal field, and E_0 is the median value of this field.

However, $\varepsilon^2/2$ equals α, namely the ratio of the random power of the signal to its average power. From elementary probability theory [see also (4.5.6)], we find that the p.d.f. of α is $q(\alpha) = e^{-\alpha}$. The problem now is to investigate the expression for $p(x)$ by using the p.d.f. of α and (4.5.3). We have

$$p(x) = \left[\frac{q(\alpha)}{df(\alpha)/d\alpha}\right]_{\alpha=\theta(x)} \quad , \tag{4.5.6}$$

where $\theta(x)$ is the reciprocal function of $f(\alpha)$.

From (4.5.3) we get

$$\theta(x) = \frac{1}{\beta}\left[e^{x \ln(1+\beta)} - 1\right] \quad . \tag{4.5.7}$$

Substituting into (4.5.6) the expression for $q(\alpha)$ and also $f(\alpha)$ from (4.5.3), we obtain

$$p(x) = \frac{\ln(1 + \beta)}{\beta} e^{1/\beta}\left[(1 + \beta)^x e^{-(1+\beta)^x/\beta}\right] \quad , \tag{4.5.8}$$

and using (4.5.4), finally

$$\eta = \frac{C_m}{C_0} = \frac{\ln(1 + \beta)}{\beta} e^{1/\beta} \int_0^\infty x(1 + \beta)^x e^{-(1+\beta)^x/\beta} dx \quad . \tag{4.5.9}$$

Introducing the variable $y = (1+\beta)^x/\beta$ and integrating (4.5.9) leads to

$$\eta = \frac{e^{1/\beta}}{\ln(1 + \beta)} \int\limits_{1/\beta}^{\infty} \frac{e^{-y}}{y} \, dy \quad . \tag{4.5.10}$$

The numerical values of the integral

$$\Lambda = \int\limits_{1/\beta}^{\infty} \frac{e^{-y}}{y} \, dy \tag{4.5.11}$$

can be calculated.

Thus we finally arrive at the formula

$$\eta = \frac{C_m}{C_0} = \frac{\Lambda e^{1/\beta}}{\ln(1 + \beta)} \quad . \tag{4.5.12}$$

If we plot $\eta = f(\beta)$, we find that this curve has a single minimum for $\beta \sim 5$ which is ~ 0.83. The result of our calculation shows therefore that the capacity of the Shannon-type channel *never* drops more than 17% from the value it has in the absence of slow random variations of its parameters, provided that the signal statistics through the turbulent medium follow the Rayleigh distribution.

4.5.2 The Lossless Channel

We next calculate the capacity of a channel which not only contains noise but the number of states y_j of the receiver is *greater* than the number of the states of the transmitter x_i (Fig.4.11). Such a channel is called a *lossless channel*, and its matrix in the specific example of Fig.4.11 is

	y_1	y_2	y_3	y_4	y_5	y_6
x_1	1/8	3/8	1/2	0	0	0
x_2	0	0	0	1/3	2/3	0
x_3	0	0	0	0	0	1

It has only one nonzero element in each column.

The remaining uncertainty is

$$S(X/Y) = - \sum_{i=1}^{3} \sum_{j=1}^{6} p(x_i, y_j) \log_2 p(x_i/y_j) \quad . \tag{4.5.13}$$

However, from the channel diagram or the channel matrix, we see that for every y_j received we know with *complete certainty* which x_i was transmitted. Hence $p(x_i/y_j)$ is either a *one* or a *zero* for every case. Thus for a lossless channel: $S(X/Y) = 0$ and so the information transfer $I(X \rightarrow Y) = S(X/Y)$, the entropy of the transmitter. The channel capacity is then found by maximizing $S(X)$ or $C = \log_2 q$, where q is the number of the states of the transmitter (here $C = \log_2 3$).

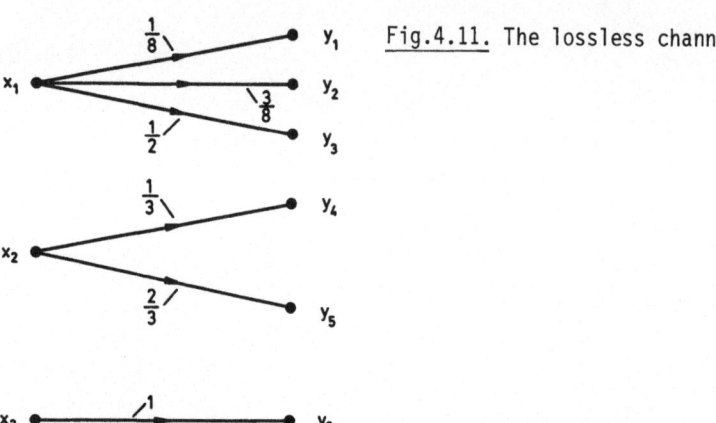

Fig.4.11. The lossless channel

4.5.3 The Deterministic Channel

In such a scheme the number of states of the transmitter x_i is *greater* than the
number of the states of the receiver y_j. The matrix of such a channel has only one
nonzero element in each row (see Fig.4.12, the channel diagram). Since each row

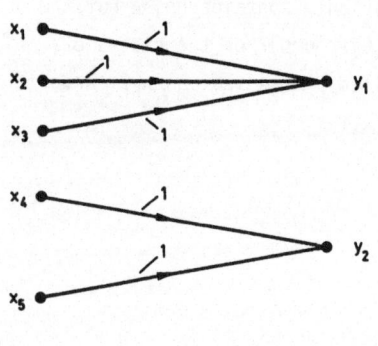

Fig.4.12. The deterministic channel

has only one nonzero element, this element must be one. The information transfer
can be written

$$I(X \rightarrow Y) = S(Y) - S(Y/X) \qquad (4.5.14)$$

and

$$S(Y/X) = - \sum_{i=1}^{6} \sum_{j=1}^{3} p(x_i,y_j) \log_2 p(y_j/x_i) \quad . \qquad (4.5.15)$$

	y_1	y_1	y_3
x_1	1	0	0
x_2	1	0	0
x_3	1	0	0
x_4	0	1	0
x_5	0	1	0
x_6	0	0	1

Since all elements $p(y_j/y_i)$ are either 0 or 1, $S(Y/X) = 0$ and $C = \max S(Y) = \log_2 r$, where r is the number of the states of the receiver (hence $C = \log_2 3$).

4.5.4 The Uniform Channel

A uniform channel is one whose channel matrix has identical rows, except for permutations, and identical columns, except for permutations. If the channel matrix is square, then every row and every column are simply permutations of the first row (Fig.4.13).

	y_1	y_2	y_3
x_1	1/2	1/4	1/4
x_2	1/4	1/4	1/2
x_3	1/4	1/2	1/4

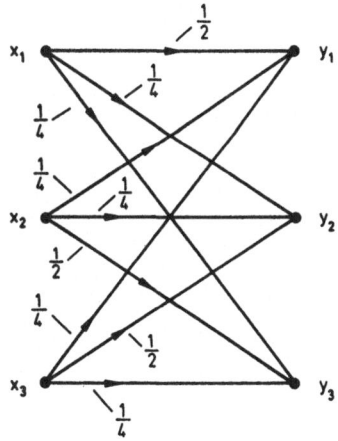

Fig.4.13. The uniform channel

For such a channel the channel entropy $S(Y/X)$ is independent of the input signal probabilities. This is easily shown by writing

$$S(Y/X) = - \sum_{i=1}^{q} \sum_{j=1}^{r} p(x_i, y_j) \log_2 p(y_j/x_i) \qquad (4.5.16)$$

$$= \left[- \sum_{i=1}^{q} p(x_i) \right] \sum_{j=1}^{r} p(y_j/x_i) \log_2 p(y_j/x_i)$$

$$= - \sum_{j=1}^{r} p(y_j/x_i) \log_2 p(y_j/x_i) \quad , \tag{4.5.17}$$

and so,

$$I(X \to Y) = S(Y) + \sum_{j=1}^{r} p(y_j/x_i) \log_2 p(y_j/x_i) \tag{4.5.18}$$

and

$$C = \log_2 q + \sum_{j=1}^{r} p(y_j/x_i) \log_2 p(y_j/x_i) \quad . \tag{4.5.19}$$

In our example $q = r = 3$. In the above summation i can take any value. So

$$C = \log_2 3 + \frac{1}{2} \log_2(\tfrac{1}{2}) + \frac{1}{2} \log_2(\tfrac{1}{4}) = \log_2 3 - 1.5 \sim 0.08 \quad .$$

4.5.5 The Binary Symmetrical Channel

This is a special and important case of the previous type of uniform channel (Fig. 4.14).

	y_1	y_2	
x_1	p	q	
x_2	q	p	p + q = 1

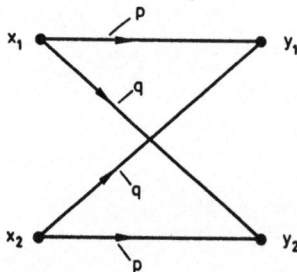

Fig.4.14. The binary symmetrical channel

The capacity of this channel is directly derived from the corresponding expression for the uniform channel (4.5.19) by putting $q = 2$, $r = 2$, $p(y_j/x_i) = p$ for $j = 1$, $i = 1$ (or 2), and $p(y_j/x_i) = q$ for $j = 2$, $i = 1$ (or 2), so

$$C = \log_2 2 + p \log_2 p + q \log_2 q \quad , \tag{4.5.20}$$

or

$$C = 1 + p \log_2 p + (1 - p) \log_2(1 - p) \quad . \tag{4.5.21}$$

If we plot C as a function of p we find that C becomes a minimum $C = 0$ for $p = 1/2$, and that $C = 1$ for $p = 1$ or $p = 0$.

4.5.6 The Binary "Erasure" Channel

	y_1	y_2	y_3	
x_1	p	q	0	p + q = 1
x_2	0	q	p	

In this channel (Fig.4.15) there is an output symbol y_2 used to detect an error in transmission. The binary erasure channel is a single-error detecting channel. Let us calculate the information transfer $I(X \rightarrow Y) = S(X) - S(X/Y)$. Let the source symbols have the following probabilities $p(x_1) = \alpha$, $p(x_2) = 1 - \alpha$. Then

$$S(X) = - \alpha \log_2 \alpha - (1 - \alpha) \log_2 (1 - \alpha) \tag{4.5.22}$$

and

$$S(X/Y) = - \sum_{i=1}^{2} \sum_{j=1}^{3} p(y_j/x_i) p(x_i) \log_2 \left[\frac{p(y_i/x_i) p(x_i)}{p(y_j)} \right] . \tag{4.5.23}$$

Here we have to take into account all six terms in the above summation. After some algebra we finally get

$$S(X/Y) = q[- \alpha \log_2 \alpha - (1 - \alpha) \log_2 (1 - \alpha)] = qS(X) \tag{4.5.24}$$

and

$$I(X \rightarrow Y) = (1 - q)S(X) = pS(X) . \tag{4.5.25}$$

The channel capacity is then

$$C = \max I(X \rightarrow Y) = p \text{ [bits/s]} . \tag{4.5.26}$$

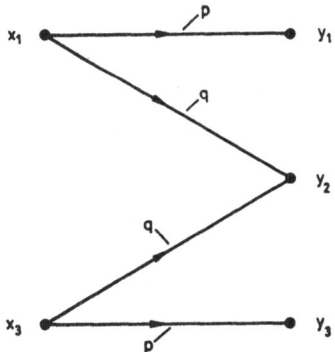

Fig.4.15. The binary erasure channel

4.5.7 Capacity of an Optical Channel

We are interested in the capacity of an optical instrument when considered as a communication channel. So far we have defined the channel capacity as the maximum number of bits that the channel can transmit per unit time. Nevertheless, since for most optical instruments the observation time is practically unlimited, it is

157

preferable, following *di Francia* [4.7], to define the capacity of an optical instrument as the number of bits that the instrument can transmit per single image, or as the greatest possible number of bits obtainable from an image formed by the instrument. We already know from Sect.3.4 the number of degrees of freedom of an image to be equal to (3.4.40)

$$F = 4\Omega \frac{\Sigma}{\lambda^2} \quad ,$$

where Σ is the "pupil area" of the instrument, λ is the wavelength of the incident light, and Ω is the total solid angle of the observation field. The illumination intensity I at each sampling point of the receiving aperture will be between 0 and I_{max}. The maximum entropy of the image will be obtained when all values of I inside the range $0 < I < I_{max}$ have equal probability. The entropy of the image will then be equal to $\log_2 I_{max}$.

Apart from thermal noise (to be treated below), in the case of an optical instrument we have also a source of "deterministic" or "static" noise coming from "stray" light, i.e., light which is incoherently scattered by inhomogeneities and/or anisotropic of the material of the receiving aperture. We will assume that the intensity of stray light I_d is uniformly distributed in the aperture, forming some sort of background noise, so that the "objective" illumination intensity I_0 will range over the interval $I_d < I_0 < I_{max} + I_d$ and $I_d = pI_{max}$. Thus

$$pI_{max} < I_0 < (1 + p)I_{max} \quad , \tag{4.5.27}$$

where p is a numerical constant depending on the degree of inhomogeneity and/or anisotropy of the material of the receiving aperture.

The entropy of the image, provided that the illuminations of different sampling points are statistically independent, will be

$$S = F \log_2 I_{max} \quad [bits] \quad . \tag{4.5.28}$$

Let us now come to the estimation of random noise — the only agent which will be responsible for a change of the entropy of the image, since stray light, being non-random, does not affect S. As quantum noise will be treated in the next section, here we will limit ourselves to identifying random noise with receptor noise. Specifically we will refer to the eye. When a portion of the retina is illuminated with an *objective* illumination I_0, the *subjective* illumination I_s, i.e., the illumination perceived by the observer, is in general somewhat different from I_0. There is an uncertainty in the observer's judgment which is related to the so-called differential threshold. This uncertainty represents noise and will be denoted by I_n. Thus we may write for the *subjective* illumination

$$I_s = I_0 + I_n \quad . \tag{4.5.29}$$

The probability distribution for I_n may be assumed to be Gaussian, so

$$p(I_n) = \frac{1}{I_N\sqrt{2\pi}} \exp\left(-\frac{I_n^2}{2I_n^2}\right) , \tag{4.5.30}$$

where I_N is the standard deviation of the observational uncertainty. It is, how-ever, experimentally known (Weber's psychological law) that $I_N = \varepsilon I_0$, where the proportionality constant ε is experimentally derived.

It is now possible to evaluate the capacity of the optical channel, calculating as usual the maximum value of the information transfer from the object X to the observer Y, i.e.,

$$\max I(X \rightarrow Y) = \max [S(Y) - S(Y/X)] , \tag{4.5.31}$$

where $\max S(Y) = \log_2 I_{max}$ bits per degree of freedom. In order to calculate the re-maining uncertainty $S(Y/X)$ about what the observer receives for a *given* object, we notice that $\max S(Y/X)$ is the entropy of $I_s = I_0 + I_n$, when I_0 is known. So $\max S(Y/X)$ is simply the entropy of I_n, i.e., of the Gaussian distribution.

Therefore, taking into account that $I_N = \varepsilon I_0$, $\max S(Y/X)$ equals the average value of $\log_2(\sqrt{2\pi e} \, I_N)$ over the range pI_{max} to $(p+1)I_{max}$, i.e.,

$$\max S(Y/X) = \frac{1}{I_{max}} \int_{pI_{max}}^{(1+p)I_{max}} \log_2(\sqrt{2\pi e} \, \varepsilon I_0)dI_0 \tag{4.5.32}$$

$$= \log_2\left[\sqrt{\frac{2\pi}{e}} \, \varepsilon \, \frac{(1+p)^{1+p}}{p^p} \right] + \log_2 I_{max} . \tag{4.5.33}$$

Substituting this expression into the formula giving $\max I(X \rightarrow Y)$, we finally get the channel capacity

$$C = \log_2\left[\sqrt{\frac{e}{2\pi}} \, \frac{1}{\varepsilon} \, \frac{p^p}{(1+p)^{1+p}} \right] \text{ bits/degree of freedom} . \tag{4.5.34}$$

If we plot C as a function of p, i.e., the percentage of stray light for a typical value of ε, say 0.1, we can see that for a usual value of $p \sim$ a few percent, ($p \sim 2$-3%), $C \sim 2.5$ bits/degree of freedom [4.7].

A good optical instrument having a pupil of $\sim$100 mm diameter and an angle of $\sim 1°$ has, in the middle of the optical region, a capacity of about $\sim 10^8$ bits.

4.5.8 Role of Quantum Noise in an Optical Channel

In Shannon's formula for the channel capacity the only type of noise preventing the value of C from becoming essentially infinite is *thermal* noise — which is, of course, preponderant for $h\nu/kT \ll 1$. What happens, however, if $h\nu/kT \gg 1$, i.e., when the car-rier frequency is high and/or the temperature is very low? According to Shannon's formula, the answer would be $C \rightarrow \infty$.

We are going to see that this is not so, because at optical frequencies, although thermal noise is negligible, *quantum* noise takes over, thereby limiting the channel capacity. Following *Lebedev* and *Levitin* [4.8], we are going to limit ourselves to the range $h\nu/kT \gg 1$. A more involved calculation of C for any value of $h\nu/kT$, again following the above authors, proves that in the limit $h\nu/kT \ll 1$ Shannon's formula is rediscovered as an asymptotic case.

We describe the signal and the noise in terms of mean occupation numbers of photons per degree of freedom of the electromagnetic field involved: $\bar{m}(\nu)$ for the signal and $\bar{n}(\nu)$ for the noise;

$$\bar{l}(\nu) = \bar{m}(\nu) + \bar{n}(\nu) \tag{4.5.35}$$

stands for the mean occupation number of photons per normal mode at frequency ν for the receiver wave form (signal plus noise).

The spectral density of photon flux (the mean number of photons per unit time and frequency) equals the mean occupation number. Thus the energy flux (power) transferred by the signal and the noise in the frequency interval $d\nu$ is

$$P_S(\nu)d\nu = \bar{m}h\nu \, d\nu \tag{4.5.36}$$

and

$$P_N(\nu)d\nu = \bar{n}h\nu \, d\nu \quad , \tag{4.5.37}$$

respectively, where

$$\bar{n}(\nu) = \frac{1}{e^{h\nu/kT} - 1} \quad . \tag{4.5.38}$$

Hence the noise power will be

$$<P_N> = \int_0^\infty \frac{h\nu \, d\nu}{e^{h\nu/kT} - 1} = \frac{\pi^2 (kT)^2}{6h} \quad . \tag{4.5.39}$$

(The zero-point energy of the field oscillators is omitted since it has no influence on the value of entropy to be calculated just below.)

The entropy flux (entropy transferred by radiation per unit time) is

$$S_N = \int_0^{<P_N>} \frac{dP}{kT} = \int_0^T \frac{1}{kT} \frac{dP(T)}{dT} \, dT \quad , \tag{4.5.40}$$

where P(T) is thermal radiation power as a function of temperature.

In the presence of a deterministic signal P, the power of total radiation will be

$$P_0 = <P_N> + P \quad . \tag{4.5.41}$$

Nevertheless, the entropy flux remains unchanged because the signal and the noise are statistically independent, and the signal is deterministic. We have already seen, however, that the maximum rate of information transmission (or the maximum entropy flux) is realized when the signal is *coded* as random noise. So if T_0 is the effective temperature for the total signal transmitted, for optimum information transfer we must have

$$\bar{I}(\nu) = \frac{1}{e^{h\nu/kT_0} - 1} \; .$$

(4.5.42)

Then the total radiation entropy flux is

$$S_0 = \int_0^{T_0} \frac{1}{kT} \frac{dP(T)}{dT} \, dT \; ,$$

(4.5.43)

and the channel capacity will simply be equal to

$$C = S_0 - S_N = \int_T^{T_0} \frac{1}{kT} \frac{dP(T)}{dT} \, dT \; .$$

(4.5.44)

From (4.5.41) we get

$$P_0(T) = \frac{\pi^2 (kT_0)^2}{6h} = \frac{\pi^2 (kT)^2}{6h} + P \; .$$

(4.5.45)

Substituting T_0 from (4.5.45) into (4.5.44), we finally get

$$C = \frac{\pi^2 kT}{3h} \left[\sqrt{1 + \frac{6hP}{\pi^2 (kT)^2}} - 1 \right]$$

(4.5.46)

or

$$C = \frac{\pi^2 T}{3h} \left[\sqrt{1 + \frac{6hP}{\pi^2 (kT)^2}} - 1 \right] \; \text{bits} \; .$$

(4.5.47)

From (4.5.47) we can see that for high signal power $6hP/\pi^2 (kT)^2 \gg 1$ the capacity of the channel is limited only by quantum effects, i.e.,

$$C_{quant} \sim \pi \sqrt{\frac{2P}{3h}} \; .$$

(4.5.48)

The bandwidth required to obtain the capacity given by (4.5.47) is approximately expressed by

$$\nu_{max} \sim \frac{kT_0}{h} \sim \frac{kT}{h} \sqrt{1 + \frac{6hP}{\pi^2 (kT)^2}} \; .$$

(4.5.49)

4.5.9 An Introduction to the "Genetic Channel" and the Genetic Code

We now will try to use the concepts of information theory in order to explain (or rather "explain away"!) a basic phenomenon occurring in biological systems, namely the control that the genes exercise on protein synthesis (see also Appendix A).

Before entering the calculation proper, some introduction to the subject matter is needed. In the first place, we are asking a question about the nature and the structure of DNA. The chemical elements involved are carbon (C), hydrogen (H), oxygen (O), phosphorus (P) and nitrogen (N). The atoms of these elements are linked in a very complicated way to form macromolecules.

About 20 years ago the genetic material was first successfully recognized and photographed with the aid of an electron microscope, i.e., it is possible to display in a photograph on half a page of a book a single DNA molecule enlarged 44 000 times (extracted from a bacteriophage for instance).

Each DNA molecule consists of two individual helices ("strands") wound around each other in the shape of a double helix; each single strand is composed of interconnected molecules of *phosphoric acid* and a type of sugar (*deoxyribose*) in regular alternation. To each sugar molecule an organic (nitrogenous) *base* molecule is connected laterally. There are *four types* of these bases, A (adenine), T (thymine), G (guanine), and C (cytosine).

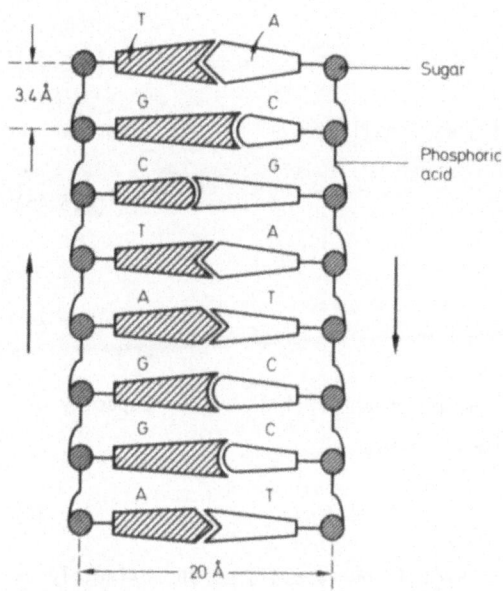

Fig.4.16. Sketch of DNA strand (see text)

Adenine is always joined with *thymine* to form a step and *guanine* is always joined with *cytosine* (Fig.4.16). But the steps may follow each other "endlessly" and in any order (i.e. randomly): AT, GC, CG, AT, AT, GC, CG, ... with the A and T or G and C lying opposite to each other and loosely connected (through hydrogen bonds). The "space wavelength" of the helix is about 34 Å per 10 steps. The diameter of the helix is about 20 Å.

The two strands of the double helix run in directions opposite to each other, and by virtue of this they are said to be of opposite chirality (one is right- and the other left-handed). An A of one strand is always directly opposite a T of the other strand, and a G is opposite a C. Each strand is also irreversible (it cannot be "read" backwards).

A DNA molecule is very long and thin. It also displays "supercoiling", that is, the initial coil is twisted again on a second level, so to speak. The molecular weight of a single molecule may be as much as 2×10^9, which corresponds to about 3×10^6 base pairs. Such a structure may be approximately 1 mm in length and 20 Å in diameter —analogous to a very thin silk string, one mile long. Molecules of DNA are therefore rather fragile and easy to break into smaller pieces. The *amount* of DNA in a living organism and the *complexity* of the organism seem to be somewhat related, i.e., in a *bacterium* the DNA might total ~0.5 mm and hold ~seven million steps. In a *virus* the length might be ~0.1 mm and contain 170 000 steps.

How is it possible that hereditary determinants are encoded in such a molecule? There is only one possible way: in the sequence of the nucleotide pairs (A, T) and (G, C). It is one of the most exciting facts in the natural process that the immense differences among the hereditary characteristics of men, animals, plants, bacteria, and viruses (as well as the hereditary differences between races, and individuals of the same race) rest almost entirely on the difference in the random sequence of (A, T) and (G, C)[2]. It is this sequence which determines also (in dynamical coupling with the specific environment) which behavioral traits will develop in the phenotype.

The connection between the information source DNA and the characteristics of the mature orgnaism involves many intermediate *hierarchical levels* (see also Sect.4.8 and Appendix A). To date, only the *first pair* of these hierarchical levels is amenable to some "biocybernetic interpretation", or modeling (Fig.4.17). This metabolic step consists of processes in which DNA directs the synthesis of *proteins*. The proteins in turn, acting via a hierarchical feedback loop, direct the synthesis of enzymes which control the switching "on" and "off" of individual portions of DNA, the *genes* (see below). Proteins are not only the basic building materials —together with

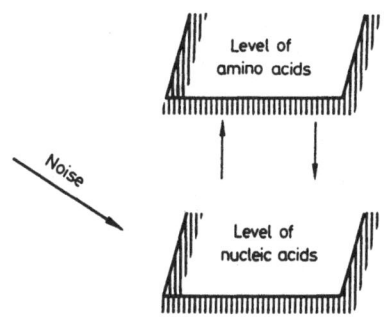

Level of
amino acids

Noise

Level of
nucleic acids

Fig.4.17. The hierarchical system of DNA and proteins

---

2 In that sense, the "transmutation of genomes", achievable perhaps through genetic engineering, should not appear in principle more strange than the "transmutation of elements" achieved through nuclear engineering. Of course, the analogy is at best superficial since transmutation of genomes requires the *compression* of a long random series —which constitutes a transcomputational problem (see also Sect. 4.6 as well as Chaps.5 and 6).

hydrocarbon and fats —of which organisms are composed (practically the only exception in higher animals being the hard substance of the bones, a mineral composed of
phosphorus and calcium) but are also the organism's chemical tools as, for example,
the enzymes through catalysis make possible all chemical reactions in the organism.
Proteins, belonging to the "second hierarchical level" of the organism, have a composition completely different from DNA. They consist of the elements C, H, O, N,
and S. Their molecules are "levo-thread" shaped, like those of DNA, but they consist of a single strand. This strand is made up of components arranged in a linear
sequence; there are not just four types of components as in the case of DNA, but
twenty different ones called *amino acids*.

The first step in the informational relationship or mapping between DNA and the
characteristics of the organism is that the order of the bases in the DNA determine
the order of amino acids in the polypeptide chain that is synthesized. (The role
of the intermediate substance, the messenger RNA, will be mentioned later.)

Here the question is, how many nucleotides are necessary in any given case in
order to determine the incorporation of a *single* specific amino acid molecule within
a protein-chain molecule? Since there are four different nucleotides, they could
determine the formation of four different amino acids. But then, what about the
others? Thus a single nucleotide is not sufficient for the direction of the formation of the amino acids. If two nucleotides were used for each amino acid then 16
(4^2) different amino acids could be directed to their appropriate places in the
protein molecule —again not enough. Thus, only combinations of three nucleotides
are sufficient to form the twenty different signals that correspond to the twenty
amino acids. Biochemistry has confirmed this conclusion. In terms of information
theory, we can say that a nucleotide carries the information quantity $\log_2 4 = 2$ bits.
Correspondingly, an amino acid carries $\log_2 20 = 4.32$ bits. If the nucleotides are
to determine the amino acids, they must contain per "word" at least as much information as a single amino acid does. Combinations of two nucleotides are not, then,
sufficient for this, because they contain $2(\log_2 4) = 4$ bits. Combinations of three
nucleotides, however, contain $\log_2 4^3 = 3 \log_2 4 = 6$ bits > 4.32 bits.

Biochemical research has found some indication as to what role the "extra"
nucleotide triplets play. To be sure, they are not sufficient to make the genetic
code a "protected" code; for that, many more combinations would be needed than are
supplied by the triplets. Today we know that the code is in part "redundant",
i.e., several nucleotide sequences give the signal for the same amino acid. On the
other hand, it is known that some of the nucleotide triplets do not contribute or
correspond to any amino acid but are free to play quite a different role: when they
occur in an RNA chain (see below), the synthesis of protein molecules that is dependent on the RNA is stopped. Thus, the code words of these "spare" triplets seem
to have a meaning similar to the symbol that signals the end of a sentence and the
beginning of a new one.

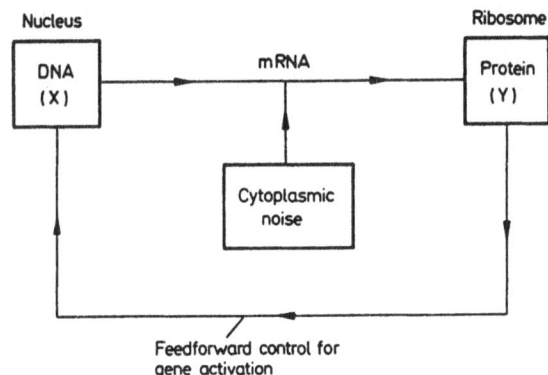

Fig.4.18. Sketch of the protein-
production loop

We have now to examine how the message from DNA (which is located in the cell's
nucleus) reaches the ribosomes in the cytoplasm of the cell where the protein synthesis takes place. In other words, we have to examine how the information transfer
from the level of nucleotides X to the level of amino acids Y takes place (Fig.4.18)
— in the presence of intracellular noise.

A typical protein is ~200 amino acids long. A typical gene (the part of DNA
responsible for the formation of a single protein) has ~600 base-pairs. Genes are
separated by "punctuation" marks, i.e., parts of DNA devoid of "expressive" poten-
tiality. The first stage in the expression of genetic information contained in the
DNA is the synthesis of RNA molecules whose nucleotide sequences are determined by
those of DNA. This takes place by DNA "unzipping" to produce messenger RNA (mRNA)
(and transfer, tRNA). Moreover, mRNA is a single-strand chain that may be looped
and folded upon itself to some extent. The pairs A, T are now substituted in RNA
by A, U (U = uracil). It is the mRNA which plays the role of messenger of triplets
(including U and not T) and carries the information for the building of proteins
from the nucleus, through the cytoplasm to the ribosomes (20 μm in diameter), where
the final synthesis of the polypeptides takes place. About 80% of all RNA in the
ribosomes is ribosomal RNA. Transfer RNA brings to the "assembly line", as it were,
the amino acids already directed to the proper position by mRNA, and rRNA is instrumental in forming the final polypeptide chain —which finally may fold spontaneously and form a three-dimensional structural protein in a way which depends on the
sequence of amino acids in the chain and the constraint of minimization of its free
energy.

Now the channel capacity of the system can, in principle, be calculated, as
usual, from the information transfer

$$I(X \rightarrow Y) = S(X) - S(X/Y) = S(Y) - S(Y/X) \tag{4.5.50}$$

between nucleus X and ribosome Y. The maximum value of S(X) occurs when all triplets
("codons") are equally probable, and $p(x) = 1/64$. This means that the DNA nucleotides
should be present in equal amounts, a condition which is found to be nearly the

165

case in *E. coli*. In other cases the situation is different so there is no *univer-sality* in the value of $I(X \rightarrow Y)$.

The present channel, however, reminds us of the *deterministic channel* treated in Sect.4.5.3. Indeed here the number of states of X is 64 and the number of states of Y is 20 (or 23 because of the punctuation mark, the start mark, and the stop mark which may involve extra amino acids).

So the channel capacity can be taken as equal to

$$S(Y)_{max} = \log_2 23 = 4.52 \text{ bits} \tag{4.5.51}$$

per amino acid and the redundancy of the channel equals *at least*

$$R = 1 - \frac{S(Y)_{max}}{S(X)_{max}} = 1 - \frac{4.52}{6} \sim 25\% \quad . \tag{4.5.52}$$

It has to be noted that during the lifetime of a cell, R goes down due to informa-tion loss caused by *random mutations* —a source of noise impossible to model for the time being. Thus more experimental results are needed in order to corroborate or discard the above simple model.

Before closing this section, it is useful to refer to some known information processes that take place not in cells but within fully developed organs. The kidney is a good example [4.9].

We know that the generation of one bit of information about a system under ob-servation halves the uncertainty of the number of possible complexions and that the entropy change with one binary decision is

$$\Delta S = -k \ln 2 \sim 10^{-21} \text{ J/(bit K)} \quad . \tag{4.5.53}$$

In biological systems there are control processes which are required to make so many observations per unit time that the energy expenditure becomes significant, pre-dictable, and measurable. The kidney is second only to the heart in oxygen consump-tion. The conventional thermodynamic calculations equating renal output with osmo-tic work give as a power output about ∼0.5 calories per minute for man. A compari-son of this value with the renal power input (about 100 calories per minute) gives an efficiency of less than 1% —an extravagance out of proportion to the efficiency of other organs.

If, however, we approach the kidney not just as a pump but rather as a "cybernetic machine" the situation changes: as in the course of metabolism the extracellular fluid (blood) tends to deviate from its "optimum" composition, the kidneys must maintain a continuous surveillance over ions and molecules in the fluid and a con-tinuous sorting out of unwanted particles. This is a matter of a very large number of "observations". Purposeful sorting out of particles requires that the kidney re-cognizes each particle sorted. This process must involve some kind of coupling be-tween the kidney and the particle, an interrogation signal, so to speak.

Now what will be the minimum cost of generating such a signal? The simplest sort of observation recognizing one of two equally probable things cannot be accomplished for less than kT ln2 units of free energy. Multiplying this (small) cost by a vast number of selections made by the kidney over a period of time gives a power requirement which is large enough to put the kidney well into a respectable efficiency (this cost is, of course, a minimum based on noiseless conditions). Regardless of the mechanism, recognition of a particle requires the generation of some sort of "interrogation signal" distinguishable from background thermal noise, which has a mean amplitude of ~kT or ~0.027 eV at "body temperature". The reliability of the selection process is dependent upon the signal-to-noise ratio. The composition of the extracellular fluid is maintained within a range of about ±5%.

A signal of strength greater than four times that of the thermal background gives a chance of less than ~2% for errors due to thermal perturbations. This is a reasonable level of reliability for a homeostatic mechanism. The above mentioned range of ±5% corresponds roughly to a signal-to-noise ratio of ~3.

This level of reliability requires then a recognition signal of at least $3 \times 0.027 = 0.08$ eV. In man the selection involves $\sim 10^{22}$ ions (mainly Na^+) per minute. The power requirement for recognition of these ions is $0.08 \times 10^{22} = 8 \times 10^{22}$ eV ~30 calories per minute. The power consumption of the human kidney being 100 calories per minute, the above calculation offers an efficiency of 30%. So we see that the evaluation of the efficiency of an organ which functions as a *control* device cannot be based strictly on input-output energy considerations.

Similarly, an electronic element like a valve or a transistor wastes nearly *all* the energy it consumes, yet it may be highly efficient as an information handling device. Such behavior is characteristic of all *dissipative structures*.

4.5.10 The Phase-Locked Loop in the Absence and Presence of Noise

We will close this section of specific examples of channels and their error-correction abilities by analytically treating the very important case of a phase-locked loop (PLL). This device is used in a receiver just before the stage of demodulation and carrier rejection, and aims at obtaining and dynamically maintaining phase coherence between an environmental fluctuating signal and a reference signal. In short, it aims at self-organization, or at achieving the greatest amount of *compressibility* on the random phase of the received signal.

Any *automatic phase control system* is generally referred to as a PLL (essentially a nonlinear feedback gadget). It consists of three major components, a *multiplier*, a *time-invariant linear filter* and a *nonlinear voltage-controlled oscillator* (Fig. 4.19).

Let the received signal power be A^2 watts. Then the peak voltage is $\sqrt{2}A$, and the received signal may be denoted by $\sqrt{2}A \sin\theta(t)$. Let the VCO ("voltage, control, os-

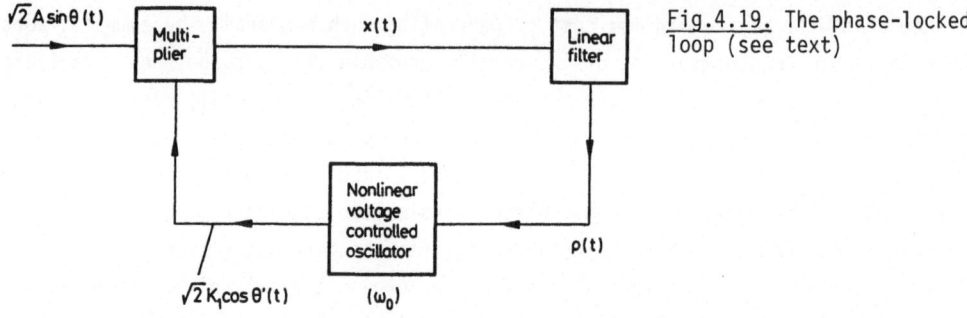

Fig.4.19. The phase-locked loop (see text)

cillator") output signal be denoted by $\sqrt{2}K_1 \cos\theta'(t)$, where K_1 is its root-mean-square amplitude. When $e(t)$, the control signal to the VCO, is removed, the oscillator generates a constant-frequency sinusoid, ω_0 rad/s. When the control signal is applied, the VCO frequency becomes $\omega_0 + K_2 e(t)$, where K_2 is the proportionality constant whose units are rads/Vs.

Thus the time derivative of the phase angle of the VCO output phase is

$$\frac{d\theta'}{dt} = \omega_0 + K_2 e(t) \quad . \tag{4.5.54}$$

The output of the multiplier will be

$$x(t) = 2AK_1 \sin\theta(t) \cos\theta'(t) = AK_1\{\sin[\theta(t) - \theta'(t)] + \sin[\theta(t) + \theta'(t)]\} \quad . \tag{4.5.55}$$

The sum-frequency term is eliminated by the filter-VCO combination and so it may be discarded.

The linear time-invariant filter operates on its input $x(t)$ to produce the output

$$e(t) = e_0(t) + \int_0^t x(t - t')f(t')dt' = e_0(t) + \int_0^t x(t')f(t - t')dt' \quad , \tag{4.5.56}$$

for $t \geqslant 0$, where it is assumed that the input is applied at time $t = 0$. Here $e_0(t)$ is the "zero-point" input response, which depends only on the initial conditions at $t = 0$. We generally set $e_0(t) = 0$ for all t. The weighting function $f(t)$ is known as the *impulse response* of the filter. We get therefore

$$\frac{d\theta'}{dt} = \omega_0 + K_2 \int_0^t f(t - t')AK_1 \sin[\theta(t') - \theta'(t')]dt' \quad , \tag{4.5.57}$$

and if we define the *phase error* $\varphi(t) = \theta(t) - \theta'(t)$ and the *loop gain* $K = K_1 K_2$, we get

$$\frac{d\varphi(t)}{dt} = \frac{d\theta(t)}{dt} - \omega_0 - AK \int_0^t f(t - t')\sin\varphi(t')dt' \quad , \tag{4.5.58}$$

or

$$\frac{d\varphi(t)}{dt} = \underbrace{\omega - \omega_0}_{\Delta} - AK \int_0^t f(t - t')\sin\varphi(t')dt' \quad,$$

or

$$\frac{d\varphi(t)}{dt} = \Delta - AK \int_0^t f(t - t')\sin\varphi(t')dt' \quad. \qquad (4.5.59)$$

For a given input frequency $\omega(t)$, the solution $\varphi(t)$ of the above integrodifferential equation describes exactly the operation of the PLL, i.e., the dynamics of phase locking.

We now assume the existence of a zero-mean, wide-band Gaussian noise, superimposed on the received signal. After passing from a narrow-band filter centered around ω_0, the noise with spectral density N_0 W/Hz is given by the stationary process

$$n(t) = \sqrt{2}[n_1(t)\sin\omega_0 t + n_2(t)\cos\omega_0 t] \quad, \qquad (4.5.60)$$

where $n_1(t)$, $n_2(t)$ are independent Gaussian processes with zero-mean and identical spectral densities.

To examine how additive "white" (Gaussian) noise affects the operation of the PLL, let the received signal be

$$\sqrt{2} A \sin\theta(t) + n(t) = \sqrt{2}\{A[\sin \omega_0 t + \theta_1(t)] + n_1(t)\sin\omega_0 t + n_2(t)\cos\omega_0 t\} \quad, \qquad (4.5.61)$$

where $\theta_1(t) = \theta(t) - \omega_0 t$.

The VCO output signal is

$$\sqrt{2} K_1 \cos\theta'(t) = \sqrt{2} K_1 \cos[\omega_0 t + \theta_2(t)] \quad, \qquad (4.5.62)$$

where $\theta_2(t) = \theta'(t) - \omega_0 t$. In this case, of course, $\theta'(t)$ is a function not only of the signal modulation $\theta(t)$ but also of the noise. The multiplier output is then

$$x(t) = AK_1 \sin[\theta_1(t) - \theta_2(t)] - K_1 n_1(t)\sin\theta_2(t) + K_1 n_2(t)\cos\theta_2(t)$$

$$+ AK_1 \sin[2\omega_0 t + \theta_1(t) + \theta_2(t)] + K_1 n_1(t)\sin[2\omega_0 t + \theta_2(t)]$$

$$+ K_1 n_2(t)\cos[2\omega_0 t + \theta_2(t)] \quad. \qquad (4.5.63)$$

All the terms centered about $2\omega_0$ rad/s can be neglected since the spectrum of the signal term is relatively narrow and since $n_1(t)$, $n_2(t)$ are negligible for $|\omega| > \omega_0$, while the VCO "quiescent frequency" or "autoperiodic" frequency ω_0 is much greater than the frequency range of the loop components.

Thus we may write

$$e(t) = K_1 \int_0^t [A \sin\varphi(t') - n_1(t')\sin\theta_2(t') + n_2(t')\cos\theta_2(t')][f(t - t')]dt' \quad, \qquad (4.5.64)$$

where $\varphi(t) = \theta(t) - \theta'(t) = \theta_1(t) - \theta_2(t)$.

If we let $K = K_1 K_2$ and $n'(t) = -n_1(t)\sin\theta_2(t) + n_2(t)\cos\theta_2(t)$, we get

$$\frac{d\varphi(t)}{dt} = \Delta - K \int_0^t [A \sin\varphi(t') + n'(t')]f(t - t')dt' \quad . \tag{4.5.65}$$

If the impulse response of the filter is deltafunction-like then (4.5.65) becomes

$$\frac{d\varphi(t)}{dt} = \Delta - AK \sin\varphi(t) - Kn'(t) \quad . \tag{4.5.66}$$

We are interested now in following up the solution of the equation from which the p.d.f. $p(\varphi)$ will emerge, describing the diffusion of the probability for locking-in, under the initial condition

$$p(\varphi, 0) = \delta(\varphi - \varphi_0) \tag{4.5.67}$$

and

$$\int_{-\infty}^{\infty} p(\varphi, t)d\varphi = 1 \quad , \tag{4.5.68}$$

of course. We assume that $\varphi(t)$ is a simple Markovian process [i.e., the probability $p(\varphi)$ of the phase assuming a given value depends only on the last previous state and not on any state (value) before that]. That is to say, the p.d.f. $p(\varphi, t)$ of the Markov process $\varphi(t)$ depends only on the initial condition $\varphi(0) = \varphi_0$. Hence we denote the conditional probability density $p(\varphi/\varphi_0;t)$. The quantity $p(\varphi/\varphi_0;t)d$ is the probability that the value of the process lies in the infinitesimal interval between φ and $\varphi + d\varphi$, given that t seconds ago its value was φ_0.

Consider the process where the time durations t and Δt are arbitrary, and let the values of the process at the three instances t_0, $t_1 = t_0 + t$, and $t_2 = t_1 + \Delta t$ be φ_0, φ', and φ respectively.

Given φ' and φ_0, one could specify the probability density of φ as $p(\varphi/\varphi';\Delta t, \varphi_0;t + \Delta t)$. However, since the transition probability at t_1 depends only on φ', it follows that the probability density at t_2 is independent of the value φ_0 at t_0. Thus

$$p(\varphi/\varphi';\Delta t, \varphi_0;t + \Delta t) = p(\varphi/\varphi';\Delta t) \quad . \tag{4.5.69}$$

From this definition there follows a fundamental integral equation for the conditional density function of a Markov process. Let us omit the relative times for the moment. The joint probability density of the three samples of the process may be denoted by $p(\varphi,\varphi',\varphi_0)$. Then

$$p(\varphi,\varphi',\varphi_0) = p(\varphi',\varphi_0)p(\varphi/\varphi',\varphi_0) \quad . \tag{4.5.70}$$

Integrating both sides of (4.5.70) with respect to φ', we obtain

$$p(\varphi,\varphi_0) = \int_{-\infty}^{\infty} p(\varphi',\varphi_0)p(\varphi/\varphi')d\varphi' \quad . \tag{4.5.71}$$

Dividing by the density function $p(\varphi_0)$ we obtain [since $p(\varphi/\varphi_0)p(\varphi_0) = p(\varphi,\varphi_0)$], inserting again the relative times,

$$p(\varphi/\varphi_0;t + \Delta t) = \int_{-\infty}^{\infty} p(\varphi'/\varphi_0;t)p(\varphi/\varphi';\Delta t)d\varphi' \quad . \tag{4.5.72}$$

Equation (4.5.72) is a fundamental relation in the conditional density function of a Markov process (Chapman-Kolmogorov equation).

From (4.5.72) and the initial condition $p(\varphi/\varphi_0;0) = \delta(\varphi - \varphi_0)$, we can derive a *partial differential equation* in terms of $p(\varphi;t)$. We begin by considering the integral

$$I = \int_{-\infty}^{\infty} R(\varphi) \frac{p(\varphi/\varphi_0;t)}{\partial t} d\varphi \quad , \tag{4.5.73}$$

where $R(\varphi)$ is an arbitrary analytic function upon whose derivatives we place certain conditions, to be stated below. The above integral can be written as

$$I = \lim_{\Delta t \to 0} \int_{-\infty}^{\infty} R(\varphi)d\varphi \frac{p(\varphi/\varphi_0;t + \Delta t) - p(\dot{\varphi}/\varphi_0;t)}{\Delta t} \tag{4.5.74}$$

$$= \lim_{\Delta t \to 0} \frac{1}{\Delta t} \left[\int_{-\infty}^{\infty} R(\varphi)d\varphi \int_{-\infty}^{\infty} p(\varphi'/\varphi_0;t)p(\varphi/\varphi';t)d\varphi' - \int_{-\infty}^{\infty} R(\varphi')p(\varphi'/\varphi_0;t)d\varphi' \right] \quad . \tag{4.5.75}$$

Interchanging the order of integration and expanding the analytic function $R(\varphi)$ in a Taylor series about φ' yields

$$I = \lim_{\Delta t \to 0} \frac{1}{\Delta t} \int_{-\infty}^{\infty} p(\varphi'/\varphi_0;t) \sum_{n=1}^{\infty} \frac{R^{(n)}(\varphi')}{n!} d\varphi' \int_{-\infty}^{\infty} (\varphi - \varphi')^n p(\varphi/\varphi';\Delta t)d\varphi \tag{4.5.76}$$

where

$$R^{(n)}(\varphi') = \frac{d^n R(\varphi')}{d\varphi'^n} \quad .$$

Now let us denote the limit of the normalized nth conditional moment of the increment $(\varphi - \varphi')$ during the time Δt by

$$A_n(\varphi') = \lim_{\Delta t \to 0} \frac{1}{\Delta t} \int_{-\infty}^{\infty} (\varphi - \varphi')^n p(\varphi/\varphi';\Delta t)d\varphi \qquad (n \geqslant 1) \quad . \tag{4.5.77}$$

Then we get

$$I = \sum_{n=1}^{\infty} \frac{1}{n!} \int_{-\infty}^{\infty} R^{(n)}(\varphi')A_n(\varphi')p(\varphi'/\varphi_0;t)d\varphi' \quad . \tag{4.5.78}$$

Under the assumption that $R(\varphi')$ and its derivatives decrease sufficiently rapidly as $\varphi' \to \pm\infty$ so that

$$R^{(n-1)}(\varphi')A_n(\varphi')p(\varphi'/\varphi_0;t)\Big|_{-\infty}^{\infty} = 0$$

$$R^{(n-2)}(\varphi') \frac{\partial}{\partial \varphi'} [A_n(\varphi')p(\varphi'/\varphi_0;t)]\Big|_{-\infty}^{\infty} = 0 \qquad (4.5.79)$$

$$\vdots \qquad\qquad \vdots$$

$$R(\varphi') \frac{\partial^{n-1}}{\partial \varphi'^{n-1}} [A_n(\varphi')p(\varphi'/\varphi_0;t)]\Big|_{-\infty}^{\infty} = 0 \ ,$$

we now integrate the nth term of the sum by parts, n times. After subtracting (4.5.78) from (4.5.73) and replacing the variable of integration φ' in the former with φ, we obtain

$$\int_{-\infty}^{\infty} R(\varphi)d\varphi \left\{ \left[\frac{\partial p(\varphi/\varphi_0;t)}{\partial t}\right] - \sum_{n=1}^{\infty} \frac{(-1)^n}{n!} \frac{\partial^n}{\partial \varphi^n} [A_n(\varphi)p(\varphi/\varphi_0;t)] \right\} = 0 \ . \qquad (4.5.80)$$

Since $R(\varphi)$ was an arbitrary analytic function except for the above conditions on its derivatives, in order for the integral to vanish, the quantity in braces must also vanish. Hence

$$\frac{\partial p(\varphi,t)}{\partial t} = \sum_{n=1}^{\infty} \frac{(-1)^n}{n!} \frac{\partial^n}{\partial \varphi^n} [A_n(\varphi)p(\varphi,t)] \qquad (4.5.81)$$

with the initial condition $p(\varphi,0) = \delta(\varphi - \varphi_0)$.

The quantity $A_n(\varphi')$ is the limit of the nth moment of the increment $\Delta\varphi'$ and can be written

$$A_n(\varphi') = \lim_{\Delta t \to 0} \frac{1}{\Delta t} \int_{-\infty}^{\infty} (\Delta\varphi')^n p(\Delta\varphi'/\varphi')d(\Delta\varphi') = \lim_{\Delta t \to 0} \frac{<(\Delta\varphi')^n/\varphi'>}{\Delta t} \ . \qquad (4.5.82)$$

We are going to verify soon that the quantities $A_n(\varphi')$ vanish for $n \geqslant 3$, which means that the changes in the process $\varphi(t)$ occur slowly enough so that moments higher than the second one vanish more rapidly than Δt as the latter approaches zero. In this case (4.5.81) becomes

$$\frac{\partial p(\varphi,t)}{\partial t} = -\frac{\partial}{\partial \varphi} [A_1(\varphi)p(\varphi,t)] + \frac{1}{2} \frac{\partial^2}{\partial \varphi^2} [A_2(\varphi)p(\varphi,t)] \qquad (4.5.83)$$

where $p(\varphi,0) = \delta(\varphi - \varphi_0)$. This equation is generally called the *Fokker-Planck equation*. Its solution gives the p.d.f. at any given time, which completely describes the process statistically.

In order now to obtain the specific partial differential equation for $p(\varphi,t)$ — the p.d.f. of the PLL — we must determine the quantities $A_n(\varphi)$. These are obtained from the differential equation (4.5.66) by integrating both sides over the infinitesimal interval from to to $t+\Delta t$; thus,

$$\Delta\varphi = \varphi(t + \Delta t) - \varphi(t) = [\Delta - AK \sin\varphi(t)]\Delta t - K \int_{t}^{t+\Delta t} n'(t)dt \ . \qquad (4.5.84)$$

The conditional density of $\Delta\varphi$, given $\varphi(t)$, is clearly Gaussian for $n'(t)$ Gaussian. Recalling that $n'(t)$ is white noise of zero-mean and one-sided spectral density N_0, we find that the first two normalized moments are

$$A_1(\varphi) = \lim_{\Delta t \to 0} \frac{<\Delta\varphi|\varphi>}{\Delta t} = \Delta - AK \sin\varphi \; ; \tag{4.5.85}$$

$$A_2(\varphi) = \lim_{\Delta t \to 0} \frac{<(\Delta\varphi)^2|\varphi>}{\Delta t} = \lim_{\Delta t \to 0} \frac{K^2}{\Delta t} \int_t^{t+\Delta t} \int_t^{t+\Delta t} <n'(t_1)n'(t_2)>dt_1 dt_2$$

$$= \lim_{\Delta t \to 0} \frac{K^2 N_0}{2\Delta t} \int_t^{t+\Delta t} \int_t^{t+\Delta t} \delta(t_2 - t_1)dt_1 dt_2 = \frac{K^2 N_0}{2} \; . \tag{4.5.86}$$

Since the square of the first term of (4.5.84) is of the order of $(\Delta t)^2$ and the cross term involves the mean value of $n'(t)$, which is zero, it can be easily shown that $A_n(\varphi) = 0$ for $n > 2$. So the Fokker-Planck equation becomes

$$\frac{\partial p(\varphi,t)}{\partial t} = - \frac{\partial}{\partial\varphi} \{[\Delta - AK \sin\varphi(t)]p(\varphi,t)\} + \frac{K^2 N_0}{4} \frac{\partial^2 p(\varphi,t)}{\partial\varphi^2} \; . \tag{4.5.87}$$

A result of great interest is the stationary distribution corresponding to $\partial p(\varphi,t)/\partial t = 0$. Such a stationary solution exists only in the limit $t \to \infty$, i.e.,

$$p(\varphi) = \lim_{t \to \infty} p(\varphi,t) \; .$$

Letting $\alpha = 4A/KN_0$ and $\beta = 4\Delta/K^2 N_0$, we obtain from (4.5.87)

$$0 = \frac{\partial}{\partial\varphi} \left[(\alpha\sin\varphi - \beta)p(\varphi) + \frac{\partial p(\varphi)}{\partial\varphi} \right] , \tag{4.5.88}$$

or

$$p(\varphi) = C \exp(\alpha\cos\varphi + \beta\varphi) \left[1 + D \int_{-\pi}^{\varphi} \exp(-\alpha\cos x - \beta x)dx \right] \tag{4.5.89}$$

for $-\pi \leqslant \varphi \leqslant \pi$.

To evaluate the constants C and D, we use the conditions

$$p(\varphi,0) = \delta(\varphi - \varphi_0) \qquad \text{and} \qquad p(\pi,t) = p(-\pi,t) \; , \qquad (\text{for} \quad -\pi \leqslant \varphi \leqslant \pi)$$

$$\int_{-\pi}^{\pi} p(\varphi,t)d\varphi = 1 \qquad (\text{for all } t) \; .$$

In the limit, as $t \to \infty$, we have $p(\pi) = p(-\pi)$ and $\int_{-\pi}^{\pi} p(\varphi)d\varphi = 1$, so

$$D = \frac{\exp(-2\beta\pi) - 1}{\int_{-\pi}^{\pi} \exp[-(\alpha\cos x + \beta x)]dx} \tag{4.5.90}$$

and C is evaluated by using $\int_{-\pi}^{\pi} p(\varphi)d\varphi = 1$.

In the special case $\omega = \omega_0$ ($\Delta = 0$), $D = 0$ and

$$C = \frac{1}{\int_{-\pi}^{\pi} \exp(\alpha\cos\varphi)d\varphi} = \frac{1}{2\pi I_0(\alpha)} \; , \tag{4.5.91}$$

so that

$$p(\varphi) = \frac{\exp(\alpha\cos\varphi)}{2\pi I_0(\alpha)} \quad .$$ (4.5.92)

The parameter α plays an important role. Since $\alpha = 4A/KN_0$, we can write it also as $\alpha = A^2/N_0(AK/4)$. However, A^2 is the received-signal power while $AK/4$ is simply the loop bandwidth B_L and $\alpha = A^2/N_0B_0$ is the signal-to-noise ratio of the bandwidth of the loop. The relation (4.5.92) resembles a Gaussian distribution for a large signal-to-noise ratio and becomes flat as α approaches zero. Indeed, as $\alpha \to \infty$, the pertinent Bessel function I_0 gives

$$I_0(\varphi) \sim \frac{e^\alpha}{\sqrt{2\pi\alpha}}$$

and

$$p(\varphi) = \frac{\exp(\alpha\cos\varphi)}{2\pi I_0(\alpha)} \quad \frac{\exp[\alpha(\cos\varphi - 1)]}{\sqrt{2\pi/\alpha}} \quad ,$$

or expanding $\cos\varphi$ in a Taylor series we get

$$p(\varphi) = \frac{\exp\{(-\alpha\varphi^2/2)[1 - (2\varphi^2/4!) + (2\varphi^4/6!) + ...]\}}{\sqrt{2\pi/\alpha}} \quad , \quad -\pi \leqslant \varphi \leqslant \pi \; ; (4.5.93)$$

when α is large, $p(\varphi)$ decreases rapidly with φ; hence the function is very small for all but very small values of φ. The higher-order terms of the series representation of $\cos\varphi$ have very little effect for moderate values of $p(\varphi)$.

The cumulative steady-state distribution is

$$P(|\varphi| < \varphi_1) = \int_{-\varphi_1}^{\varphi_1} p(\varphi)\,d\varphi \quad , \quad 0 \leqslant \varphi_1 \leqslant \pi$$ (4.5.94)

and gives $\Big($for $\omega = \omega_0$ always, and since

$$p(\varphi) = \frac{\exp(\alpha\cos\varphi)}{2\pi I_0(\alpha)} = \frac{1}{2\pi I_0(\alpha)}\left[I_0(\alpha) + 2\sum_{n=1}^{\infty} I_n(\alpha)\cos n\varphi\right]\Big):$$

$$P(|\varphi| < \varphi_1) = 2\int_0^{\varphi_1} p(\varphi)\,d\varphi = \frac{\varphi_1}{\pi} + \frac{2}{\pi}\sum_{n=1}^{\infty} \frac{I_n(\alpha)\sin(n\varphi_1)}{nI_0(\alpha)} \quad .$$ (4.5.95)

This series converges very rapidly.

The variance of φ can be similarly obtained

$$\sigma_\varphi^2 = \frac{1}{2\pi I_0(\alpha)}\int_{-\pi}^{\pi} \varphi^2 \exp(\alpha\cos\varphi)\,d\varphi = \frac{\pi^2}{3} + 4\sum_{n=1}^{\infty} \frac{(-1)^n I_n(\alpha)}{n^2 I_0(\alpha)} \quad .$$ (4.5.96)

This series converges even more rapidly and as $\alpha \to 0$, $\sigma_\varphi^2 \to \pi^2/3$ (which is the variance of a random variable that is uniformly distributed between $-\pi$ and π).

For the general case now, $\omega \neq \omega_0$ ($\Delta \neq 0$), a computer is needed to evaluate the pertinent integrals. In these cases a steady-state solution does not exist necessarily. (For computing simulation of phase locking, see Appendix C.)

Finally we have to treat the following important issue: since the random process considered here is a *phase error*, the principal concern is that the absolute value $|\varphi|$ is maintained small at *all times* if possible, below a given "limit" value φ_1. A valuable statistic in that respect is the expected time required for the phase error to reach $+\varphi_1$ for the first time, given that it was initially at φ_0, where $\varphi_0 < \varphi_1$. Closely related to this *first-passage* problem is the *frequency of skipping cycles*. (For the mechanical analogue it is the inverse of the expected time for the pendulum to swing a complete cycle in either direction.) For the PLL, it represents the frequency with which the loop VCO *gains* or *drops* a cycle relative to the received signal. In either case, it corresponds to letting $\varphi_1 = 2\pi$ in the determination of expected time.

We will treat only the above case of the first-order loop for which the VCO quiescent frequency is tuned to the received frequency ($\omega = \omega_0$), so that the equilibrium phase-error position is $\varphi = 0$. For the first-order loop, the same approach can be used when $\omega \neq \omega_0$, but the results are in the form of integrals which require numerical calculations. We assume then that $\omega = \omega_0$, and that the loop is initially in lock so that $\varphi_0 = 0$.

As long as the loop phase error (or the "pendulum" angle) remains within the limits $|\varphi| < \varphi_1$, the density function of φ, which we denote here by $q(\varphi,t)$, satisfies the Fokker-Planck equation

$$\frac{\partial q(\varphi,t)}{\partial t} = \frac{\partial}{\partial \varphi} [(AK \sin\varphi)q(\varphi,t)] + \frac{K^2 N_0}{4} \frac{\partial^2 q(\varphi,t)}{\partial \varphi^2} \quad \text{for all} \quad |\varphi| < \varphi_1 \quad (4.5.97)$$

within the initial conditions $q(\varphi,0) = \delta(\varphi)$. We have denoted the density function of the phase error by $q(\varphi,t)$ to distinguish it from the corresponding function $p(\varphi,t)$ for the *unbounded* case. As soon as $|\varphi|$ reaches φ_1 for the first time, the pendulum is instantly removed from action so that $q(\varphi,t) = 0$ for all $|\varphi| \geq \varphi_1$ and all t.

Thus, in addition to the initial condition, we have the boundary conditions $q(\varphi_1,t) = q(-\varphi_1,t) = 0$ for all t. Solution of the Fokker-Planck equation over the interval $-\varphi_1 < \varphi < \varphi_1$ with these boundary conditions would yield $q(\varphi,t)$. Its integral form over the interval

$$\Psi(t) = \int_{-\varphi_1}^{\varphi_1} q(\varphi,t)d\varphi \quad (4.5.98)$$

gives the probability that φ has not *yet* reached φ_1 at time t. Note that as a consequence of the condition $q(\varphi,t) = 0$ for $|\varphi| \geq \varphi_1$, the limits on the above integration could just as well be infinite, so that

$$\int_{-\infty}^{\infty} q(\varphi,t)\,d\varphi = \Psi(t) \leqslant 1 \quad . \tag{4.5.99}$$

This is the fundamental difference between $q(\varphi,t)$ and $p(\varphi,t)$, for $\int_{-\infty}^{\infty} p(\varphi,t)\,d\varphi = 1$ for *all* t. In other words, $q(\varphi,t)$ is not, strictly speaking, a p.d.f.

Since $\Psi(t)$ is the probability that $|\varphi|$ is less than φ_1 at time t and has never lost lock in the time interval from 0 to t, it must be a monotonically non-increasing function of t. Thus, the p.d.f. of the time required for $|\varphi|$ to reach φ_1 for the first time is

$$\lim_{\Delta t \to 0} \left[\frac{\Psi(t) - \Psi(t + \Delta t)}{\Delta t} \right] = -\frac{\partial \Psi(t)}{\partial t} \quad , \tag{4.5.100}$$

and the expected time to reach the position φ_1 for the first time is

$$T = \int_0^{\infty} -t\,\frac{\partial \Psi(t)}{\partial t}\,dt = -\left[t\Psi(t) \right]_0^{\infty} - \int_0^{\infty} \Psi(t)\,dt \quad . \tag{4.5.101}$$

If the non-increasing function $\Psi(t)$ approaches zero faster than $1/t$, the first term on the right side of the above equation is zero. (This must be the case if $\int_0^{\infty} \Psi(t)\,dt$ is to exist.)

We have, then

$$T = \int_0^{\infty} \int_{-\varphi_1}^{\varphi_1} q(\varphi,t)\,d\varphi\,dt \quad , \tag{4.5.102}$$

and by integrating the Fokker-Planck equation (4.5.97) for $q(\varphi,t)$, we find

$$q(\varphi,\infty) - q(\varphi,0) = \frac{d}{d\varphi}\left[(AK\,\sin\varphi)Q(\varphi) \right] + \frac{N_0 K^2}{4}\,\frac{d^2 Q(\varphi)}{d\varphi^2} \quad , \tag{4.5.103}$$

where

$$Q(\varphi) = \int_0^{\infty} q(\varphi,t)\,dt \quad .$$

Clearly, $q(\varphi,\infty) = 0$, and since φ is assumed to be initially zero, $q(\varphi,0) = \delta(\varphi)$. Therefore, we get

$$-\delta(\varphi) = \frac{d}{d\varphi}\left[(AK\,\sin\,)Q(\varphi) \right] + \frac{N_0 K^2}{4}\,\frac{d^2 Q(\varphi)}{d\varphi^2} \quad , \tag{4.5.104}$$

with the boundary conditions:

$$Q(\varphi_1) = \int_0^{\infty} q(\varphi_1,t)\,dt = 0 \quad ,$$

$$Q(-\varphi_1) = \int_0^{\infty} q(-\varphi_1,t)\,dt = 0 \quad .$$

The solution of (4.5.104) may then be integrated with respect to φ over the interval $(-\varphi_1, \varphi_1)$ to obtain T, the expected time to first passage. Taking the indefinite integral of both sides of (4.5.104) we have

$$C - u(\varphi) = (AK \sin\varphi)Q(\varphi) + \frac{N_0 K^2}{4} \frac{dQ(\varphi)}{d\varphi} \quad , \tag{4.5.105}$$

where $u(\varphi)$ is the unit step function and C is a constant to be evaluated from the boundary conditions.

The solution to this first-order differential equation is

$$Q(\varphi) = D \exp(\alpha\cos\varphi) + \exp(\alpha\cos\varphi) \int_{-\varphi_1}^{\varphi_1} \frac{\exp(-\alpha\cos x)}{\gamma} [C - u(x)] dx \quad , \tag{4.5.106}$$

where

$$\alpha = \frac{A^2}{N_0 AK/4} = \frac{A^2}{N_0 B_L}$$

and

$$\gamma = \frac{N_0 K^2}{4} = \frac{AK}{\alpha} = \frac{4B_L}{\alpha} \quad .$$

Applying the boundary conditions $Q(\varphi_1) = Q(-\varphi_1) = 0$ yields the values of the constants as $D = 0$ and $C = 1/2$. Thus

$$Q(\varphi) = \frac{\exp(\alpha\cos\varphi)}{\gamma} \int_{-\varphi_1}^{\varphi_1} \exp(-\alpha\cos x)\left[\frac{1}{2} - u(x)\right] dx \quad , \tag{4.5.107}$$

and integrating with respect to φ over the interval $(-\varphi_1, \varphi_1)$, we obtain the expression for the mean first-passage time

$$T = \int_{-\varphi_1}^{\varphi_1} Q(\varphi) d\varphi$$

$$= \frac{1}{\gamma} \int_{-\varphi_1}^{\varphi_1} d\varphi \int_{-\varphi_1}^{\varphi_1} \exp[\alpha(\cos\varphi - \cos x)]\left[\frac{1}{2} - u(x)\right] dx$$

$$= \frac{1}{\gamma} \int_{0}^{\varphi_1} d\varphi \int_{-\varphi_1}^{\varphi_1} \exp[\alpha(\cos\varphi - \cos x)] dx \quad , \tag{4.5.108}$$

Taking into account the expansion

$$\exp(\pm\alpha\cos\varphi) = I_0(\alpha) + 2 \sum_{m=1}^{\infty} (\pm 1)^m I_m(\alpha)\cos m\varphi \quad ,$$

(4.5.108) becomes

$$T = \frac{1}{\gamma} \left\{ \frac{I_0^2(\alpha)\varphi_1^2}{2} + 2I_0(\alpha) \sum_{n=1}^{\infty} I_n(\alpha) \int_{0}^{\varphi_1} d\varphi \int_{-\varphi_1}^{\varphi_1} [\cos n\varphi + (-1)^n \cos nx] dx \right.$$

$$+ 4 \sum_{m=1}^{\infty} \sum_{n=1}^{\infty} (-1)^n I_m(\alpha) I_n(\alpha) \int_0^{\varphi_1} \cos m\varphi \, d\varphi \int_{-\varphi_1}^{\varphi_1} \cos nx \, dx \Big\} \quad . \tag{4.5.109}$$

The most important result is, of course, the expected time between skipping cycles, i.e., $T(2\pi)$. So for $\varphi_1 = 2\pi$, from the general expression (4.5.109) we get

$$T(2\pi) = \frac{2\pi^2}{\gamma} I_0^2(\alpha) = \frac{\pi^2 \alpha I_0^2(\alpha)}{2B_L} \quad , \tag{4.5.110}$$

where $B_L = AK/4$, so that

$$\text{frequency of skipping cycles} = \frac{2B_L}{\pi^2 \alpha I_0^2(\alpha)} = \frac{AK}{2\pi^2 \alpha I_0^2(\alpha)} \quad . \tag{4.5.111}$$

For $\alpha \gg 1$ (large signal-to-noise ratio), since

$$I_0(\alpha) \sim \frac{e^\alpha}{(2\pi\alpha)^{1/2}} \quad ,$$

we get the frequency of "intermittent behavior" or the

$$\text{frequency of skipping cycles} \sim \frac{AK \, e^{-2\alpha}}{\pi} \quad . \tag{4.5.112}$$

4.6 Modeling of Stochastic Time Series

We next consider the following problem. We are observing a (discrete) time series of 0's and 1's as the *behavior* of a certain dynamical system—the complexity of which we intend to *compress* or to *simulate*. What is the number of *minimal-state* (finite-state) deterministic automatons that can simulate the observed time series? How does the average *number* of the states in the optimal model (i.e., the expected complexity) increase with the number of observations, i.e., with the (ever-increasing) length of the received time series? The observer does not know whether the time series he receives comes from a "coin-flipping" (or "dice-rolling") process or whether it derives from a causal process.

Here we are interested in the case in which the modeler, prompted by some psychological bias, assumes a priori that the observed behavior is generated by a *causal* system and therefore forms a deterministic model of it. For any finite length N of observed time series, there will be, of course, an unbounded number of finite-state machines whose input-output behavior, starting from a specific state, is identical to that observed; out of these possible models there will be some such that no other machine has a smaller number of states. An optimal *causal* modeler is therefore defined as one who always chooses such a minimal-state machine as a model.

We would like, following *Gaines* [4.10], to derive a relationship between the length N of the sequence of observations and the expected number of states in the model formed by the optimal causal modeler (that is, the average complexity). In other words, we are interested in the behavior of the ratio

$$R_N = \frac{\text{expected number of states of the model}}{\text{number of observations}}$$

as a function of the length of the observed sequence. One perhaps would expect that for $N \to \infty$, $R_N \to 0$. In fact, as $N \to \infty$, $R_N \to 1$ which means that the number of the states of the optimal (causal) model increases with the length of the observed time series, or the complexity of the optimal causal model increases with the number of observations.

We shall see that the slightest introduction of probabilistic acausality in the behavior of the dynamical system we intend to simulate via a deterministic finite-state machine gives rise to models whose complexity is basically proportional to the length of observation. Suppose the behavior of the system about to be simulated is expressed as a discrete time series of 0's and 1's with equal a priori probability $P(0) = P(1) = 1/2$ [in general, we have a Bernoulli series with $P(0) \neq P(1)$]. This is what one expects as behavior from a two-state stochastic automaton.

There are 2^N possible sequences of N observations; all of these sequences are equiprobable with probability 2^{-N}. Now let us enumerate the different S-state deterministic automata that are available as models of these sequences. We note that a given automaton, starting in a given state, can act as a model of only one of the sequences, and we are interested in finding an upper bound on the variety of models available for a given S.

Suppose for example that S = 6. One model will be 001010, another will be 010001, etc. The general form of a model will be a transient chain followed by a cycle, for example:

$$0 \;\; 0 \;\; 1 \;\; 0 \;\; 1 \;\; 0 \;\; 1 \;\; .$$

If there are S states to be filled with two symbols 0, 1 and the cycle can commence in any state, there are at most $S2^S$ models (some of which may generate the same sequence, making this an *upper bound* on the number of different models available), so

$$M = S2^S > 2^N \;\; . \tag{4.6.1}$$

(Note that this number includes also models with less than S states.)

Now we search for the *mean number* of states in the minimal forms of these M models. There are at most $R2^R$ models with R states; hence a lower bound of the *mean* (expected) number of states in the set whose maximum number of states is S is calculated as follows. Let M(R) be the probability of finding models with states from R = 1 to R:

$$M(R) = \int_1^R \frac{dM(R)}{dR} \, dR \;\; , \tag{4.6.2}$$

where $dM(R)/dR$ is the corresponding p.d.f. Since, however, R is a discrete variable, the average value of $M(R)$ for R between 1 and $S-1$ equals

$$\overline{M(R)} = \frac{1}{S-1} \sum_{R=1}^{S-1} \frac{\Delta M(R)}{\Delta R} R \quad . \tag{4.6.3}$$

But

$$\frac{\Delta M(R)}{\Delta R} = 2^R + R2^{R-1} = R2^R - (R-1)2^R + R2^{R-1}$$

$$= R2^R - (R-2)2^{R-1} \quad R2^R - (R-1)2^{R-1} \quad ,$$

so

$$\overline{M(R)} > \frac{1}{S-1} \sum_{R=1}^{S-1} R[R2^R - (R-1)2^{R-1}] \quad . \tag{4.6.4}$$

Since the average number μ_S of state S gives

$$\mu_S 2^S > \overline{M(R)} > \frac{1}{S-1} \sum_{R=1}^{S-1} R[R2^R - (R-1)2^{R-1}] \quad , \tag{4.6.5}$$

we have

$$\mu_S > \frac{1}{(S-1)2^S} \frac{1}{2} \sum_{R=1}^{S-1} (R^2 2^R + R2^R) \tag{4.6.6}$$

or, after some algebra,

$$\mu_S > S - 2 + \frac{2}{S-1} - \frac{2}{(S-1)2^{S-1}} > S - 2 \quad . \tag{4.6.7}$$

From the inequality (4.6.1), we also get

$$S + \log_2 S > N \tag{4.6.8}$$

so that

$$S > N - \log_2 S \quad , \tag{4.6.9}$$

but since $S < N$, $S > N - \log_2 N$. Hence

$$\mu_S > S - 2 > N - \log_2 N - 2 \quad , \tag{4.6.10}$$

that is, the mean number of states in the ensemble of models necessary to account for all possible behaviors is greater than $N - \log_2 N - 2$, where N is the length of the received time series. Therefore

$$R_N = 1 - \frac{\log_2 N + 2}{N} \tag{4.6.11}$$

which asymptotically approaches unity as $N \to \infty$.

The conclusion is that if a modeler insists on simulating a stochastic behavior through a deterministic finite-state automaton, he gets an over-complex view of the system under observation. In Chap.6 we will see how systems possessing strange attractors effectively *compress* the behavior of a system whose number of degrees of freedom equals the dimensionality of the state space in which the strange attractor is embedded as a compact subset.

4.7 Communication Between Two Hierarchical Systems Modeled by Controlled Markov Chains

We now consider in some detail a specific example dealing with the interaction between *two* self-organizing systems, each possessing two hierarchical levels. The dynamics at the lower levels Q, Q' of the systems involved are emulated as finite-state controlled Markov chains; their transitional probability matrices are parametrized by control variables which are related to the probabilities of "payoff" in an underlying two-agent game which simulates collectively all hierarchical levels below Q and Q', respectively. The higher levels W, W' are modeled by semi-Markov chains, whose mean holding times are inversely proportional to the degree of organization (or redundancy) of the respective lower levels Q, Q'.

The higher hierarchical levels W, W' receive afferently from the respective lower levels Q, Q' collective properties which measure (a) the percentage of occupancy of a state selected a priori [on the lower level (s)] as a "homeostatic" one, and (b) the cross-correlation(s) between the state sequences of the levels (Q, W') or (Q', W), respectively. The transition probabilities for the semi-Markov chains at the higher levels W, W' are functions of the above-mentioned quantities (a) and (b).

The communication process between the two hierarchical systems is pursued as a bidirectional information transaction where the lower levels play the roles of receivers and the higher levels, transmitters —the levels standing for "experience" and "behavior", respectively, or "hardware" and "software" levels.

The intersystem communication becomes adaptive by built-in control, efferent (feedforward) mechanisms exercised intrasystemically from the higher levels W, W' to their respective lower levels Q, Q' via the controlling parameters of the underlying games. In this example the development of an extensive computer program is undertaken in order to simulate such an adaptive communication process. The evolution with time of the behavioral mode (state), switching at the levels W and W', is computed for specified control laws, randomly selected from all possible feedforward mechanisms —which increase exponentially with the number of quantization levels of the control variables. As far as the selection of the control algorithm is concerned, the criterion is the maximization of the "joint figure of merit" having to do with the long-term average of an appropriately weighted sum of the conflicting terms: probabilities of the homeostatic states and the intersystemic cross-correlations between the levels (W, Q') and (W', Q).

The present discussion is inspired by contemporary psychophysiological and biological research. The simulation on the computer of the mathematical model presented here may contribute to a better understanding of certain aspects of communication between organisms and suggest the design of new neurophysiological experiments. The purpose of this discussion is to propose cybernetical models for a comprehensive presentation of a large number of phenomena related to communication between biological, i.e., self-organizing, systems.

4.7.1 Introduction: Elaboration of the Nature of Hierarchical Systems

Self-organizing systems are hierarchical structures [4.11] which simultaneously undergo a variety of distinguishable activities; different sets of variables and parameters are appropriate to state-space descriptions pertaining to these several activities taking place at the individual levels. In this example we model some aspects of the communication procedure which goes on between two self-organizing systems —each possessing a pair of hierarchical levels Q, W and Q', W', respectively (Fig.4.20). The discussion, although cybernetic-mathematical in nature, is inspired by psychophysiological research and evidence. In particular, we have been encouraged and supported in our deliberations by the recent impact of biochemistry and neurophysiology on modern psychiatry [4.12] involving, for instance, some factual experience regarding the biology of mental illness [4.13], as well as by modern trends in psychosomatic medicine [4.14].

Inversely, starting from higher hierarchical levels concerning the pragmatics of human communication [4.15], we have been intrigued by the correlative way in which the "neurotic" behavior, especially concerning the misuse of the hierarchical codes of the symbolic language (the "double bind" mode of communication [4.16,17]

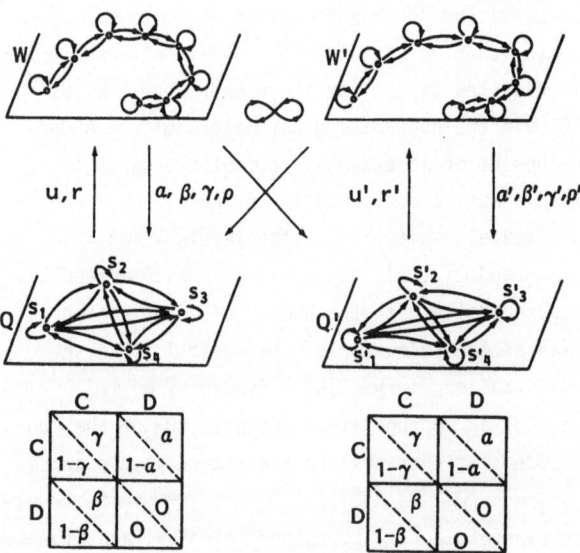

Fig.4.20. The general layout of the communication process between two hierarchical systems

is certainly an appropriate example), is instrumental in breaking the communication and occasionally precipitating organic disturbances in the person(s) involved.

In all the references above, the point worth noting is the correlation (albeit sometimes weak) between *physiological* and *psychological variables*. In this discussion these stochastic dependences are isomorphically incorporated in our model, i.e., they determine the way that successive hierarchical levels in the individual organism(s) interact. But what exactly defines *physically* a "hierarchical level" and in what respects do "hierarchical levels" differ from each other? Before proceeding further it is constructive, we think, to clarify these two important points. *A hierarchical level is determined by the number, location, nature* (i.e., structure or complexity) as well as the *degree of coherent cooperativeness* among functionally similar, although anatomically perhaps quite different, "modules" of the system concerned (or "tissues" in the case of a biological organism). By "functionally similar" modules we imply a dynamical set of elements which have to cooperate (i.e., keep a number of flexibly bounded, ordered relationships among themselves) for the successful performance of the given activity.

In respiration, for example, the object is the supplying of oxygen to the alveoli of the lungs and its diffusion through the walls of the alveoli into the blood. For this to be achieved, a complex muscular apparatus is necessary, incorporating the diaphragm and the intercostal muscles capable of expanding and contracting the chest, as well as a complex system of nervous structures in the brainstem and higher centers to control the apparatus. It is obvious that the above processes are carried out as *functions* embodying many components belonging to different quarters of the secretory, motor, and nervous apparatus.

In the brain also, all "higher" processes such as perception, speech, writing, reading, and feeling, cannot be regarded as isolated or even indivisible "faculties" which can be presumed to be the direct functions of limited cell groups or to be localized in specific areas in the brain. Instead of "localized areas" of dynamical activity, we have rather pathways of information transfer (afferent pathways), and storage and integration processes. So "mental" functions are to be considered here as complex functional hierarchical systems which cannot be localized in particular zones of the cortex or in isolated cell groups, but must be organized in groups of coherently working zones, each of which plays the role of a nonlinear dynamic (random) variable and which may belong to completely different and often very distant areas of the brain. Such groups will define the hierarchical level, performing each time the appropriate function.

From the above examples, one could perhaps propose for a higher organism the following set of hierarchical levels in ascending order: The levels of (a) nucleic acids, (b) protein, (c) individual cells, (d) individual organs, and (e) groups of organs (performing somatic functions like digestion, etc.). Higher hierarchical levels are characterized by an increasing proportion of participation of cerebral tissues of progressively ascending order and increasing complexity (from the brain-stem, along the reticular formation all the way up to the cerebral cortex). These last levels are termed "cognitive" levels and most of them, unlike the lower "soma-tic" levels, are formed and broken down not only by evolution but also by *learning* taking place during the limited life-span of the individual organism.

Let it be understood, nonetheless, that the material or hardware "canvas" for all these highest cognitive levels is the *same*, namely groups of neurons. What makes these cognitive levels different from each other is the nature and combination of the cooperating tissues and the statistical averaging processes performed or the resulting degree of abstraction (compressibility).

At each hierarchical level we can introduce a state-space description involving a number of dynamical variables and a set of parameters pertaining to the particu-lar level. For reasons of simplicity, in this work (see also Appendix B) we quantize the state space at each hierarchical level and admit only a finite number of states which stand for the salient subtraits at the level involved. Thus the stochastic nonlinear differential equations corresponding to the continuous state description are now replaced at all hierarchical levels by discrete-time Markov chains charac-terized by appropriate transition matrices $\{P_{ij}\}$, fully describing the transitions between the possible states of the systems at the level involved.

The lower levels Q, Q', are modeled as four-state Markov chains which are deduced as outcomes of respective antagonistic processes or "games" going on between pairs of dynamic agents (see Chap.5 and [4.18]); these games simulate collectively all hierarchical levels below Q and Q', respectively. At the levels Q, Q', one of the available four states will be considered as the "homeostatic" state(s); these are the states towards which the activities at the levels Q and Q' should evolve pre-ferentially, but not exclusively, in order to achieve optimization of the intrinsic regulation and stability of the organism at the levels concerned. On the other hand, in order to ensure *adaptability* via learning the organism(s) at the same level(s) Q, Q' should communicate adequately with each other, i.e., form high cross-correla-tion(s) between the incoming set of triggers induced by the higher level(s) of the partner W', W and prestored dynamical patterns at the base level(s) Q, Q'.

For any given set of triggers emanating from the partner, the above requirement for high cross-correlation implies a rather extensive use of the repertoire of the states available at the level(s) Q, Q'. However, such an extensive "wandering" be-tween individual states of the systems at the levels Q, Q' should certainly jeopar-dize the requirement for homeostasis. Therefore, in general, the two basic deliber-ations for each organism, namely the homeostatic necessity and the tendency for continuous adaptability to the partner's cues, *seem to be in conflict*.

The higher levels W, W' are modeled by *semi*-Markov chains of eight states whose transitional probabilities are functions of (a) the percentage of occupancy of the homeostatic state of the underlying level u, and (b) the correlation r between the state sequences of the levels (Q, W') or (Q', W), respectively. The *holding times* at the states of the levels W, W' follow a geometrical distribution with a mean value which is inversely proportional to the "degree of organization" or the redundancy of the underlying levels (as will be discussed in the following subsections).

Upon receiving the above data u, r, u', r', the highest levels hold the existing state for some time and then a transition occurs. The new state is emitted (as "behavior") to the partner in a three-bit binary symbolic code. On the other hand, depending on the received values r and u, r' and u', efferent control commands are precipitated to the underlying game(s) which alter(s) the amplitudes and phases of their slowly and rhythmically varying parameters α, β, γ, p, α', β', γ', p', respectively, according to some control algorithm (to be revealed in the next section). The *new* game parameters will influence the transition probabilities at the level(s) Q, Q', thereby altering the homeostatic probability as well as the cross-correlations (Q', W) and (Q, W'). This procedure repeats itself in successive cycles, asserting the communication between the partners.

There are many regimes of communication. In this discussion we restrict ourselves to the cooperative regime of communication whose goal is a control with two objectives, namely the maximization of the joint "figure of merit" which weighs in preselected proportions the homeostasis probabilities and the cross-correlations (W, Q'), (W', Q). If someone were naively to take as a measure of the quality of communication the degree of synchronization between the behavioral state turnover at the levels W, W' as evolving with time, he certainly would be disappointed. Synchronization may be good for a simple device like the phase-locked loop but says virtually nothing about a time-delayed evolutionary phenomenon like a "dialogue", which constitutes essentially a *chaotic process* (see Sects.6.5.6 and 7.2 for more details).

In the next section we give a description of the communication process including the underlying games, the resulting dynamics at the levels Q, Q', as well as the dynamics at the levels W, W'. In Sect.4.7.3 we elaborate on the semi-Markov chain structure of the higher levels as well as the statistics of the holding times. In Sect.4.7.4 the control problem is laid out and the criteria of optimization are explained. In Sect.4.7.5 an extensive computer simulation of the communication-control processes is presented. We close the example (Sect.4.7.6) with some concluding remarks concerning the relevance of the presented model to some current research on human communication.

4.7.2 Dynamics at the Base Levels Q, Q' and the Underlying Game

Let us now try to deduce the dynamical activity at the base levels Q, Q', as an "offshoot" of an underlying game. In each case we postulate the existence of two dynamical agents I, II involved in alternating plays. Concerning, for instance,

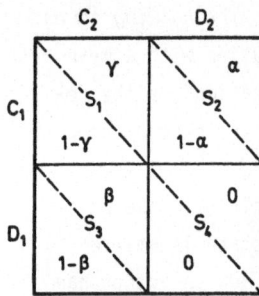

Fig.4.21. Diagram of the two-agent underlying "game"

the first partner, i.e., the one with levels (Q, W), on every trial of the game each agent makes either a "cooperating" (C) or a "defecting" (D) move (Fig.4.21).

If both agents defect, their probabilities of winning are zero. If both agents cooperate, agent II "wins" with probability γ and agent I "wins" with probability $1 - \gamma$, where "winning" corresponds to dominance of the agent concerned of the system's deliberations at the level Q. If agent I cooperates and agent II defects, the latter has a probability α of winning and the former a probability $1-\alpha$ of winning. If, finally, agent I defects and agent II cooperates, the former wins with probability β and the latter with probability $1-\beta$.

We further assume that:

1) If an agent makes a winning move, then with probability *one* he repeats that move on the next trial.
2) If an agent makes a losing move, then with probability p he will make the alternative move on the next trial.

We therefore understand that the first system at the base level Q has four states:

State S_1: Both agents cooperate.
State S_2: Agent II defects, agent I cooperates.
State S_3: Agent I defects, agent II cooperates.
State S_4: Both agents defect.

It should be noted that in the above we have considered an *asymmetrical* matrix in which the states (C, D) and (D, C) are distinguishable. Identical criteria hold for the second partner (Q', W') with α', β', γ', p' being the corresponding parameters.

What could be a plausible physiological interpretation of the parameters α, β, γ, p? The probabilities of winning, α, β, γ, may be considered as correlatives to the probability distributions of the instantaneous concentration of relevant substances such as hormones or neurotransmitters —which are rhythmically varying entities and whose concentration is correlated to the degree of excitability or arousal of the "tissues" involved[3]. (Footnote see next page)

The parameter p could be considered as analogous to the probability of turnover C→D or D→C for either agent I or II; it is derived therefore from the distribution of the *differential* concentration of the appropriate substances (hormones or neuro-transmitters) responsible for the above turnover. In physiology there are excellent cases of asymmetrical "games", such as the above, taking place, for example, between the two branches of the autonomous nervous system, sympathetic and parasympathetic (mediated by noradrenaline and acetylcholine, respectively). There is also the basic and, at the moment, rather poorly understood antagonism between the degree of arousal of the ascending branch of the reticular formation and the thalamocortical pacemaking activity. The first of these examples is a classic one; in what follows we comment briefly on some contemporary tentative views concerning the second example, since the "game" involved is of capital importance since the level Q is associated with "higher" activities such as orientation, motivation, and commitment.

During the last 20 years a number of investigations concerning the electrical activity of the brain have been performed, based on the statistical analysis of recordings obtained simultaneously with several external electrodes and microelectrodes. Such measurements appear to indicate that the rhythmic part of an electroencephalogram (the "alpha" rhythm) is characterized by a highly organized *circulation* of activity via the networks of the cortex and the thalamus [4.19]. Since this circulating activity follows along *curved* lines and shows frequent reversals of direction and progressive shifts in the location of the groups of the neurons involved in it, it has been concluded that the pathways of circulation extend along a series of loops whose locus advances progressively through the neuronal network. Morphological studies have also indicated that large cortical areas possess reciprocal connections with a number of thalamic nuclei [4.20]. This suggests that corticothalamic feedback loops, which conduct impulses to the thalamic nuclei, may be important in the coordination of thalamocortical communication.

Further investigations by *Verzeano* et al. [4.21] have shown that the interaction between the circulating neuronal activities of several thalamic nuclei forms the cellular basis of the pacemaker mechanism that the thalamus exercises in the cortex, thereby sustaining a rhythmic activity such as the alpha rhythm. In all cases, however, even though the pacemaking mechanism may be based on the activity of thalamic nuclei, it is probable that its stability and coordination are greatly influenced by corticothalamic feedback loops.

From the investigations reported above it may be concluded that the so-called synchronized regime in the brain which characterizes mainly *relaxed wakefulness* is based on the activity of a neuronal system distributed over all thalamic nuclei,

---

3 By the "excitability" or "arousal" of a certain aggregate of coupled neurons we mean the percentage of simultaneously firing ("on") elements, or the percentage of simultaneously depolarized elements below the threshold.

communicating with the cortex and coordinated by the nonspecific nuclei. The co-ordinating mechanism is the highly organized circulation of neuronal activity through cortical and thalamic networks and through cortical feedback loops.

On the other hand, there also exists in the brain the desynchronized or chaotic regime, which characterizes mainly *alertness* and *attentive wakefulness*: it is based on the activity of the ascending reticular activating system whose action in the brain is mediated in part by "relay" stations located in the thalamic nuclei [4.22]. When these relay stations are activated by reticular stimulation (resulting from *arousal* caused by information transfer from lower to neocortical levels or from the environment via the peripheral nervous system), thalamocortical neuronal activity is polarized and disrupted for periods of time which increase with the degree of excitation, with the degree of arousal of the reticular formation ascending branch, or with the *rate* of information transmission along this channel.

Thus, it is concluded, *two antagonistic* neuronal systems operate in the thalamus: one concerned with circulation of neuronal activity and synchronization and the other concerned with the *disruption* of circulation of neuronal activity and desynchroniz-ation. Moreover, it is known that different neurotransmitters operate in the two systems; *serotonin* in the synchronizing system and *acetylcholine* in the desynchroniz-ing system. "Normally" the activities of these two agents should exist in some sort of dynamical balance. An increased influence of the first system leads to a state of relaxation and drowsiness. An increased influence of the second leads to arousal and alertness.

In the context of our present discussion, it is perhaps relevant to touch upon some "pathological" situations where one of the two agents may be *secularly over-driven* (e.g., by an excessive action of its own neurotransmitter or by depletion of the neurotransmitter of the other system) leading to addition of noise and in extreme cases to convulsive states. Under such circumstances, the communication be-tween the hierarchical level(s) involved is progressively disrupted.

From the above example one may obtain some idea about the plausibility and rele-vance of the model, including the distinguishability between the states (C, D) and (D, C) (for further discussion see Sect.4.7.4).

We next intend to express the 16 elements P_{ij} of the transition matrix at Q as functions of the parameters α, β, γ, p. This has been done in a recent work [4.18]. In Figs.4.22-25 we display the corresponding (self-explanatory) tree diagrams.

From a knowledge of the transition matrix we can further compute the likelihood of being in the state S_i ($i = 1,2,3,4$) at the start of the trial $n + 1$.

Let this probability be $u_{i,n+1}$; it follows immediately that

$$u_{i,n+1} = \sum_{j=1}^{4} u_{j,n} P_{ji} \ . \tag{4.7.1}$$

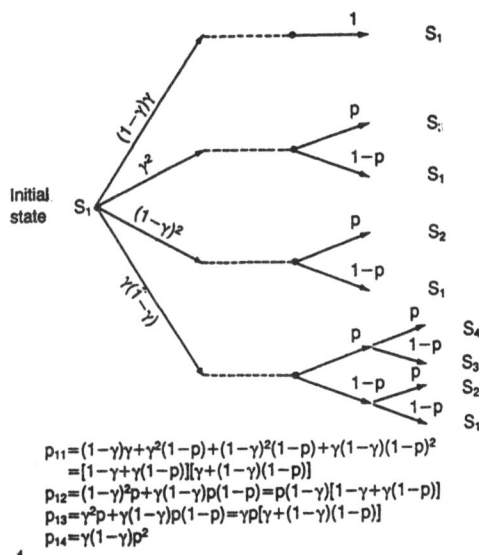

$p_{11}=(1-\gamma)\gamma+\gamma^2(1-p)+(1-\gamma)^2(1-p)+\gamma(1-\gamma)(1-p)^2$
$\quad=[1-\gamma+\gamma(1-p)][\gamma+(1-\gamma)(1-p)]$
$p_{12}=(1-\gamma)^2p+\gamma(1-\gamma)p(1-p)=p(1-\gamma)[1-\gamma+\gamma(1-p)]$
$p_{13}=\gamma^2p+\gamma(1-\gamma)p(1-p)=\gamma p[\gamma+(1-\gamma)(1-p)]$
$p_{14}=\gamma(1-\gamma)p^2$

$\sum\limits_{j=1}^{4} p_{ij}=1$

Fig.4.22. Tree diagram for transition
probabilities with initial state S_1

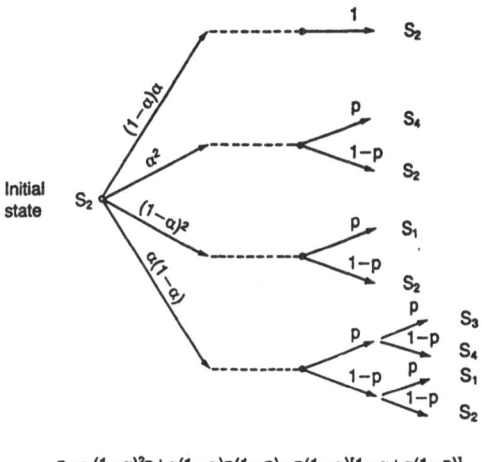

$p_{21}=(1-\alpha)^2p+\alpha(1-\alpha)p(1-p)=p(1-\alpha)[1-\alpha+\alpha(1-p)]$
$p_{22}=(1-\alpha)\alpha+\alpha^2(1-p)+(1-\alpha)^2(1-p)+\alpha(1-\alpha)(1-p)^2$
$\quad=[1-\alpha+\alpha(1-p)][\alpha+(1-\alpha)(1-p)]$
$p_{23}=\alpha(1-\alpha)p^2$
$p_{24}=\alpha^2p+\alpha(1-\alpha)p(1-p)=\alpha p[\alpha+(1-\alpha)(1-p)]$

$\sum\limits_{j=1}^{4} p_{ij}=1$

Fig.4.23. Tree diagram for transition
probabilities with initial state S_2

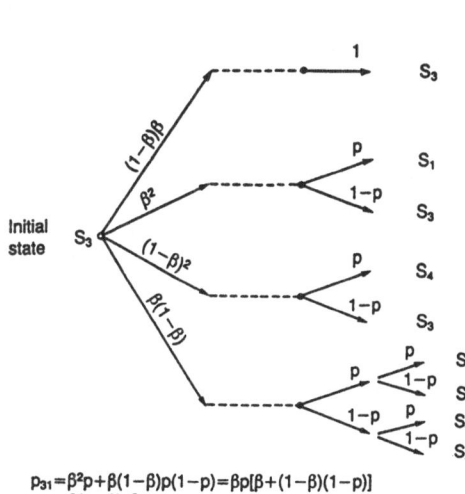

$p_{31}=\beta^2p+\beta(1-\beta)p(1-p)=\beta p[\beta+(1-\beta)(1-p)]$
$p_{32}=\beta(1-\beta)p^2$
$p_{33}=\beta(1-\beta)+\beta^2(1-p)+(1-\beta)^2(1-p)+\beta(1-\beta)(1-p)^2$
$\quad=[1-\beta+\beta(1-p)][\beta+(1-\beta)(1-p)]$
$p_{34}=(1-\beta)^2p+\beta(1-\beta)p(1-p)=p(1-\beta)[1-\beta+\beta(1-p)]$

$\sum\limits_{j=1}^{4} p_{ij}=1$

Fig.4.24. Tree diagram for transition probabilities with initial state S_3

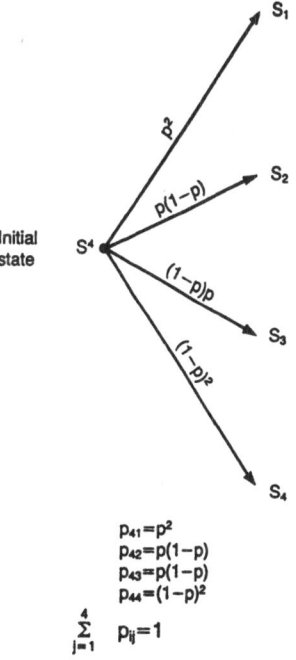

$p_{41}=p^2$
$p_{42}=p(1-p)$
$p_{43}=p(1-p)$
$p_{44}=(1-p)^2$

$\sum\limits_{j=1}^{4} p_{ij}=1$

Fig.4.25. Tree diagram for transition probabilities with initial
state S_4

[In the case of an aperiodic Markov chain there is, as we already know, an asymptotic distribution of (steady) states u_i, independent of the initial conditions, i.e., $u_i = \Sigma_{j=1}^{4} u_j P_{ji}$.]

So, one way or the other, we can deduce the values of u_1, u_2, u_3, u_4 as functions of the parameters α, β, γ, p. (Likewise, for the second partner/system we deduce the values of u_1', u_2', u_3', u_4' from the parameters α', β', γ', p'.)

$$P_{11} = (1 - \gamma)\gamma + \gamma^2(1 - p) + (1 - \gamma)^2(1 - p) + \gamma(1 - \gamma)(1 - p)^2$$

$$= [1 - \gamma + \gamma(1 - p)][\gamma + (1 - \gamma)(1 - p)] \qquad (4.7.2)$$

$$P_{12} = (1 - \gamma)^2 p + \gamma(1 - \gamma)p(1 - p) = p(1 - \gamma)[1 - \gamma + \gamma(1 - p)] \qquad (4.7.3)$$

$$P_{13} = \gamma^2 p + \gamma(1 - \gamma)p(1 - p) = \gamma p[\gamma + (1 - \gamma)(1 - p)] \qquad (4.7.4)$$

$$P_{14} = \gamma(1 - \gamma)p^2 \qquad (4.7.5)$$

$$\sum_{j=1}^{4} P_{ij} = 1$$

$$P_{21} = (1 - \alpha)^2 p + \alpha(1 - \alpha)p(1 - p) = p(1 - \alpha)[1 - \alpha + \alpha(1 - p)] \qquad (4.7.6)$$

$$P_{22} = (1 - \alpha) + \alpha^2(1 - p) + (1 - \alpha)^2(1 - p) + \alpha(1 - \alpha)(1 - p)^2$$

$$= [1 - \alpha + \alpha(1 - p)][\alpha + (1 - \alpha)(1 - p)] \qquad (4.7.7)$$

$$P_{23} = \alpha(1 - \alpha)p^2 \qquad (4.7.8)$$

$$P_{24} = \alpha^2 p + \alpha(1 - \alpha)p(1 - p) = \alpha p[\alpha + (1 - \alpha)(1 - p)] \qquad (4.7.9)$$

$$\sum_{j=1}^{4} P_{ij} = 1$$

$$P_{31} = \beta^2 p + \beta(1 - \beta)p(1 - p) = \beta p[\beta + (1 - \beta)(1 - p)] \qquad (4.7.10)$$

$$P_{32} = \beta(1 - \beta)p^2 \qquad (4.7.11)$$

$$P_{33} = \beta(1 - \beta) + \beta^2(1 - p) + (1 - \beta)^2(1 - p) + \beta(1 - \beta)(1 - p)^2$$

$$= [1 - \beta + \beta(1 - p)][\beta + (1 - \beta)(1 - p)] \qquad (4.7.12)$$

$$P_{34} = (1 - \beta)^2 p + \beta(1 - \beta)p(1 - p) = p(1 - \beta)[1 - \beta + \beta(1 - p)] \qquad (4.7.13)$$

$$\sum_{j=1}^{4} P_{ij} = 1 .$$

We next introduce the second collective variable of our problem, namely the *degree of cross-correlation* between the "environmental" triggers, as they arrive from the partner, at the level Q, and the dynamical pattern stored at Q. In order to proceed, however, we now have to describe the dynamics at the higher levels W and W' of the two partners.

4.7.3 A Semi-Markov Chain Model for the Hierarchical Levels W and W'

The higher hierarchical levels W, W' are endowed with dynamics simulated, for example, by an eight-state Markov chain model. Each of the above dynamical processes —displaying the behavior of the first system —now plays the role of the "environment" for the partner/system.

We associate with each state W_i, W_i' of these levels a three-digit "word", i.e., 000, 001, 011, 111, 100, 010, 110, and 101. A (different) subset of four of these "words" is associated with the four states S_1, S_2, S_3, S_4 and S_1', S_2', S_3', S_4' of the two partners at the levels Q and Q', respectively. In this discrete form, the degree of "similarity" between synchronous strings of individual states $W' \otimes Q$, $W \otimes Q'$, in each system at the level Q or Q', respectively, is given by the expression

$$r_i = 1 - \frac{1}{3} D_i \quad \text{or} \quad r_i' = 1 - \frac{1}{3} D_i' \ , \tag{4.7.14}$$

where D_i (or D_i') is the distance between the corresponding "words", i.e., the number of digits by which these two words/states differ. The communication between the two systems is therefore envisaged as the sequential formation of cross-correlations between three-digit "words" corresponding to the states belonging to the levels/pairs (W, Q') and (W', Q). The results, concerning the individual cross-correlations r_i, r_i' and the probabilities u_i, u_i' (measured by the frequency of appearance) of the preselected homeostatic states at the levels Q and Q', respectively, are reported to the (higher) levels W and W'.

According to these values, the levels W, W' (which also play the role of the controller for their own lower levels Q and Q') have to precipitate (feedforward) efferent control commands resulting in modification of the values of the basic parameters $(\alpha, \beta, \gamma, p)$, $(\alpha', \beta', \gamma', p')$. Thus the "joint figure of merit" [the long-term average of an appropriately weighted sum of the conflicting terms for each system or the probabilities of the homeostatic states and the intersystem cross-correlations between the levels (W, Q') and the (W', Q)] will be maximized.

The modeling processes at the levels W, W' have to do with the parametrization of the transition elements P_{ij}. We envisage them as follows. Let us consider the Markovian chain(s) at the level(s) W (or W'), connected as in Fig.4.26. Only successive states (labeled 1 to 8) communicate. Let us try to assign appropriate transitional probabilities as functions of u and $r = 1 - D/3$, so that certain intuitive postulates relative to the nature of the level(s) W (or W') are met.

In the human communication context, for instance, in the case where W, W' stand for "emotionally driven" behavioral levels, the string of states from 1 to 8 may

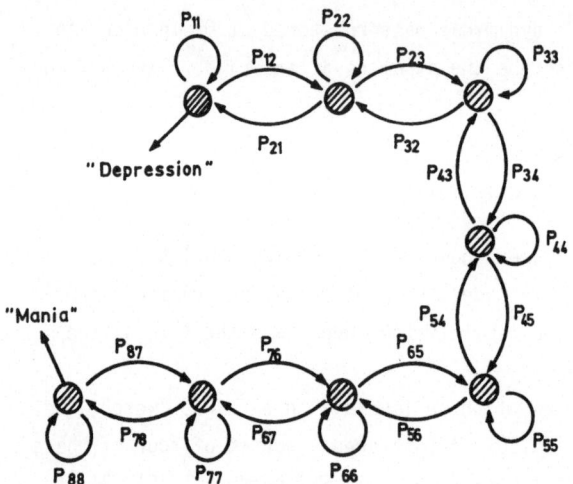

emulate a scale ranging from "depression" to "euphoria" or, behaviorally speaking, from *catatonia* to *hyperactivity*. In such a paradigm it seems appropriate that "ascending" transitional probabilities $P(k, k+1)$ $(k=1,...,7)$ (i.e., sequential shifting $1 \to 8$ from catatonia to hyperactivity) should be *increasing* functions of the cross-correlations (r) with the partner/environment and *decreasing* functions of the homeostasis level (u). For example, we may put

$$P(k, k+1) = P_\uparrow(r,u) = C_k\left[1 - e^{-\nu_k r}\right]e^{-\xi_k u} + \pi_k \quad , \tag{4.7.15}$$

where $k=1,2,...,7$, $0 \leqslant \pi_k \leqslant 1$, $0 \leqslant C_k \leqslant 1 - \pi_k$, and ν_k, ξ_k are positive numbers. Here π_k corresponds to an "intrinsic" spontaneous jump to the next state, in the absence of triggers transmitted from the partner.

Likewise, we may take for the "descending" transition probabilities the following r,u-parametrization:

$$P(k, k-1) = P_\downarrow(r,u) = d_k\left[1 - e^{-\mu_k u}\right]e^{-\lambda_k r} + q_k \quad , \tag{4.7.16}$$

where q_k has a similar meaning to π_k above, $k=2,3,...,8$, $0 \leqslant q_k \leqslant 1$, $0 \leqslant d_k \leqslant 1 - q_k$, and μ_k, λ_k are positive numbers.

Finally, the probabilities of remaining in the same state are calculated as

$$P(k,k) = 1 - P(k,k+1) - P(k,k-1)$$

$$= 1 - \pi_k - q_k - C_k\left[1 - e^{-\nu_k r}\right]e^{-\xi_k u} - d_k\left[1 - e^{-\mu_k u}\right]e^{-\lambda_k r} \quad , \tag{4.7.17}$$

for $k=2,...,7$, and

$$P(1, 1) = 1 - P(1, 2) \tag{4.7.18}$$

$$P(8, 8) = 1 - P(8, 7) \quad . \tag{4.7.19}$$

Similar conditions hold for the deliberations on the level W'.

We now introduce a *second* kind of parametrization of the transition elements P_{ij} at the levels W and W'. The Markov chains with which the dynamical behavior at the hierarchical levels Q or W has been simulated thus far have been implicitly endowed with the property that a transition is called for at regular intervals. The transition might return the system to the state previously occupied but a transition occurs nevertheless. Now, regarding the higher levels W, W' we examine a more general class of processes (*semi*-Markov chains), where the time between transitions may follow a given probability density function (p.d.f.) and where this transition time may depend on the transition under consideration. When the process at the level W enters a state i we know that it determines the next state j to which it will move according to state j's conditional probabilities $P_{ij}(r,u)$ —as described above. However, after j has been selected, but *before* making this transition from state i to state j, the process, we imagine, "holds" or "gets stuck" for a time τ_{ij} in state i. The holding times are positive, integer-valued random variables, each governed by a probability mass function $h_{ij}(\tau)$, called the "holding time density function".

If we do not know its successor state, the probability that the system at W will spend τ time units in state i is

$$\Lambda_i(\tau) = \sum_{j=1}^{N} P_{ij} h_{ij}(\tau) \quad , \tag{4.7.20}$$

where N is the number of all successor states. We call τ the "waiting" time in state i and $\Lambda_i(\tau)$ the waiting time probability mass function. Thus a waiting time is merely a holding time that is *unconditional* on the destination state. The "mean" waiting time $\langle\tau_i\rangle$ is related to the mean holding time $\langle\tau_{ij}\rangle$ as follows:

$$\langle\tau_i\rangle = \sum_{j=1}^{N} P_{ij}\langle\tau_{ij}\rangle \quad . \tag{4.7.21}$$

In the model just revealed, the process at the level W *first* selects its next state using the transitional probabilities P_{ij}, and *then* selects its time of transition from a holding time mass function $h_{ij}(\tau)$, which is conditional on the transition made. But we can equivalently *reverse* the process to allow the system to select its time of transition *first* and *then* select a new state —conditional on the transition time. In the present model this sequence of events is more natural.

The waiting time distribution, $\Lambda_i(\tau)$, already defined, is the distribution of holding time unconditional on destination state. In the present alternative description of the process, $\Lambda_i(\tau)$ would have to be specified for all states. Then we have to supply the probability of making a transition to each state, given the time interval that the system has held in its present state before making the transition.

Let $P_{ij}(\tau)$ be the probability that a process which is now in state i and which will make a transition out of state i at time τ will make that transition to state j. Thus the $P_{ij}(\tau)$ are transitional probabilities *conditional* on holding time; we call them "conditional" transitional probabilities. The waiting time distribution $\Lambda_i(\tau)$ and the set of conditional transition probabilities $P_{ij}(\tau)$ provide a complete alternative definition of the semi-Markov process.

So, if the process is originally described (as in the present case) in terms of P_{ij} and $h_{ij}(\tau)$, we compute $\Lambda_i(\tau)$ from (4.7.20) and then we obtain

$$P_{ij}(\tau) = \frac{P_{ij}h_{ij}(\tau)}{\Lambda_i(\tau)} = \frac{P_{ij}h_{ij}(\tau)}{\Sigma_{j=1}^{N} P_{ij}h_{ij}(\tau)} \quad . \tag{4.7.22}$$

What is the reason for including in the dynamics of the *higher* levels W, W' this holding time distribution dependence? The holding time concept has to do, operationally speaking, with the "jerkiness" of the "clocking" mechanism activating the Markov chain(s) involved, i.e., the mechanism responsible for the turnover of behavioral modes (states). We postulate that this clocking mechanism (some sort of a master pacemaker) becomes inactivated within time intervals (holding times) which increase with the degree of disorganization at the lower levels Q and Q' for the systems concerned, respectively. More precisely, we postulate that any (random) imbalance among the concentrations of key substances (hormones, neurotransmitters) $\sim\alpha$, β, γ, α', β', γ' or different concentrations $\sim p$, p', externally or internally induced, may change the elements of the transition probabilities P_{ij} at the lower levels Q or Q' in such a way that the redundancy or the degree of organization at that level will fluctuate concomitantly.

The redundancy at the level Q is defined as

$$R_Q = 1 - H_Q/H_{max} \quad , \tag{4.7.23}$$

where

$$H_Q = \frac{1}{\tau} \sum_{v=1}^{\tau} H_{iv} \tag{4.7.24}$$

is the time-average entropy or uncertainty per state; at the level Q during the previous holding time,

$$H_i = - \sum_{k=1}^{4} P_{ik} \log_2 P_{ik} \tag{4.7.25}$$

is the uncertainty of the state i (i = 1,2,3,4); and $H_{max} = \log_2 4 = 2$; i_v stands for the index of the state at the particular moment v.

The same imbalance mentioned above between the concentrations of key substances (or "game parameters") may be responsible for or concomitant to the temporal de-synchronization or inactivation of the master pacemaker, thereby introducing holding times at the higher level(s) W, W' *correlative* with the drop of redundancy at the lower levels. In Sect.4.7.6 we will discuss recent neurophysiological work in favor of this argument.

Let us now define the holding time probability mass function as a geometrical distribution

$$h_{ij}(\tau) = n_{ij}(1 - n_{ij})^{\tau-1} \quad , \tag{4.7.26}$$

where $0 \leqslant n_{ij} \leqslant 1$ stands for the conditional probability of staying in the state i at the higher level W exactly one unit time before commuting, given that the transition is from state i to state j. The mean value of the above distribution equal $1/n_{ij}$, and its variance equals $(1 - n_{ij})/n_{ij}^2$.

We now calculate the elements $P_{ij}(\tau)$ of the Markovian dynamics at the level W (similar things hold for the level W'). We find it reasonable to postulate — as far as the holding time dependence is concerned — that $P_{ij}(\tau)$ should be a decreasing function of τ as one moves from catatonia to hyperactivity, and an increasing function of τ as one moves from hyperactivity to catatonia. This means that the longer the system at W stays in a depressive state, the greater the probability that it will continue to remain in this regime; and the longer the system stays in a euphoric state, the greater the probability that it will switch to a less euphoric or more depressive state.

We write

$$n_{i(i+1)} = n_d \qquad \text{for} \qquad i = 1,2,\ldots,7 \quad ,$$

$$n_{i(i-1)} = n_\mu \qquad \text{for} \qquad i = 2,3,\ldots,8 \quad , \tag{4.7.27}$$

and

$$n_{ii} = n_0 \qquad \text{for} \qquad i = 1,2,\ldots,8 \quad ,$$

assuming $n_\mu < n_0 < n_d$.

Therefore the relationship between the corresponding conditional holding times will be

$$\frac{1}{n_\mu} > \frac{1}{n_0} > \frac{1}{n_d} \quad . \tag{4.7.28}$$

The holding time dependence on the redundancy of the lower hierarchical level R_Q is introduced into our example as follows:

$$n_\mu = \frac{1 + R_Q}{36} \quad , \qquad n_0 = \frac{1 + R_Q}{30} \quad , \qquad n_d = \frac{1 + R_Q}{24} \quad . \tag{4.7.29}$$

We write now, from (4.7.26,27),

$$h_{i(i+1)}(\tau) = n_d(1 - n_d)^{\tau-1} \qquad \text{for} \qquad i = 1,2,\ldots,7 \quad ,$$

$$h_{i(i-1)}(\tau) = n_\mu(1 - n_\mu)^{\tau-1} \qquad \text{for} \qquad i = 2,3,\ldots,8 \quad , \tag{4.7.30}$$

$$h_{ii}(\tau) = n_0(1 - n_0)^{\tau-1} \qquad \text{for} \qquad i = 1,2,\ldots,8 \quad ,$$

and

$$h_{ij}(\tau) = 0 \quad \text{for} \quad (j-1)(j-i-1)(j-i+1) \neq 0 ,$$

which, together with (4.7.22), lead to the following expressions for the holding time conditional transitional probability elements:

$$P_{i(i+1)}(\tau) = \frac{P_{i(i+1)}h_{i(i+1)}(\tau)}{P_{i(i-1)}h_{i(i-1)}(\tau) + P_{ii}h_{ii}(\tau) + P_{i(i+1)}h_{i(i+1)}(\tau)}$$

$$= \frac{P_{i(i+1)}}{P_{i(i-1)} \dfrac{n_\mu}{n_d}\left[\dfrac{1-n_\mu}{1-n_d}\right]^{\tau-1} + P_{ii}\dfrac{n_0}{n_d}\left[\dfrac{1-n_0}{1-n_d}\right]^{\tau-1} + P_{i(i+1)}} , \qquad (4.7.31)$$

for $i = 2,3,\ldots,7$ and $\tau = 1,2,3,\ldots$ This is, as postulated above, a decreasing function of τ for $n_\mu < n_0 < n_d$.

We also have

$$P_{12}(\ ) = \frac{P_{12}h_{12}(\tau)}{P_{11}h_{11}(\tau) + P_{12}h_{12}(\tau)} = \frac{P_{12}}{P_{11}\dfrac{n_0}{n_d}\left[\dfrac{1-n_0}{1-n_d}\right]^{\tau-1} + P_{12}} , \qquad (4.7.32)$$

which is a decreasing function of τ.

For descending transitions, we have

$$P_{i(i-1)}(\tau) = \frac{P_{i(i-1)}h_{i(i-1)}(\tau)}{P_{i(i-1)}h_{i(i-1)}(\tau) + P_{ii}h_{ii}(\tau) + P_{i(i+1)}h_{i(i+1)}(\tau)}$$

$$= \frac{P_{i(i-1)}}{P_{i(i-1)} + P_{ii}\dfrac{n_0}{n_\mu}\left[\dfrac{1-n_0}{1-n_\mu}\right]^{\tau-1} + P_{i(i+1)}\dfrac{n_d}{n_\mu}\left[\dfrac{1-n_d}{1-n_\mu}\right]^{\tau-1}} . \qquad (4.7.33)$$

for $i = 2,3,\ldots,7$. This is an increasing function of τ for $n_\mu < n_0 < n_d$.

We also have

$$P_{87}(\tau) = \frac{P_{87}h_{87}(\tau)}{P_{87}h_{87}(\tau) + P_{88}h_{88}(\tau)} = \frac{P_{87}}{P_{87} + P_{88}\dfrac{n_0}{n_\mu}\left[\dfrac{1-n_0}{1-n_\mu}\right]^{\tau-1}} . \qquad (4.7.34)$$

Also

$$P_{ii}(\tau) = \frac{P_{ii}h_{ii}(\tau)}{P_{i(i-1)}h_{i(i-1)}(\tau) + P_{ii}h_{ii}(\tau) + P_{i(i+1)}h_{i(i+1)}(\tau)}$$

$$= \frac{P_{ii}}{P_{i(i-1)} \dfrac{n_\mu}{n_0}\left[\dfrac{1-n_\mu}{1-n_0}\right]^{\tau-1} + P_{ii} + P_{i(i+1)}\dfrac{n_d}{n_0}\left[\dfrac{1-n_d}{1-n_0}\right]^{\tau-1}} , \qquad (4.7.35)$$

for $i = 2,3,\ldots,7$ and $\tau = 1,2,3,\ldots$ Finally we have

$$P_{11}(\tau) = \frac{P_{11}h_{11}(\tau)}{P_{11}h_{11}(\tau) + P_{12}h_{12}(\tau)} = \frac{P_{11}}{P_{11} + P_{12}\dfrac{n_d}{n_0}\left[\dfrac{1 - n_d}{1 - n_0}\right]^{\tau-1}} \quad , \qquad (4.7.36)$$

$$P_{88}(\tau) = \frac{P_{88}h_{88}(\tau)}{P_{87}h_{87}(\tau) + P_{88}h_{88}(\tau)} = \frac{P_{88}}{P_{87}\dfrac{n_\mu}{n_0}\left[\dfrac{1 - n_\mu}{1 - n_0}\right]^{\tau-1} + P_{88}} \quad , \qquad (4.7.37)$$

for $\tau = 1,2,3,\ldots$.

4.7.4 The Control Problem

a) Biological Rhythms Underlying the Games

We take as expressions for the game parameters harmonic (circadian) components of periodic functions, simulating basic rhythms. Let them be

$$\alpha(t) = \frac{\alpha_{max} + \alpha_{min}}{2} + \frac{\alpha_{max} - \alpha_{min}}{2} \cos(\omega t + \varphi) \quad ,$$

$$\beta(t) = \frac{\beta_{max} - \beta_{min}}{2} - \frac{\beta_{max} - \beta_{min}}{2} \cos(\omega t + \varphi) \quad , \qquad (4.7.38)$$

$$\gamma(t) = \frac{\gamma_{max} + \gamma_{min}}{2} + \frac{\gamma_{max} - \gamma_{min}}{2} \cos(\omega t) \quad ,$$

where $p = \text{const}$, $\alpha_{max} \geq \alpha_{min}$, $\beta_{max} \geq \beta_{min}$, $\gamma_{max} \geq \gamma_{min}$ are given constants expressing the extreme values of the fluctuations concerned.

We take as one control variable the phase difference φ between $\alpha(t)$ and $\gamma(t)$, and we arrange for $\alpha(t)$ and $\beta(t)$ to be $180°$ out of phase. This emulates the case of phase rearrangement via control action among individual biological rhythms. The other control parameter is p. The control vector is $\mathbf{V} = (p,\varphi)$.

We also consider the less severe control mechanism which consists of changing p and the initial phase φ of the three rhythms, which are otherwise in step, i.e.,

$$\alpha(t) = \alpha_1 + \alpha_2 \cos(\omega t + \varphi) \quad ,$$

$$\beta(t) = \beta_1 + \beta_2 \cos(\omega t + \varphi) \quad , \qquad (4.7.39)$$

$$\gamma(t) = \gamma_1 + \gamma_2 \cos(\omega t + \varphi) \quad ,$$

where α_i, β_i, γ_i $(i = 1,2)$ are constants.

b) Description of the Communication and Control Processes

Let us consider the deliberation from the viewpoint of partner A(Q,W). For a given value of the control vector $\mathbf{V} = (p,\varphi)$, immediately after a state transition at the level W, the time variation of α, β, and γ is completely specified. Having $(\alpha, \beta,$

γ, p), the 16 elements for $P_{ij}(\tau)$ at the level Q are determined. Let τ be the holding time at the current state of W. The succession of states S_{i1}, S_{i2}, ..., $S_{i\tau}$ at the level Q (where $i = 1,2,3$ or 4) during this holding time is governed by the above time-varying transitional probabilities and is simulated on the computer by a Monte Carlo method described below.

For a preselected homeostatic state S_h (h = 1,2,3, or 4), we calculate the relative frequency of occurrence u of that state during the holding time τ. We also evaluate the cross-correlation

$$r = 1 - \frac{D}{3} = \frac{1}{\tau} \sum_{i=1}^{\tau} \left[1 - \frac{D^{(i)}}{3} \right] \qquad (4.7.40)$$

between the state sequences of the level Q of partner A and the higher level W' of partner B(Q', W'), during the same time interval τ. During this time interval a number of transitions may have occurred at the level W'.

The pair of values (u, r) is reported afferently at the end of the holding time to the level W, thereby fixing the transition probabilities P_{ij} at that level as described in Sect.4.7.3. The higher level W plays the dual role of "transmitter" toward the level Q' of partner B and "controller" of its own underlying game at level Q. In this last role it has as a mission the modification — via efferent (feedforward) control commands — of the control vector $\mathbf{V} = (p, \varphi)$ on the basis of the received signals (u, r).

The objective of the control is to satisfy a properly defined criterion. The situation calls for a two-objective control procedure. Namely, from the point of view of partner A, the objective could be the maximization of a weighted sum of the average values E(u) and E(r) of u and r respectively, i.e., maximization of the "figure of merit"

$$F = \lambda_1 E(u) + \lambda_2 E(r) \quad , \qquad (4.7.41)$$

where λ_1 and λ_2 are non-negative constants with unit sum.

From the point of view of partner B, we should have as an objective the maximization of another figure of merit, i.e.,

$$F' = \lambda_1' E(u') + \lambda_2' E(r') = \max \quad , \qquad (4.7.42)$$

where $0 \leqslant \lambda_1' \leqslant 1$ and $\lambda_2' = 1 - \lambda_1'$.

In such an antogonistic two-objective control problem, the partners involved could compromise by seeking a regime of *mutual adaptability* or *coexistence* which amounts to maximizing a joint figure of merit:

$$F_j = \sigma F + \sigma' F' \quad , \qquad (4.7.43)$$

where σ, σ' are non-negative constants with unit sum.

c) Selection of Control Mechanisms

The values of the control variables (p, φ) as well as (p', φ') are selected by the hierarchical levels W and W', respectively, according to the collective observables (u, r), (u', r') immediately after the respective transitions, i.e.,

$$p = f_1(r, u) \quad , \qquad p' = f_1'(r', u') \quad ,$$
$$\varphi = f_2(r, u) \quad , \qquad \varphi' = f_2'(r', u') \quad . \tag{4.7.44}$$

The joint selection of the above functions [or mappings: $(r, u) \to (p, \varphi)$ and $(r', u') \to (p', \varphi')$], in a way which will maximize the joint figure of merit F_j, constitutes the control problem. This is, in general, a very difficult stochastic control problem. In the present discussion we assume that the control vector $\mathbf{V}$ can take on only a finite number of values: $v_1, v_2, \dots, v_N$. We also consider that the rectangular region $0 \leqslant r \leqslant 1$, $0 \leqslant u \leqslant 1$ of the "received" vector (r, u) or (D, u) is partitioned into M regions: $R_1, R_2, \dots, R_M$, as shown in Fig.4.27, where for computer simulation purposes we have taken $M = N = 8$. In this specific case we have considered that p can take on two values and φ four values:

$$p \in \{p_1, p_2\} \quad ; \quad \varphi \in \left\{0, \frac{\pi}{2}, \pi, \frac{3\pi}{2}\right\} \quad . \tag{4.7.45}$$

The regions R_i ($i = 1, 2, \dots, 8$) are defined as shown in Fig.4.27.

In the absence of heuristics a priori excluding a number of maps clearly nonoptimal or incompatible with the nature of the problem, one should consider all $(N^M)^2 = N^{2M}$ pairs of mappings. Clearly this excludes an exhaustive search for this combinatorial optimization, even for a small number of quantization steps, as, for example, in the case of Fig.4.27. Under such conditions, we confine our search to a randomly selected subset of the $8^{16} = 2^{48}$ possible "joint behavioral modes".

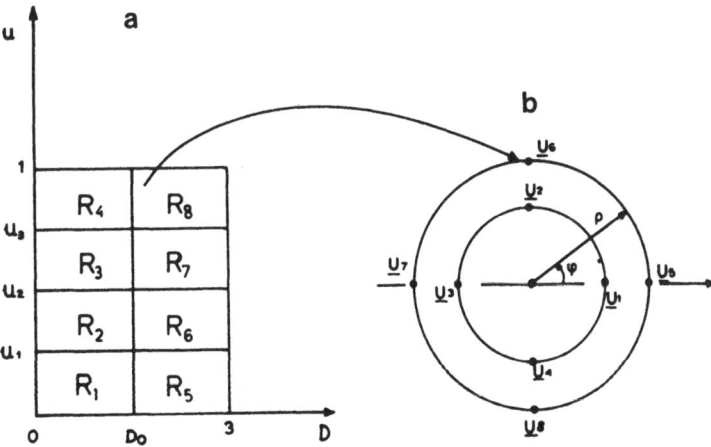

Fig.4.27. **a** Quantization on the plane (D, u). **b** The discrete values of the control vector $\mathbf{V}(\rho, \varphi)$

4.7.5 Computer Simulation

We first select a random map of the eight points of the (D, u) plane onto the eight points of the control vector (p, φ) plane (Fig.4.27). The one-to-one map is characterized by the 8! permutations of the numbers 1 to 8 $(n_1, n_2, \ldots, n_8)$, where $n_i = 1, 2, \ldots, 8$, in the sense that the region R_i of the (D, u) plane is mapped into the point v_{n_i} of the control variable plane $V = (p, \varphi)$. The selection of the map (i.e., of the random permutation) is made by the SMAP subroutine. In this sub-routine we rearrange the numbers 1 to 8 in a Monte Carlo fashion as follows. Using an appropriate subroutine we can produce random numbers, uniformly distributed in the interval (0, 1). We multiply one such number by eight and the integral part of the result is the first number n_1 of the permutation. We then produce another ran-dom number; we multiply it by 7. Let m be the integral part of the result; then the mth largest number among the remaining seven numbers is the second number n_2 of the permutation. We continue in a similar fashion. This way we put the eight numbers in a random sequence, which corresponds to a specific map.

After the selection of two random maps, one for each partner, we calculate the holding times for both. The absolute time of the transition points at the higher level W for each partner is determined as the sum of the corresponding successive holding times.

For the partner with the earlier transmission (Fig.4.28) we calculate, using the CLQ subroutine, the next state of level W and the next holding time. To obtain these parameters we first evaluate the distance D between the level Q of the one partner and the level W of the other. To do this, we determine the succession of states S(1), S(2), S(3), S(4) of the Markov chain at the hierarchical level Q, in

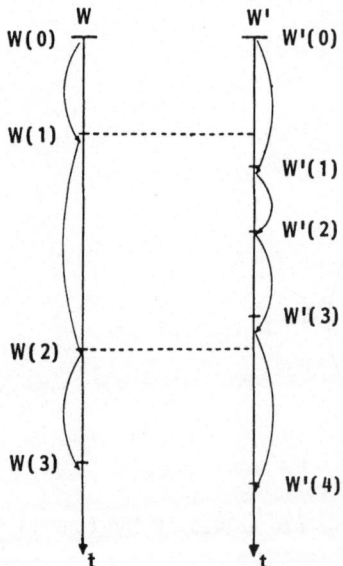

Fig.4.28. Computer simulation: state transitions at the levels W and W'

200

a Monte Carlo fashion as follows. Let S(i), i =1,2,3,4, be the present state at Q. The next state S(j) is produced by generating, with the use of the appropriate sub-routine, random numbers uniformly distributed in the interval (0, 1), which is di-vided into four subintervals according to the transition probabilities P_{ij}. The in-dex of the next state S(j) of the Markov chain, at Q, is determined from the inter-val in which the particular random number falls.

The states for the levels Q and W are represented by three-bit binary numbers from 000 to 111. Specifically, for the levels W, we use as a code for the state W_i (or W_i') the three-bit binary representation of the index i. For the four states of hierarchical levels Q and Q' of the two partners, respectively, we use the codes S(1) = 001, S(2) = 010, S(3) = 011, S(4) = 100, S'(1) = 011, S'(2) = 100, S'(3) = 101, S'(4) = 110. The distance between two elements α_1, α_2, α_3, β_1, β_2, β_3, where α_i and β_i are equal to 0 or 1 (i =1,2,3), is calculated as

$$D = |\alpha_1 - \beta_1| + |\alpha_2 - \beta_2| + |\alpha_3 - \beta_3| \quad . \tag{4.7.46}$$

A sequence of N states for level Q is produced in the way described above, where N is the number of time intervals during each holding time. The distances $D^{(i)}$ be-tween the corresponding elements are calculated and the average distance during the holding time of the corresponding higher level is

$$D = \frac{1}{N} \sum_{i=1}^{N} D^{(i)} \quad . \tag{4.7.47}$$

The value of u is the percentage of appearance of the homeostatic state during the holding time. Having obtained the values of u and D (for the selected homeostatic state), we determine in what region R_i (i =1,...,8) in Fig.4.27, the point (D, u) falls. With the quantized values of D and u, we select a new value for the control vector (p, φ), depending on the map we produced at the beginning. The functions α(t), β(t), γ(t) and the parameter p are now completely determined, so that the transition probabilities for the lower level can be calculated.

Also with the calculated values of u and r = 1 - D/3, we evaluated the transition probabilities P(k, k - 1), P(k, k), P(k, k +1) for the level W (or W'), using the previously presented expressions. Th next state for level W is obtained in a Monte Carlo fashion similar to that for the Q level.

For each state of level Q (or Q') we calculated (4.7.25)

$$H_i = \sum_{j=1}^{4} P_{ij} \log_2 P_{ij} \quad ,$$

and at the end of each holding time τ (including N time intervals)

$$H_Q = \frac{1}{N} (H_{i1} + \ldots + H_{iN}) \quad , \qquad i_k = 1, 2, 3, 4 \quad , \tag{4.7.48}$$

and, from (4.7.23),

$$R_Q = 1 - \frac{H_Q}{\log_2 4} = 1 - \frac{H_Q}{2} \ . \tag{4.7.49}$$

From R_Q and the next state of W, we determine the parameters of the geometric distribution that the holding time follows:

i) for "upward" transitions, $n = n_d = (1 + R_Q)/24$;

ii) for remaining in the same state, $n = n_0 = (1 + R_Q)/30$;

iii) for "downward" transitions, $n = n_\mu = (1 + R_Q)/36$.

For the computer generation of the geometrically distributed holding time T with parameter n, we take

$$T = \left[-\frac{1}{n} \ln R \right] + 1 \ , \tag{4.7.50}$$

where [A] is the integral part of A, i.e., the greatest integer not exceeding A, and R is a random number a series A of which are uniformly distributed in the interval (0, 1).

Outputs of the program are the joint figures of merit for 30 randomly selected maps along with the corresponding redundancies at the lower levels for the two partners. The program also selects the two maps (one for each partner) which produce the best (suboptimal) joint figure of merit among the tried control laws.

For the determination of the "best" pair of homeostatic states (i.e., the pair for which the maximum of the maxima of the joint figure of merit is obtained) we run the program 16 times corresponding to the 4×4 combinations of the selected homeostatic states. In Figs.4.29,30 (see pages 203,204) we display the behavioral W, W' state transitions as functions of time for the extreme values (out of $16 \times 30 = 480$ values) of the joint figure of merit. We also present in Figs.4.31,32 (see pages 205,206) some results for the case where the rhythms α, β, γ keep constant phase relationships.

4.7.6 Biological Relevance of the Model

A common objective in a common *person-to-person* communication is the revelation (or the concealing) of one's experience [perception, cognition — level(s) Q, Q'] through concomitant behavior [performance, symbolic acting out — level(s) W, W']. This process is implemented in a time series of discrete modes (states) at the levels W and W'.

The adoption of an optimum *common* code which will correlate each partner's experience and behavior is therefore of the utmost importance: such a code would maximize the joint figure of merit, or it would optimize the communication between the two partners under the constraints of the special situation. To find such a code one should either go through an exhaustive search of all possible mappings $\{u, r\} \rightleftharpoons \{\alpha, \beta, \gamma, p\}$, or find heuristics[4] (footnote see page 203) which will allow

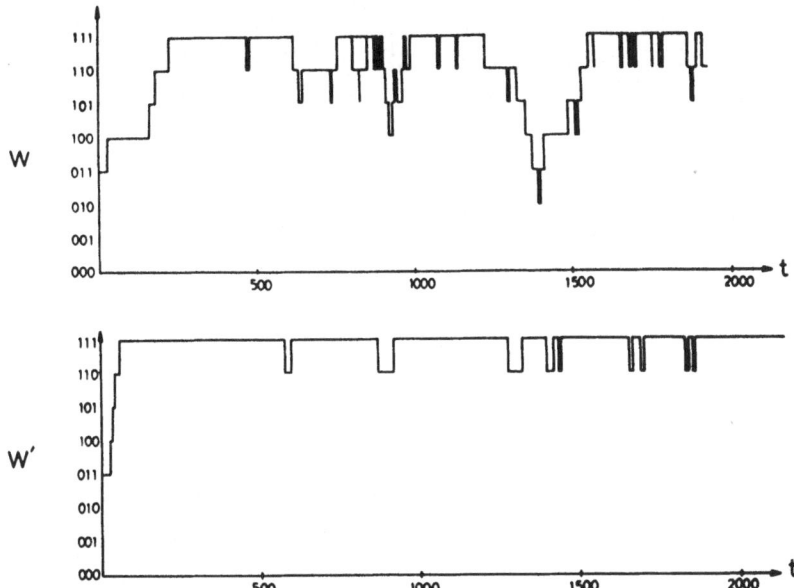

Fig.4.29. Evolution of the states of the higher levels W and W' when the underlying rhythms are expressed by $\alpha(t) = [(\alpha_{max} + \alpha_{min}) + (\alpha_{max} - \alpha_{min})\cos(\omega t + \varphi)]/2$, $\beta(t) = [(\beta_{max} + \beta_{min}) - (\beta_{max} - \beta_{min})\cos(\omega t +)]/2$, and $\gamma(t) = [(\gamma_{max} + \gamma_{min}) + (\gamma_{max} - \gamma_{min})\cos(\omega t)]/2$. The units of t are arbitrary. Values of the parameters for the computer simulation: $\alpha_{max} = 0.9$, $\alpha_{min} = 0.1$, $\beta_{max} = 0.8$, $\beta_{min} = 0.2$, $\gamma_{max} = 0.7$, $\gamma_{min} = 0.3$; $\alpha'_{max} = 0.7$, $\alpha'_{min} = 0.3$, $\beta'_{max} = 0.8$, $\beta'_{min} = 0.2$, $\gamma'_{max} = 0.9$, $\gamma'_{min} = 0.1$; $\omega = \pi/10$. Higher level transition probability parameters: $\pi_K = 0.1$, $q_K = 0.2$, $c_K = d_K = 0.8$, $\pi'_K = 0.2$, $q'_K = 0.1$, $c'_K = d'_K = 0.8$, $\lambda_K = \mu_K = \nu_K = \xi_K = \lambda'_K = \mu'_K = \nu'_K = \xi'_K = 2$ for $K = 1,2,\ldots,8$. Values of control variables $p \in \{0.2; 0.8\}$, $p' \in \{0.1; 0.5\}$, φ and $\varphi' \in \{0, \pi/2, \pi, 3\pi/2\}$. The figure corresponds to the maps represented by the permutations 47312586 for the "left" partner and 48726153 for the "right" partner, which yielded the maximum joint figure of merit (F = 0.29) among 30 computer runs with randomly selected maps

the a priori exclusion of incompatible or unpromising mappings and limit the exhaustive search among the members of a very small subset of allowable codes.

Such heuristics exist at present in only a handful of cases. Suppose, for the sake of argument, that the levels Q, Q' stand for nuclei acid dynamics whereas the levels W, W' stand for enzyme (protein) dynamics. In such a case it is valid to recall the well-known *one-to-one* correspondence gene →protein, and exclude all other mappings. The adoption of such a heuristic to codes involving higher hierarchical levels should limit in our problem the possible number of maps from 8^8 to 8! — still a very large number, practically excluding exhaustive search.

---

4 We call "heuristical" any method, constraint, or guidance principle that helps eliminate unfruitful possibilities in a search process.

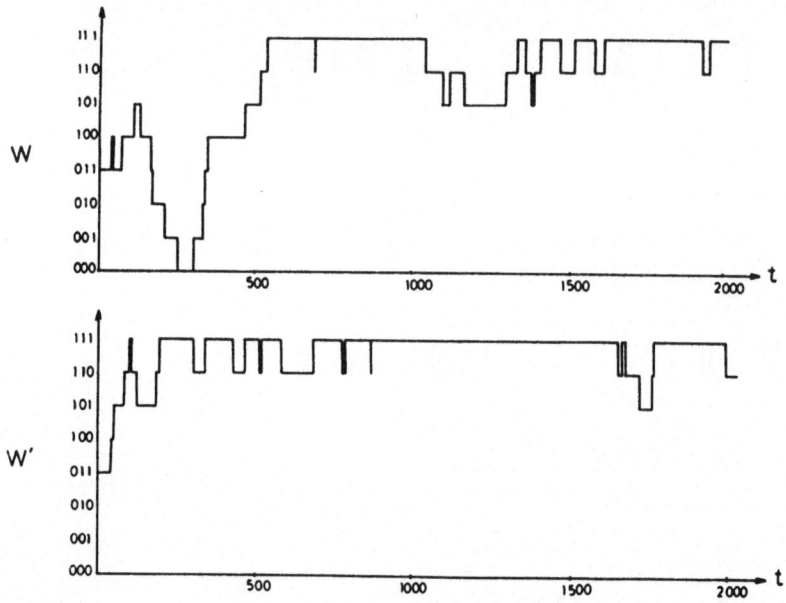

Fig.4.30. Evolution of the states of the higher levels W and W' with parameters as in Fig.4.29. The figure corresponds to the maps represented by the permutations 74628135 for one partner and 87436125 for the other, which yielded the minimum joint figure of merit (F = 0.19) among 30 computer runs with randomly selected maps

One possible way of detecting heuristics will perhaps be opened up through the study of neurophysiological mechanisms underlying (in higher vertebrates) the behavioral mode turnover. It has been hypothesized by *Kilmer* and *McCulloch* [4.23]—on the basis of the physiological work of the *Scheibels* [4.24] on the "poker chips" model of the reticular formation—that the RF in vertebrates may be the "embodiment" of the function mediating behavioral selection, and commitment (to a single mode at a time), from a multitude of competing synchronous experiences impinging externally and internally upon the organism. The "selection rules" through which the organism commits itself to a single behavioral mode (state) at a time appear to depend on the "arousal" of the RF. What we call arousal or excitability in a given tissue (or "module") of the RF is the percentage of simultaneously firing (or subthreshold depolarized) neuronal elements. We point out also that this degree of electrical activity is concomitant with the metabolic activity of the neurons involved—in terms of manufacturing and releasing hormones and neurotransmitters. (This is how we imagine a relationship between arousal and the game parameters α, β, γ, p in our model). Such an activity may be due in part to endogenous reasons (i.e., to permanent cortisol or adrenalin implantation in a number of pertinent tissues as a result of adrenocortical or sympathetic overactivity). Aside from spontaneous, endogenous reasons, arousal may also occur as a result of preceding *cognitive appraisals*, i.e., a number of cross-correlations between a set of impinging stimuli and a set of dynamically prestored patterns.

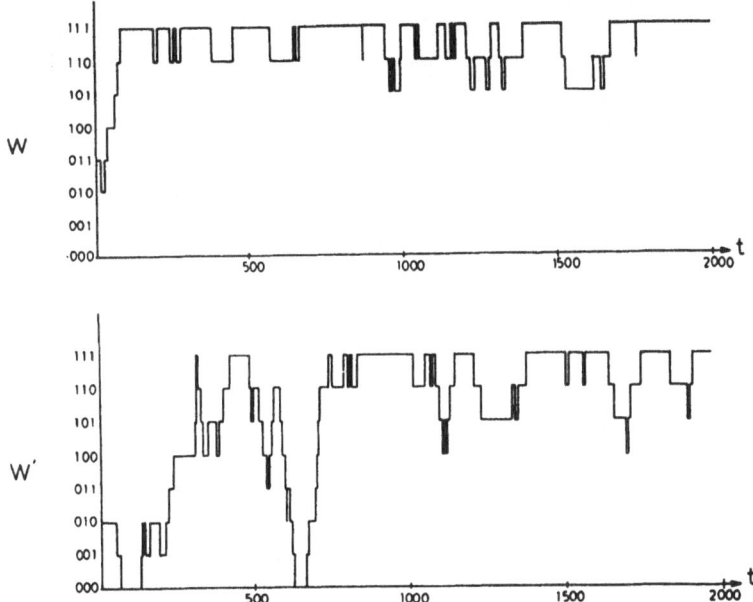

Fig.4.31. Evolution of the states of the higher levels W and W' when the underlying rhythms are expressed by $\alpha(t) = [(\alpha_{max} + \alpha_{min}) + (\alpha_{max} - \alpha_{min})\cos(\omega t + \varphi)]/2$,

$\beta(t) = [(\beta_{max} + \beta_{min}) + (\beta_{max} - \beta_{min})\cos(\omega t + \varphi)]/2$, and $\gamma(t) = [(\gamma_{max} + \gamma_{min}) + (\gamma_{max} - \gamma_{min})\cos(\omega t + \varphi)]/2$. The units of t are arbitrary. Values of the parameters for the computer simulation: $\alpha_{max} = 0.9$, $\alpha_{min} = 0.1$, $\beta_{max} = 0.8$, $\beta_{min} = 0.2$, $\gamma_{max} = 0.7$,

$\gamma_{min} = 0.3$; $\alpha'_{max} = 0.7$, $\alpha'_{min} = 0.3$, $\beta'_{max} = 0.8$, $\beta'_{min} = 0.2$, $\gamma'_{max} = 0.9$, $\gamma'_{min} = 0.1$;
$\omega = 50/\pi$. Higher level transition probability parameters: $\pi_K = q_K = 0.2$, $c_K = d_K = 0.8$,

$\pi'_K = q'_K = 0.3$, $c'_K = d'_K = 0.7$, $\lambda_K = \mu_K = \nu_K = \xi_K = \lambda'_K = \mu'_K = \nu'_K = \xi'_K = 2$ for $K = 1,2,\ldots,8$.
Values of control variables $\rho \in \{0.2; 0.8\}$, $\rho' \in \{0.1; 0.5\}$, φ and $\varphi' \in \{0, \pi/2, \pi, 3\pi/2\}$. The figure corresponds to the maps represented by the permutations 86741253 for the "left" partner and 65128374 for the "right" partner, which yielded the maximum joint figure of merit (F = 0.41) among 30 computer runs with randomly selected maps

Now each "module" of the RF constitutes a semiautonomous neuronal unit, perpendicular to the ascending-descending streamlines of the RF: the hypothesis is that such a module receives most if not all simultaneously formed cross-correlations ("cognitive appraisals") between the set of operating stimuli (external or internal) and the set of stored patterns but weights them differently, i.e., it assigns different degrees of arousal to each of them. The various modules are loosely coupled to each other. However, it is hypothesized [4.23] that in order to form an "independent opinion", upon the presentation of a *new* stimulus each module should briefly uncouple from the rest of the modules to a degree that increases with both the significance (i.e., cross-correlation) of the signal received and the degree to which the preceding signal —associated with the previously selected behavioral mode —has been entrenched.

205

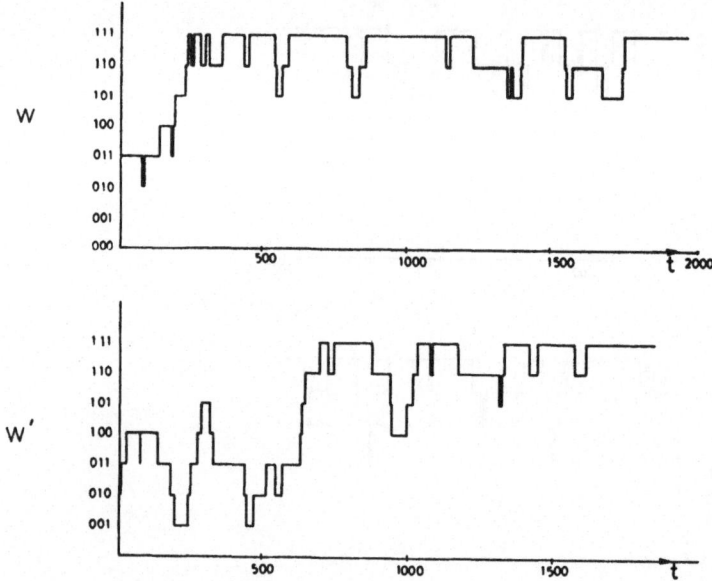

Fig.4.32. Evolution of the states of the higher levels W and W' with parameters as in Fig.4.31. The figure corresponds to the maps represented by the permutations 12756348 for one partner and 51683427 for the other, which yielded the minimum joint figure of merit (F = 0.23) among 30 computer runs with randomly selected maps

Through a set of still unknown rules it appears that a *consensus* is *rapidly* reached among the different modules as to which single behavioral mode will be selected and launched at a given time. It is believed that the predominant mode is the one corresponding to the cognitive appraisal (experience) which has been voted "highly arousing" —above a certain threshold—by a distinct majority of the modules. Nevertheless, there are indications that the reticular formation activating system does not reach the above decision in absentia of the cerebral context. According to the work of *Verzeano* [4.19] for example, it appears that the rate of information processing in the cortex (normally of the order of 10 Hz, the basic frequency of the thalamocortical pacemaker) can be modulated over a wide range by the degree of arousal of the ascending branch of the reticular formation activating system (RFAS).

We may consider therefore (see Sect.4.7.2) these two dynamical agents (RFAS and the thalamocortical pacemaker) as involved in an antagonistic process (game) mediated by the instantaneous concentrations of key substances such as neurotransmitters and hormones. The outcome of this game determines the statistics of the modes (states) of experience of the system(s) [level(s) Q, Q'], as well as, in a correlative way, the modes (states) of behavior of the system(s) [level(s) W, W'].

It remains unknown how one could deduce heuristic suggesting optimum codes mediating experience and behavior out of such a dynamical scheme, which incorporates as an *intensive factor* the degree of arousal of the RF. Much more experimentation is needed in the neurophysiology of the RF and the corticothalamic neural dynamics before one could perhaps exclude a number of one-to-one $Q \rightleftharpoons W$ mappings as incompatible with the experimental evidence.

In conclusion, in this long example we have attempted a plausible modeling of certain aspects of (human) communication. The key issue which has remained unresolved has to do with the heuristics of the selection of control laws between *behavior* and *experience* which give the best compromise in the intrinsic conflict between self-preservation ("homeostasis") and *transcendence* (via learning how to best simulate the partner).

4.8 Emergence of New Hierarchical Levels in a Self-Organizing System

This section can be read independently of the previous one.

We finally make an attempt to sketch the conditions under which an evolving self-organizing system may enrich the number of its hierarchical levels. After a general formulation of the problem we focus our attention on the specific topic of the emergence of a *new*, higher cognitive level functioning via coherent interactions between hitherto noncooperating components (cerebral tissues in a higher organism). We finally put forth some speculative remarks related to the possible outcome of a dynamical *conflict*, caused by inadequate communication, between two hierarchical levels of the organism. These last remarks also pave the way to a general treatment of conflict —couched in terms of "game theory" —in Chap.5.

4.8.1 Formulation of the Problem

A self-organizing system is a hierarchical structure (Fig.4.33). Graphically such a system may be symbolized by an "inverted tree" with branches spreading laterally as one moves downwards. For the sake of simplicity we depict in Fig.4.33 a single branch. Contributions concerning information arrivals at a given hierarchical level along this branch and other branches from more than one "platform" underneath are lumped together.

The system as a whole simultaneously displays a variety of distinguishable activities; for different groups of activity, different types of description (i.e., different sets of variables and parameters) may be appropriate. Which hierarchical level we are dealing with depends therefore on which aspect of the behavior of the system we are interested in, and this will generally depend on the way in which we interact with the system.

But what exactly defines a "hierarchical level" *physically*" and in what respects do "hierarchical levels" differ from each other? Before proceeding further, it is constructive to clarify these two important points.

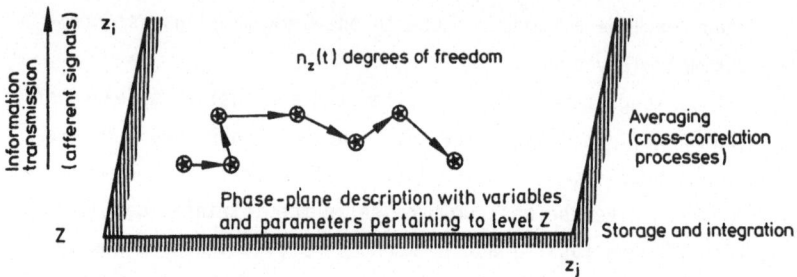

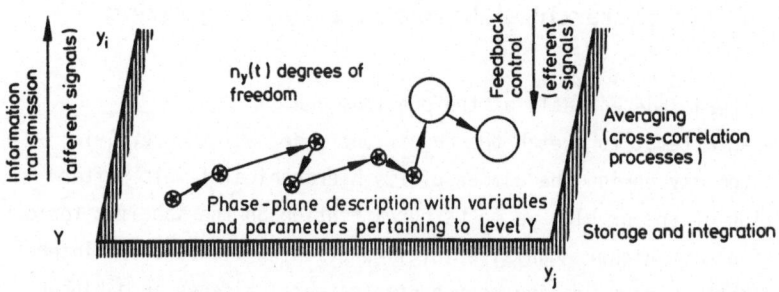

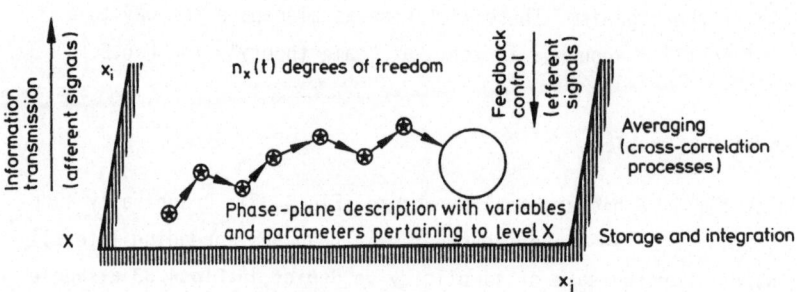

Fig.4.33. Successive hierarchical levels X,Y,Z of a self-organizing system. The individual levels are correlative to each other. At each level z_i a set of averaging
processes takes place between the converging signals from the lower level (y_j) and
the environmental fluctuations (w_{ij}). The "information transmission" and "feedback
control" signals are mediated at the neuronal level, for example, by pulses consisting of sequences of action potentials and/or hormonal and neurotransmitter releases,
whereas the "storage and integration" process is mediated by the postsynaptic subthreshold membrane potentials —cross-correlated with the incoming set of triggers
via synaptic neurotransmission. A given hierarchical level can be "recognized" only
by a higher one. For example $z_i = z_{0_i} + \Sigma_j <\alpha_{ij}(t)w_{ij}^K(t)y_j^\xi(t)(t-\tau)>$, where z_{0_i} stands
for the self-sustained activity at the hierarchical level z and $\alpha_{ij}(t)$ are coupling
coefficients. The large circles sketch possible periodic trajectories (limit cycles)

A hierarchical level is determined by the number, the location, and the nature (i.e., structure and complexity) of, as well as the degree of coherent cooperativeness among, functionally similar (although anatomically perhaps quite different) "elements" of the system concerned (or "tissues" in the case of a living organism). By "functionally similar" elements we imply a set of elements which have to cooperate for the successful performance of the given task or activity.

In respiration, for example, the object is the supplying of oxygen to the alveoli of the lungs and its diffusion through the walls of the alveoli into the blood. For this to be achieved, however, a complex muscular apparatus is necessary, incorporating the diaphragm and the intercostal muscles capable of expanding and contracting the chest, as well as a complex system of nervous structures in the brainstem and higher centers. It is then obvious that the totality of the above functions embodies many components belonging to different quarters of the secretory, motor, and nervous apparatus.

In the brain also, all "higher" processes such as perception, speech, writing, reading, and feeling, cannot be regarded as isolated or indivisible "faculties" which can be presumed to be the direct functions of limited cell groups or to be strictly localized in a specific area in the brain. Instead of localized areas, we have rather a number of pathways of information transfer (afferent pathways), feedback control (efferent pathways) and storage and integration processes. So "mental" functions are viewed here as complex functional systems organized in groups of coherently working zones, each of which may belong to completely different and often very distant areas of the brain. Such groups will define the hierarchical level, performing each time the appropriate function.

In the framework of a macroscopic description, the activity of each working zone will be described by a limited number of dynamical variables, which will generally be coupled via nonlinear feedback processes.

Higher hierarchical levels are characterized by an increasing proportion of participation of cerebral tissues of progressively ascending order and increasing complexity (from the brainstem, along the reticular formation all the way up to the cerebral cortex). These last levels are termed "cognitive levels" and most of them, unlike the lower "somatic" levels, are formed and broken down not only by evolution but also by learning taking place during the limited life-span of the individual organism. Let it be understood, nonetheless, that the material "canvas" for all these higher cognitive levels is the same, namely the neurons. What makes these cognitive levels different from each other is the number, the location, and the combination or connectedness of the cooperating tissues.

At each hierarchical level we now introduce a phase-space description involving the type and number of dynamical variables and parameters, and the resulting type and number of states pertaining to the particular level. The motion and transactions that the system undergoes in this phase space at a fixed hierarchical level (such

as a sequential passage from steady state to steady state, "time of rest" or stability properties of each state, as well as the flight time between successive
steady states) depend essentially on the parameters of the system which are perturbed, thereby leading to a series of transitions associated with the branching,
or bifurcation of the solutions of the equations of evolution. Such perturbations
are the result of a concerted influence from (a) the information transmission (afferent signals) coming from the level immediately below, (b) the information impinging from the external environment (i.e., via the peripheral nervous system),
and (c) the feedback control (efferent signals) coming from the level immediately
above.

As a result of these critical changes, the (hitherto) stable reference state(s)
in which the system resided become unstable. The system at the level considered
(say the level Z) has to move or accommodate itself into another steady state provided by the same set of coupled nonlinear differential equations modeling the system at Z, supplemented by the new boundary conditions (e.g., the new values of the
control parameters). Under some conditions, the latter state may display a greater
or lesser degree of coherence than the previous one(s).

In any case, the permanence of the system at the *specific* level Z during these
interactions is ensured by the "sameness" of the variables and of the major characteristics of the resulting states: the system at Z moves from steady state to steady
state by breaking down and resynthesizing a set of ordered relationships between
dynamical variables (e.g.. nonlinear oscillators), the nature of which does *not*
change during these transactions.

As an example, consider a homogeneous mixture of n chemically reacting substances
with relative composition z_i. The homogeneous mixture can find itself at many homogeneous steady states, depending on the number of possible stable combinations among
the z_i's. All these states belong to the same hierarchical level. Suppose now that
an external perturbation gives rise to a particular bifurcation as a result of which
the system moves to a steady state which is inhomogeneous, i.e., the concentrations
of some of the components $z_i(r)$ are now functions of space variables. This new
state does not belong to the previous hierarchical level Z; the system has moved
itself into another level where its nature differs fundamentally from the previous
one (as far as, for instance, the establishment of new, long-range correlations
among the variables is concerned).

Such symmetry-*breaking* transitions can be caused by diffusion, as first pointed
out by *Turing* [4.25]. *Nicolis* and *Prigogine* [4.26] have demonstrated the generality
of this type of phenomena in nonlinear systems far from equilibrium and introduced
the term *dissipative structure* to describe these new hierarchical levels. The biological implications of such transitions are nowadays widely recognized. (For a complementary approach see [4.27].)

Our aim in the present section is to propose a general formulation for the emergence of new hierarchical levels in a complex system. In a quite general way, we shall consider each hierarchical platform as a dynamical element where a "storage and integration" takes place. As we are primarily interested in the activity of the nervous system, we will represent storage and integration by a cross-correlation. This means that the signals $x_\lambda(t)$ transmitted sequentially from a lower level (each one corresponding to a certain state of the system at the level say x, see Fig.4.33) and the ones arriving from the external environment $w_\lambda(t)$ are (for example, time-) averaged upon arrival at the "receiving platform" Y (e.g., at a particular post-synaptic membrane aggregation in the cortex). The result is that some of the variables x_λ are completely "washed out" at the higher level, while others are non-linearly cross-correlated or matched as

$$<x_\lambda^\nu(t)w_\lambda^\rho(t + \tau)> \quad , \quad 0 < \tau < T \quad , \tag{4.8.1}$$

where T is the transmission time interval either from the lower level X or from the external environment. Events taking place simultaneously at different hierarchical levels have, therefore, correlative rather than causal correspondence.

It should already be clear that the description we adopt here is global, in the sense that the dynamical variables x_λ, w_λ entering the analysis are quantities averaged over a large number of physicochemical configurations of the elementary units constituting the system. Such global descriptions are necessary in dealing with complex situations, such as the central nervous system.

4.8.2 Creation of a New Hierarchical Level

As we pointed out in the previous section, on an existing hierarchical level the system may proceed in time from steady state to steady state via a sequence of bi-furcations which do not change the nature of the dynamical variables involved. On the other hand, *across* a given level (i.e., moving from this hierarchical level to the next higher one) a cross-correlation operation of the type already described in the preceding section is necessary. The two procedures seem at first glance quite distinct. One could think, however, of a generalized bifurcation scheme *built into the dynamics of the system*, which might incorporate a stochastic operator responsible for the formation of cross-correlations between a collective property at the actual level X with some process representing the fluctuating environment. This operator should be inactive or "masked" as long as the system moves on the level X, and should manifest itself as soon as a bifurcation leading to a change of the nature of the variables is inevitable, in much the same way as diffusion manifests itself in the bifurcation involving a symmetry-breaking instability leading to an inhomogeneous steady state.

By what sort of constraint imposed by the environment could such a bifurcation take place? In the present context, by appropriate environmental fluctuations with variance exceeding a certain threshold, fluctuations which could seriously disturb the system at X.

Facing the prospect of irreversible disorganization at the level X, the system might use its only alternative, i.e., "activate" the cross-correlation operator or change variables and operate on a higher level Y —from which, by sending back efferent control commands, it may also inhibit the excessive fluctuations at the level X.

Once this stabilization at the level X is secured, the new states may be abandoned and the level Y cease to exist altogether; it is beneficial nevertheless for the organism to "keep" the new level as long as possible, initially as a biasing agency against possible recurrence of similar perturbations and later on as an functional (software) element of a more sophisticated "self". (Please take the last sentence as sheer metaphor.)

In order now to model the system at the "disturbed" level X, we adopt a description whereby the internal dynamics of the system at this level is expressed by suitable nonlinear "rates" $f_i(x_j,t)$ standing for all possible couplings between the pertinent variables. In addition to the phenomena arising from these couplings, however, it is conjectured that a collective behavior of a new kind may establish itself via a "weighting procedure", described by a stochastic operator as follows: the individual dynamical variables x_i are weighted with a stochastic matrix of environmental fluctuations $w_{ij}(t)$. Subsequently, this action feeds back on the evolution of the variables x_i and constrains the dynamics at the level X, in a way that reflects explicitly the properties of the environment. The existence of numerous feedback pathways between neocortical and limbic levels adds credence to this view [4.28].

We may now write down a set of coupled nonlinear integral differential equations for the macroscopic dynamical variables x_i, which in the absence of spatial differentiation ($\nabla^2 x_i = 0$) read

$$\frac{dx_i}{dt} = f_i(x_j,t) + w_i(t) + \sum_{j=1}^{K} \int_0^t x_j(t')w_{ij}(t' + \tau)dt' \quad . \tag{4.8.2}$$

Here $w_i(t)$ stands for the fluctuations induced by the environment and the third term on the right represents the influence exercised by the cross-correlation operator on the hierarchical level X. For simplicity, it has been assumed that the weighting action of this operator involves a linear dependence on x_j's, although this need not be true in the most general case. Obviously, for a noisy environment modeled as a random process with zero-mean and root-mean-square deviation (r.m.s.) equal to σ, this term is zero when x_i is constant (steady states at the level X) or when it oscillates with a random phase, or thirdly, when the frequencies ω_k of the k (for k sufficiently large) oscillators are uncorrelated (or irrational to each other).

Dynamically speaking, the "cross-correlation operator" can be imagined as a manifestation of a "master hard nonlinear oscillator" (MHNLO) potentially existing at the level X with two stable states —one steady state "at rest" (i.e., when non-functioning) and one limit cycle when in the position "on". This MHNLO in a pool of environmental (white) noise w(t) with zero-mean and power spectral density σ [W/Hz] can be represented, for example, as a "hard" van der Pol relaxation oscillator

$$\ddot{x} - 2\epsilon(1 - 4\alpha x^2 + 8\beta x^4)\dot{x} + \omega_0^2 x = w(t) \quad . \tag{4.8.3}$$

It has been calculated ([4.11] and references therein, also Appendix A) that the ratio of the (mean) times during which such an oscillator is in the nonexcited versus the excited state is given by

$$p \sim \frac{(\alpha^2 - 4\beta)^{\frac{1}{4}}}{2 \sqrt{\pi\epsilon}} \sqrt{\sigma} \, e^{c/\sigma} \quad , \tag{4.8.4}$$

where

$$c = \frac{\epsilon}{12\beta} (\alpha + \sqrt{\alpha^2 - 4\beta}) \left[1 - \frac{\sqrt{\alpha^2 - 4\beta}}{4\beta} (\alpha + \sqrt{\alpha^2 - 4\beta})\right] \quad . \tag{4.8.5}$$

For α, β of the same order of magnitude (strong "hard" nonlinearity) we may have either $c > 0$ or $c < 0$.

The above result indicates then that for given parameters α, β, and ϵ, $p \sim \sqrt{\sigma} \exp(\pm c/\sigma)$. In the case $p \sim \sqrt{\sigma} \exp(c/\sigma)$, p initially decreases with increasing σ, passes through a minimum, and then increases. In the case $p \sim \sqrt{\sigma} \exp(-|c|/\sigma)$, p increases monotonically with σ. In the first case we can always determine a "favorable" set of parameters α, β, ϵ, so that $c > 0$, and for a given (moderate) level of environmental fluctuations σ, $p \ll 1$. This means that the oscillator will be activated with practical certainty whenever the r.m.s. of the environmental fluctuations exceeds σ. The (probabilistically) activated MHNLO will subsequently move into its limit cycle with a basic frequency ω_0 and an appreciably large amplitude.

What are the conditions under which the above described hard oscillator can influence the behavior of the system? Suppose that initially some of the variables in (4.8.2) are excited in such a way that they perform almost synchronous oscillations. Then it is possible that the MHNLO will cause further frequency entrainment among a number k of such oscillating components at the level X, so that

$$x_i \sim A_i(t)\cos[\omega_0 t + \varphi_i(t)] \quad . \tag{4.8.6}$$

The individual members of the entrained group of oscillators can subsequently enter into a sequential phase-locking relationship with spectral components of the environmental random phasor $w_i(t)$ and establish a set of ordered relationships

$$\varphi_{i+1} - \varphi_i < \omega_0 \tau \quad . \tag{4.8.7}$$

Under the above-postulated dynamics, we will eventually obtain a set of nonzero terms

$$\int_0^t x_j(t')w_{ij}(t' + \tau)dt' \tag{4.8.8}$$

(due to phase coherence) and a nonzero sum

$$\sum_{j=1}^k \int_0^t x_j(t')w_{ij}(t' + \tau)dt' \tag{4.8.9}$$

(due to frequency entrainment).

In other words, the correlation operator appearing in (4.8.2) will be "activated" very much as the diffusion operator is "switched on" if the initial condition corresponds to an inhomogeneous distribution of matter. Therefore we readily visualize a concrete case where, in principle at least, a bifurcation implied by the last term on the right-hand side of (4.8.2) can be implemented. The quantity

$$\int_0^t x_j(t')w_{ij}(t' + \tau)dt' \sim y_j \tag{4.8.10}$$

will be regarded as the variable pertaining to a new hierarchical level Y. The term $\sum_{j=1}^k y_j$ then gives the total amount of efferent feedback control exercised by Y on X. For

$$\sum_{j=1}^k y_j < 0 \tag{4.8.11}$$

we get negative feedback, i.e., a tendence to restrain the excessive fluctuations on level X.

In order to substantiate these conjectures it would be necessary to establish, at least on simple models, the occurrence of bifurcations in the system of the stochastic equation (4.8.2). We should, in other words, be able to estimate in such cases the probability density functions for the real parts $\text{Re}\{\lambda_i\}$ of the eigenvalues of the cross-correlation operator and establish the conditions under which $\text{Re}\{\lambda_i\} > 0$ can take place at least for one i with an appreciable probability. A solution to this extremely complicated mathematical problem will not be attempted here.

The number of cognitive levels man can "build" is equal essentially to the number of abstractions he is capable of; in principle, there is no "ceiling". Interestingly enough, this symbolic hierarchy seems also to be bottomless. One should think perhaps of some a priori existence of a "lowest" hierarchical level —de facto an irreducible one —constituting a set of "axioms" both complete and self-consistent. Gödel's theorems prove that there is no such thing.

Even more devastating is the fact that the hierarchical "pyramid" seems to fold and "bite its own tail", as it were, by joining the occasionally highest with the occasionally lowest level. Indeed, the "elementary-particle" level constitutes in fact a very complex conceptual artifact. To "constrain" an elementary particle one needs a mind —which in turn emerges as an evolutionary by-product of interacting particles —subjected to primordial initial and boundary conditions. This "bootstrap" scheme may give rise to a host of "paradoxes of self-reference", real or illusory. For instance, one's image of his own mind is itself an item present in the mind. So the image includes an image that includes an image, and so on, in an infinite nonconverging regress. A dynamical approach to the problem is suggested in Chap.6.

4.8.3 A Comment on Typical Cases of "Psychosomatic Disturbances"

Two given hierarchical levels of an organism are in "conflict" whenever they do not communicate adequately[5]. This happens when the afferent and/or efferent routes by which information is transmitted "upwards" and by which feedback control commands are received "downwards" are impaired. One cause of such a conflict might be the presence of excessive noise (i.e., unconstrained fluctuations) arising at the inter-face between the hierarchical levels concerned.

The conflict (lack of adequate communication) between a *neocortical* hierarchical level and a lower level belonging to the *limbic system* may have as a result the precipitation of an efferent (feedback) secular disturbance which may lead to an overt perturbation at a specific organ of the body or an impairment of an otherwise homeostatic somatic activity (through an imbalance, for example, of the parasym-pathetic-sympathetic nervous system). Such a conflict may also be attributed, in some cases, to a relatively immature cortical level, unable to constrain properly the limbic afferent signals. This profile of "infanilism" may be corrected if the system, with the help of an adequate environment, builds itself a new cognitive hierarchical level. By appropriate negative feedback control exerted from this new level, the required dynamical balance between signal *variety* and *reliability*, in-dispensable for the coordinated evolution of the organism, will again be ensured.

We finally suggest a possible biological mechanism of implementing via learning the creation of a new hierarchical level. This may be achieved by activating the neuronal genome through, for example, massive neurotransmitter releases which modu-late the postsynaptic membrane potential fluctuations, thereby creating random ionic imbalances which in turn may influence the nucleus and excite hitherto silent genes [4.11]. In this way new types of proteins will be manufactured, and subse-

---

5 The interesting case where conflict *persists* even in the presence of perfect com-munication will be dealt with in Chaps.5 and 6: it has to do with paradoxical games and the paradox of self-reference.

quently coat the postsynaptic dendrites, and bind to specific membrane loci —simultaneously in a large number of cerebral tissues.

As a result, new patterns of postsynaptic receiving sites for the individual species of neurotransmitters are randomly created. Such "receptor sites" are presumably highly specialized for receiving the pertinent neurotransmitter's molecules in a very specific "lock-and-key" arrangement. Therefore the new neurotransmitter-receptor interactions can effectively trigger a systematic change in the fluctuations of the active transport currents across the neuronal membrane. These currents are produced as a result of cross-correlations performed by the neuronal membranes between the fluctuations of the postsynaptic membrane potentials (X) corresponding to the new membrane coating and the fluctuations of the membrane trans-conductances, corresponding to the flux of neurotransmitters (W).

Thus a coupling takes place among hitherto uncorrelated or nonexistent dynamic variables. These new variables, mutually interacting, may form the new hierarchical level. (For more details see also Appendix A.)

5. Elements of Game Theory, with Applications

Mutual simulation between two hierarchical systems —in the present case two biological organisms —seldom proceeds in a straightforward way. Usually the conflicting interests of each of the two partners (each wishes to simulate successfully, i.e., to *predict* and ultimately *control* the opponent and at the same time, present the opponent with a random behavior which does not allow *him* to simulate in return) force them to wage a *game*.

A game has thus to do with an algorithm of decision making under *conflict* and *uncertainty* —uncertainty about the motives and moves of the partner. Each partner strives under the given constraints to maximize his "gains" and/or to minimize his "losses". In this chapter we intend, with one exception, to describe and formulate in dynamical language *two*-partner game theory. The mathematical treatment will therefore be couched in terms of two coupled linear or nonlinear, first-order ordinary differential equations. The state-space description will therefore be two dimensional. Our interest is in the shape of integral curves as well as the kind of singularities —if any —present under "realistic" assumptions, and the way bifurcations may give rise not only to *evolution* but also *change* and ultimately *resolution* of the conflict. The variables in our equations will be the probabilities of staying at a certain move or strategy.

After the description of the abstract models, we come to some applications: *Animal conflicts, conflicts in production*, and similar high-level deliberations.

Fascinating applications also include some examples from epidemiology where the antagonism revolves around the issue of who is going to infect (simulate) whom *first*, and furthermore how the dynamics of "host-parasite" interacting groups evolves in time. The circulation and propagation of rumours —be it political propaganda, religious dogma, or plain advertisement —constitutes finally a very interesting special case of "epidemiology" which will be given here in stochastic albeit heuristic terms, namely by bypassing the rigorous formulation of a master equation, concerning the probability of spreading information via dyadic interactions in a given number of individuals at a given time. Rumor propagation theory constitutes a game in the sense that some individuals resist the adoption and refuse to be "used" as carriers of information, while others are more susceptible (conformists) and therefore become more easily "infected".

We deal below with a classification and the treatment of each category of abstract games starting with constant-sum games.

5.1 Constant-Sum Games

In a constant-sum game the 2×2 pay-off matrix (Fig.5.1) has elements ("payoffs") whose sum is the same for all strategy combinations. We have assumed here that each partner has just two moves in his repertoire, (A_1, B_1), (A_2, B_2) respectively, and that the "payoffs" are conventional numbers which represent symbolically rewards (when positive) or losses (when negative).

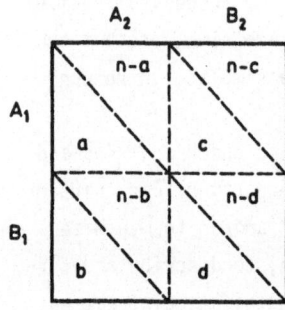

Fig.5.1. The 2×2 matrix of a constant-sum game. The terms in the lower-left and upper-right halves of each square give the payoff for the contestants I and II, respectively

A "solution" of a game in general and a constant-sum game in particular is envisaged as a (stable) apportionment of payoffs (or expected payoffs) resulting from "rational choices" by the contestants: neither player can do better *if* the opponent chooses the prescribed strategy. In turn, rational choices are those that tend to maximize the players' payoffs (or expected payoffs) under the constraint that each is attempting to maximize his own payoff (or expected payoff) — knowing that the opponent is attempting to do the same. (In our context a payoff is positive or negative regardless of whether *compressibility* of the code of the opponent is achieved. In that respect every game among equals contains an element of futility.)

In constant-sum games there is an existence theorem which essentially proves that there is an optimum pair of strategies for the contestants. (In non-constant-sum games we simply do not know: such games, as we shall see later, require "bargaining" abilities since optimal strategies leading to stable equilibria cannot be found.) Four distinct cases have to be distinguished in constant-sum games.

5.1.1 Both Players Have a Dominant Strategy

A dominant strategy or move is one which gives a "player" the maximum possible payoff, regardless of what the opponent does. In our notation this would be exemplified by a game matrix in which $a > b$, $c > d$, $a < c$, $b < d$. Clearly in this situation a rational player I would choose A_1 and a rational player II would choose A_2. The outcome would accord a to player I and n-a to player II.

5.1.2 Only One Player Has a Dominant Strategy

This can be realized for $a > b$, $c > d$, $a > c$, $d > b$. In such a case player I would choose A_1; player II, *knowing this*, would choose B_2 because $n-c > n-a$. The pay-offs would be c and n-c respectively.

5.1.3 Neither Player Has a Dominant Strategy

Still we can stay in the regime of "pure" or "deterministic" strategies (that is, accept that the game can be resolved in only *one* play) by invoking possible equilibria in the form of "saddle points". Let us refer to two examples to make the point clear. Consider first the 2×3 matrix in Fig.5.2 where each player has three moves at his disposition. (Here $n = 0$ so the constant-sum game turns into a "zero"-sum game.) Here neither player has a dominant strategy, i.e., neither player can choose a move disregarding what the other intends to do.

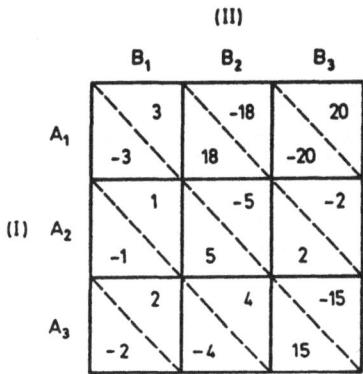

Fig.5.2. A 2×3 (2 players $\times 3$ moves) matrix of a zero-sum game

In the above example it is a good tactic for each partner to try to *minimize his maximum losses*. For example, player I thinks as follows: "If I choose move A_1 the worst result for me will come if my opponent chooses B3— then my payoff is -20. If I choose move A_2 my worst is -1, and if I choose A_3 my worst is -4." So he plays A_2. Player II argues similarly: "If I choose B_1 I cannot drop below 1. If I choose B_2 I cannot drop below -18 and if I choose B_3 I cannot drop below -15." Thinking in this conservative way, player I chooses A_2, player II chooses B_1, and they meet at the $(-1,1)$ square —which is a saddle point in the sense that the values of the payoff(s) are the minimum of the row and the maximum of the column for player I and the maximum of the row and the minimum of the column for player II. Moreover, this saddle point is "stable" in the sense that neither player can do better by unilaterally defecting from it.

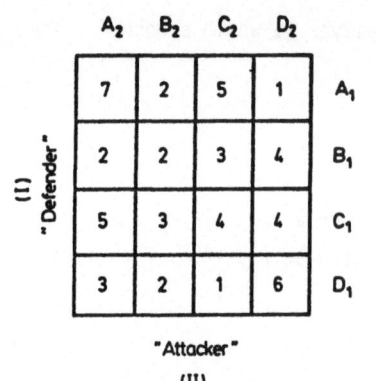

Fig.5.3. The matrix of a "war game"

	A$_2$	B$_2$	C$_2$	D$_2$	
	7	2	5	1	A$_1$
	2	2	3	4	B$_1$
	5	3	4	4	C$_1$
	3	2	1	6	D$_1$

(I) "Defender"

"Attacker"

(II)

Take another example from a "war game" (Fig.5.3). The numbers in the matrix represent days between encounters of the opponent armies. The attacker knows that time works against him and that it is in his interest to wage a "blitzkrieg", thereby *minimizing* the time between encounters with the enemy. The defender, on the other hand, adopts a strategy of "scorched earth", namely he retreats, trying to maximize the time between encounters in order to build up resistance and wear out the attacker. (An example is the early phase of the Second World War conflict between Germany and the Soviet Union. Surprisingly, another example is the courtship of exceptional animal species — including man.)

What does each contestant decide, on the basis of the available strategies in Fig.5.3? The defender will try to maximize his minimum winnings, and the attacker will try to minimize his maximum losses.

The attacker thinks as follows: "If I adopt strategy A$_2$ the worst thing which can happen to me is to wait seven days. If I choose B$_2$ the worst thing is to wait three days. If I move to C$_2$ in the worst case I'll wait five days, and if I go to strategy D$_2$ I'll have at most to wait six days. These are my *maximum* losses. To *minimize* them I choose strategy B$_2$ (three days)."

The defender thinks as follows: "I must exhaust the attacker. If I adopt strategy A$_1$ my worst case will be one day. If strategy B$_1$, two days; for strategy C$_1$, three days; and for strategy D$_1$, one day. These are my *minimum* winnings. To *maximize* them I choose strategy C$_1$ (three days)."

So the contestants move to the square C$_1$B$_2$ which contains the number which is the minimum number of its row and the maximum of its column. This pair of strategies is *stable* because either contestant who tried unilateral deviation from it would lose *more*.

5.1.4 Mixed Strategies

In a case where neither player has a dominant strategy and no saddle point is allowed in the payoff matrix, the game cannot be resolved in a single play. In our 2×2 general matrix of Fig.5.1 this amounts to having $a > d > b > c$. Under such circumstances the players adopt mixed strategies, that is, they perform many plays. Let x be the probability that player I will choose the move A_1 and y the probability that player II will choose the move A_2. Under such circumstances the expected (average) payoff of player I will be

$$G_x(x,y) = axy + b(1 - x)y + cx(1 - y) + d(1 - x)(1 - y) \qquad (5.1.1)$$

and that of player II will be

$$G_y(x,y) = n - G_x(x,y) \quad . \qquad (5.1.2)$$

The logic of the adoption of strategy A_i or B_j $[i,j \in (1,2)]$ by the players has obviously to do with their expected payoffs. Specifically, the rate of x or y (dx/dt, dy/dt) will be proportional to the differential expected profit that each player realizes by adopting the strategy A_1 or A_2, B_1 or B_2 respectively. Hence we may write

$$\frac{dx}{dt} = \frac{\partial G_x}{\partial x} \qquad \text{and}$$

$$\qquad (5.1.3)$$

$$\frac{dy}{dt} = \frac{\partial G_y}{\partial y} \quad .$$

We have thus transformed (or "reduced") the logic of the contest to a pair of coupled linear differential equations. Substituting for G_x and G_y we get

$$dx/dt = y(a - b - c + d) + c - d \quad ,$$

$$\qquad (5.1.4)$$

$$dy/dt = x(- a + b + c - d) - b + d \quad ,$$

from which the steady state follows:

$$x^* = \frac{d - b}{a - b - c + d} \quad , \qquad y^* = \frac{d - c}{a - b - c + d} \quad . \qquad (5.1.5)$$

It is of interest to examine the *stability* of the above steady state.

Let us introduce perturbations $x' = x - x^*$ and $y' = y - y^*$. Then the differential equations become

$$dx'/dt = y'(a - b - c + d) \qquad \text{and}$$

$$\qquad (5.1.6)$$

$$dy'/dt = x'(- a + b + c - d) \quad .$$

Therefore

$$\frac{dy'}{dx'} = - \frac{x'}{y'} \quad , \qquad (5.1.7)$$

which determines a circular trajectory for the perturbations given by the parametric equations $x' = \cos\theta$, $y' = \sin\theta$ around the origin as the center of the circle.

The integral

$$\int_0^{2\pi} G(x',y')d\theta = 0 \quad,$$

because $G(x',y')$ contains only terms in x', y' and $x' \cdot y'$, i.e., terms in $\cos\theta$, $\sin\theta$, and $\sin\theta \cdot \cos\theta$, whose integrals all vanish. The conclusion is that the focus (x^*,y^*) is *marginally stable* —since the differential equations we started with are essentially those of the undamped harmonic oscillator.

5.2 Non-Constant-Sum Games

We have to distinguish between two-person games in which the interests of the players are diametrically opposed (constant-sum or zero-sum games) and those where the interests are partially opposed and partially coincident (non-constant-sum games). In constant-sum games the sum of the payoffs of the two players is *the same* regardless of how the game ends: the larger the payoff to one player the smaller the payoff to the other. This is the meaning of "diametrically" opposed interests.

The solution of a two-person constant-sum game is a pair of pure or mixed strategies (one available to each player) that are in equilibrium: neither player can do better *if* the other uses the prescribed strategy.

In non-constant-sum games the players have in general partially common, partially opposed interests. Such games are sometimes called mixed-*motive* games. In non-constant-sum games, it is still possible to prove the existence of equilibria. Nevertheless, it is no longer possible to prescribe "optimal" strategies in terms of these equilibria because the choice of strategies containing equilibria by *each* of the players need *not* result in an equilibrium outcome. Therefore in non-constant-sum games optimal strategies (i.e., solutions which *guarantee* stability of a pair of choices in the sense that unilateral defection or even bilateral defection is not encouraged) do not exist. In such games the concept of *rationality* must be still further refined and generalized. It turns out that the moves prescribed by *individual* rationality may differ essentially from those of *collective* rationality. "Paradoxes" arise when the concepts of "rational decision" that are adequate on one level of conflict are applied on another level. Therefore, in the context of a two-person non-constant-sum game, the concept of rationality bifurcates, so to speak, into *individual* and *collective* rationality —the two being often at variance —and the concept of *bargaining* as a possible suboptimal resolution arises. We divide non-constant-sum games into two subcategories:

a) Non-constant-sum *negotiable* games.

b) Non-constant-sum *paradoxical* games.

Let us deal with each group separately.

5.2.1 Non-Constant-Sum "Negotiable" Games

We start with the specific example illustrated in Fig.5.4. Here we essentially follow *Rapoport* [5.1].

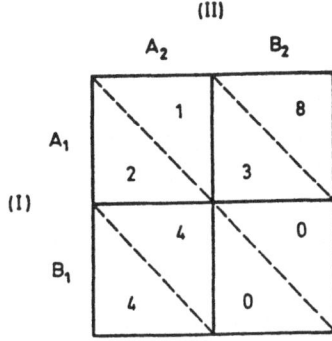

Fig.5.4. The 2 ×2 matrix of a non-constant-sum "negotiable" game

Player I must choose between the rows A_1, B_1. If he knew player II's choice, his decision would be easy: if player II were to choose A_2, player i would choose B_1 for a gain of 4. If player II were to choose B_2, he would choose A_1 for a gain of 3. However, player II faces the same sort of decision problem: if he were sure that player I would try for his biggest reward (B_1) then player II could do no better than choose A_2. But he cannot be sure that player I would not rather "play it safe" and choose A_1 (in the expectation that player II will try for *his* biggest gain of 8). In that case player II could get his largest gain. We see that this game has *no* optimal pure strategy for either player. Let us see how it stands with *mixed* strategies.

Ignoring for the moment the opponent's payoff, each player may select a strategy mixture which guarantees a *minimum* payoff *regardless* of the opponent's strategy. There are many algorithms for that, but let us suppose that the actual mixture for player I is determined by obtaining the difference in payoffs for each row (i.e., for row 1, 3-2 =1; for row 2, 4-0 =4), and then *invert* the obtained values and choose on a random basis A_1 four times for every B_1 choice. Similarly player II assigns himself a probability of 8/11 to A_2 and 3/11 to B_2.

Thus player I, by assigning a probability of 4/5 to A_1 and 1/5 to B_1, guarantees himself again

$$\frac{1}{2}\frac{4}{5}(2 + 3) + \frac{1}{2}\frac{1}{5}(4 + 0) = \frac{12}{5}$$

— by assuming full ignorance (1/2, 1/2) about player II's probabilities of playing A_2, B_2.

Player II likewise guarantees himself an amount

$$\frac{1}{2} \frac{8}{11} (1 + 4) + \frac{1}{2} \frac{3}{11} (8 + 0) = \frac{32}{11} \;.$$

At first glance we might consider this pair of mixed strategies (which guarantee a minimum "security level", so to speak, for each player) as a "solution". Nevertheless, this solution is *not* satisfactory: both players could get more than these guaranteeed minima if, for example, player I chose B_1 and player II A_2. Had they been able to *coordinate* their strategies, they could have obtained these larger gains. To be sure, they would still have had to face the problem of agreeing on *one* strategy pair or the other (or possibly a mixture of the two), because the strategy pair (B_1,A_2) favors player I while (A_1,B_2) favors player II. However, whatever strategy pair of whatever mixture of the two they agreed upon, it would have given each of them more than the guaranteed levels of 12/5 and 32/11 respectively.

The inability to coordinate strategies must be attributed here to *lack of communication* between the two partners. Suppose for a moment that the players somehow did agree on the strategy-pair A_1B_2 (or B_1A_2). Then there would be no *motive* for either player to break the agreement because either would not impair his payoff if he chose the other strategy while the other player stuck to the agreement. So each of the above two outcomes is an equilibrium: neither player can improve his payoff (and will, in general, impair it) if he moves away from an equilibrium while the other stays with the same strategy (*minimax* principle: unilateral defection is *discouraged*).

If, however, in the game under consideration the circumstances of the situation permit communication and coordination of strategies, the situation has changed. For in that case the players may be able to agree on an outcome that benefits them both. Clearly, they cannot do so in a constant-sum game since in such a game whatever outcome is better for one of the players must be worse for the other. In non-constant-sum games, on the other hand, one may meet outcomes that are preferred to other outcomes by *both* players.

The mathematical problem is, then, to find a single point G_x, G_y on the curve of possible outcomes of agreement (the "negotiation set") which can be defined as a "rational solution" or a "rational conflict resolution" (Fig.5.5).

The point on the negotiation set which is the corner of the rectangle with the *largest* area can be proved to be the solution under two conditions: first that the negotiation set has been determined in advance *and* second that the "other corner" of the rectangle, the "status quo", has also been predetermined.

In general, it is not obvious what the "status quo" point in different games should be. It might be argued, for example, that the status quo point should indicate what each player can guarantee for himself (regardless what the other does).

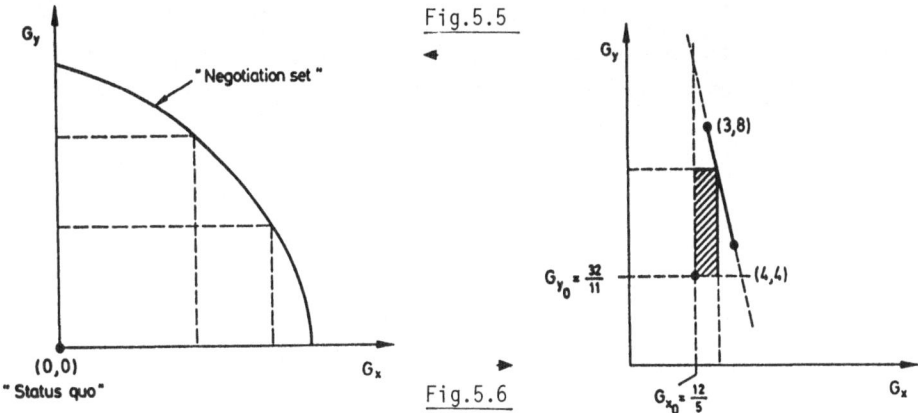

Fig.5.5

Fig.5.6

Fig.5.5. The negotiation set

Fig.5.6. The negotiation set for the specific game with the matrix given in Fig.5.4

We have already seen that in our game the coordinates of that point are $G_{x_0} = 12/5$, $G_{y_0} = 32/11$. If so, let us try to calculate the optimal payoffs by taking as a nego-tiation set the linear segment: $(3,8)$, $(4,4)$ (Fig.5.6). (Indeed it is between the above two limits that the players have to make a choice.)

Under the above conditions, the optimum point G_x^*, G_y^* will be calculated from the relation

$$F = (G_x^* - G_{x_0})(G_y^* - G_{y_0}) = \max , \tag{5.2.1}$$

but

$$G_y = 20 - 4G_x , \tag{5.2.2}$$

from the analytical expression for the line segment from $(3,8)$ to $(4,4)$, and $G_{x_0} = 12/5$, $G_{y_0} = 32/11$, so we get

$$\left(G_x^* - \frac{12}{5}\right)\left(20 - 4G_x^* - \frac{32}{11}\right) = \max$$

or

$$\frac{\partial F}{\partial G_x} = 0 ,$$

$$20 - 4G_x^* - \frac{32}{11} - 4\left(G_x^* - \frac{12}{5}\right) = 0 .$$

From this

$$G_x^* = \frac{1}{8}\left(20 - \frac{32}{11} + 4\frac{12}{5}\right) \quad \text{or}$$

$$G_x^* = 3.34 , \quad G_y^* = 6.65 . \tag{5.2.3}$$

Questions arise, however, about *the stability* of the chosen status quo $G_{x_0} = 12/5$, $G_{y_0} = 32/11$.

To guarantee himself the security level payoff, player I must play his two stra-tegies with probabilities $x = 4/5$ and $1 - x = 1/5$ respectively. If he does this, how-ever, player II can get considerably *more* than *his* security level: by playing B_2 as a *pure* strategy, player II can get $8 \times 4/5 = 32/5$ —rather than his security level $32/11$. Similarly, if player II plays his security level mixture $(8/11, 3/11)$, player I can get, by playing B_1, $4 \times 8/11 = 32/11$, that is, *more* than his security level, which is $12/5$.

It follows then that each player would be *tempted* to *depart* from his security mixed strategy if he knew that the other would *stay* with it. On the other hand, if both do so, they determine a new status quo, namely the point $(0,0)$. This clearly comes about as a result of a failure on the part of the players to agree, each in-sisting on the strategy most favorable to himself.

Let us see what the optimal outcomes are for the *same* negotiation set but the status quo point at the origin. We have now $G_{x_0} = 0$, $G_{y_0} = 0$, so (5.2.1) becomes

$$F = (G_x^* - 0)(20 - 4G_x^* - 0) = \max \quad ,$$

from which, by asking $\partial F / \partial G_x = 0$, we get

$$G_x^* = \frac{1}{8} 20 \sim 3 \quad \text{and} \quad G_y^* \cong 8 \quad . \tag{5.2.4}$$

Now in general it is assumed that to determine the status quo point, each player chooses a "threat strategy" (pure or mixed) independently; it is understood that if no agreement is reached, the pair of strategies x, y so chosen shall determine the status quo point. If agreement is reached, that is, if a point on the negoti-ation set is to be determined, this determination shall be in accordance with the solution of the bargaining game described above with the pair of threat strategies constituting the status quo point.

Suppose player I chooses for his "threat" strategy some mixed strategy accord-ing to which A_1 is chosen with probability x and, accordingly, B_1 with probability $1 - x$. Similarly player II chooses for his "threat" strategy a mixed strategy accord-ing to which A_2 is chosen with probability and B_2 with probability $1 - y$.

If these strategies go into effect (i.e., if agreement is *not* reached) the expec-ted payoff to player I will be

$$G_{x_0} = 2xy + 4(1 - x)y + 3x(1 - y) \quad . \tag{5.2.5}$$

The corresponding payoff to player II will be

$$G_{y_0} = xy + 4(1 - x)y + 8x(1 - y) \quad . \tag{5.2.6}$$

Let us now consider the bargaining game in which the point (G_{x_0}, G_{y_0}) in (5.2.5,6) *is* the status quo point and whose negotiation set lies, as previously, along the line connecting the points $(3,8)$ and $(4,4)$. The solution will be the point (G_x^*, G_y^*) on the bargaining set such that G_x maximizes the value of the expression

$$F = (G_x - G_{x_0})(G_y - G_{y_0}) = (G_x - G_{x_0})(20 - 4G_x - G_{y_0}) \quad . \tag{5.2.7}$$

The maximization value, derived from the expression $\partial F/\partial G_x = 0$, is found to be

$$G_x^* = \frac{1}{8}(20 - G_{y_0} + 4G_{x_0}) = \frac{1}{8}(20 - 9xy + 4x + 12y) \quad . \tag{5.2.8}$$

It is clearly in player I's interest to choose a threat strategy x which will maximize his final payoff G_x^*, i.e., (5.2.8).

Let us now transform the original game by multiplying player I's payoffs by 4, as shown in Fig.5.7. This transformation amounts merely to choosing a different unit of "currency" for player I which certainly does not change the "strategic structure" of the game.

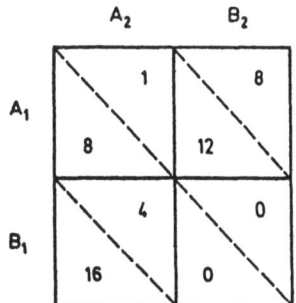

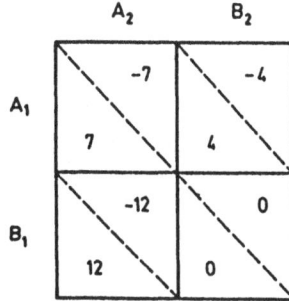

Fig.5.7

Fig.5.8

Fig.5.7. Transforming the game of Fig.5.4

Fig.5.8. Further transformation of the game of Fig.5.4 to a constant-sum game having a saddle point

As a second step we now form a *zero-sum* game (Fig.5.8) in which the payoffs are the algebraic differences of the payoffs in the above transformed game.

Let us calculate the payoff to player I in this last constant-sum game if he uses a mixed strategy x while his opponent, player II, uses a mixed strategy y. The expected payoff to player I will be

$$7xy + 4x(1 - y) + 12y(1 - x) = -9xy + 4x + 12y$$

— which is exactly the quantity he wishes to maximize in the original game [see (5.2.8)]. However — and this is the crux of the matter — the last game above is a zero-sum game with a *saddle-point* at (A_1B_2). By playing A_1, player I maximizes his minimum winnings, and by playing B_2 player II minimizes his maximum losses.

Therefore in this game a *pure best strategy* is available to player I, namely A_1. Similarly, for player II this last game is strategically equivalent to the original one, and so his best strategy is *in any case* the pure strategy B_2. The above thoughts may perhaps convince us that the negotiation set should after all be the line $G_x + G_y = 11$ (Fig.5.9), with the status quo point $G_{x_0} = 12/5$, $G_{y_0} = 32/11$.

One therefore is interested in finding out what possible "compensations" player II must offer to player I in order to attract him to this negotiation set. With this set the two players get *jointly* an amount larger than in any other outcome, hence larger than in any mixture of outcomes.

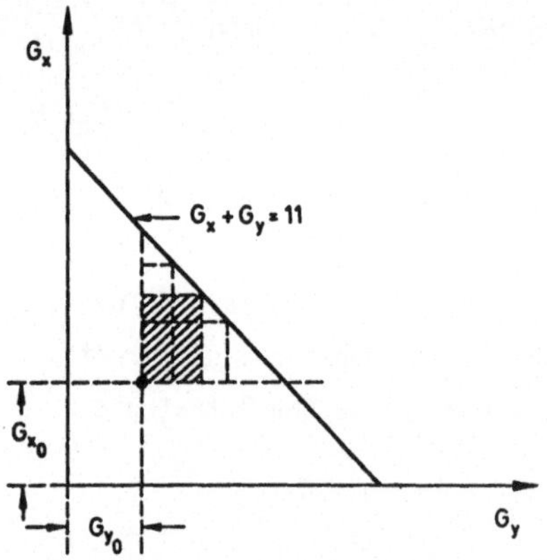

Fig.5.9. Another negotiation set for the same specific game (Fig.5.4)

Should not two *rational* players immediately agree to coordinate their moves so as to obtain the outcome (A_1B_2), and only *then* bargain about how the joint payoff(s) is (are) to be divided between them? Let us find out by maximizing the quantity

$$F = \left(G_x - \frac{12}{5}\right)\left(11 - G_x - \frac{32}{11}\right) \quad .$$

From $\partial F/\partial G_x = 0$, we get

$$11 - G_x^* - \frac{32}{11} - G_x^* + \frac{12}{5} = 0 \quad ,$$

from which

$$G_x^* \sim 5.25 \quad \text{and} \quad G_y^* \sim 5.75 \quad .$$

The concessions that player II has to make to player I in order to attract him toward the negotiation set $G_x + G_y = 11$ seem to be very heavy indeed. Yet there is an alternative way to see that these concessions are as heavy "as they should be". We can envisage the total payoff that the two-player "*coalition*" can obtain being apportioned according to the following principle: each partner gets the amount that he brings to the coalition by joining it.

Suppose the coalition is built up by player I joining the "empty" coalition first. The empty coalition, of course, gets no payoff. Player I in a coalition with *himself* gets simply his "security" level, which is 12/5. If player II joins the empty coalition first, the one-man coalition can get again 32/11 (player II's security level).

When player I joins, he literally and actually *enables* the two-man coalition to get 11. Consequently, he has brought $11 - (32/11) = 89/11$ into the coalition that he has joined, and he is entitled to *that* amount, period. We assume, however, that the order in which the two-man coalition is built up involves pure chance. (Each order is assumed to occur with probability 1/2). So the amount to be accorded to player I should be the even-weighted average of 12/5 and 89/11, namely $(1/2)(12/5) + (1/2)(89/11) \cong 5.25$. Player II gets the remainder, namely $11 - 5.25 = 5.75$.

5.2.2 Non-Constant-Sum, Nonnegotiable "Paradoxical" Games

Consider the following non-constant-sum game —widely known as the "prisoner's dilemma" (PD) (Fig.5.10). Here the diagonal payoffs can in general be assigned the
value ξ, $|\xi| \geqslant 1$. From the point of view of each player, it seems quite clear what
he should choose: Player I should choose B_1, *quite regardless* of how player II
chooses. For should player II choose A_2, player I is better off with B_1, which
gives him 10, whereas A_1 gives him only 1. Should player II choose B_2, player I
is still better off choosing B_1 and accepting a loss of 1, whereas A_1 leads him to
a loss of 10. For exactly the same reason, it is "rational" for player II to choose
B_2. The "rational" strategy-pair $(B_1 B_2)$ leads to a loss of 1 by *both* players. Had
they chosen $(A_1 A_2)$ respectively, they would each have *gained* 1. Note that in this
game it is *not* the inability to *coordinate* strategies that prevents the contestants
from getting as much as they could. There is only *one* strategic pair that leads to
a gain for *both*, namely $(A_1 A_2)$. Thus there is *no* ambiguity about the choice that is
in their *collective* interest (as there was in the previous category of negotiable
games).

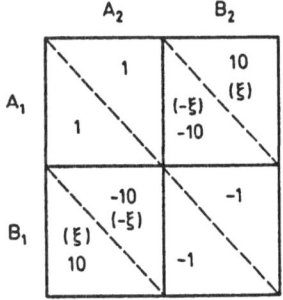

Fig.5.10. The 2×2 matrix of the "prisoner's dilemma" game

What then prevents them from achieving $(A_1 A_2)$? Evidently it is "acting in accordance with the individual interest" of each that leads them to $(B_1 B_2)$ instead of
to $(A_1 A_2)$. Also, the present game differs from the previous one in that instead of
two equilibria, as in the case of Sect.5.2.1, the present game has only one equilibrium $(B_1 B_2)$. However this would not be the outcome on which a pair of "rational"
players would agree.

If they agreed at all they would rather settle for $(A_1 A_2)$. This outcome, however,
is not an equilibrium: each player can get *more* by moving away from it *alone* (i.e.,
by breaking the agreement), but, if they *both* do so, both lose.

The situation seems desperate from the point of view of "pure" strategies. Let
us see what might happen if the partners choose mixed strategies (i.e., they get
involved in a statistically significant number of plays) x for A_1, 1 - x for B_1,
y for A_2, 1 - y for B_2.

The "equations of motion" for x, y will be, as before,

$$\frac{dx}{dt} = \frac{\partial G_x}{\partial x} \quad,$$

$$\frac{dy}{dt} = \frac{\partial G_y}{\partial y} \quad, \quad \text{where} \tag{5.1.3}$$

$$G_x = xy - \xi x(1 - y) + \xi(1 - x)y - (1 - x)(1 - y)$$

and

$$G_y = xy + \xi x(1 - y) - \xi(1 - x)y - (1 - x)(1 - y) \quad. \tag{5.2.9}$$

We get therefore

$$dx/dt = 1 - \xi \quad,$$

$$dy/dt = 1 - \xi \quad \text{or} \tag{5.2.10}$$

$$x(t) = x(0) \, e^{(1-\xi)t} \quad,$$

$$y(t) = y(0) \, e^{(1-\xi)t} \quad, \tag{5.2.11}$$

and since $|\xi| \geqslant 1$, x, y decay quickly to the steady-state values $(0,0)$, i.e., the game soon ends up at the state $(B_1 B_2)$. The situation therefore does *not* improve at all by using mixed strategies.

Let us try next to introduce mixed strategies with *memory*: this means that the moves the opponents adopt for the next play are somehow conditioned by the previous outcome. What we attempt here, then, is the introduction of a simple (one-step memory) Markov chain for the algorithm of each player. Let us consider moves A_1, A_2 as "synergetic moves" C_1, C_2, and moves B_1, B_2 as "defecting moves" D_1, D_2. Our matrix has four states $S_1(C_1C_2)$, $S_2(C_1D_2)$, $S_3(C_2D_1)$, $S_4(D_1D_2)$ as in Fig.5.11.

Let us define then for players I and II separately, taking into account also the "spirit" of the players (namely intelligence and ruthlessness), a set of *conditional*

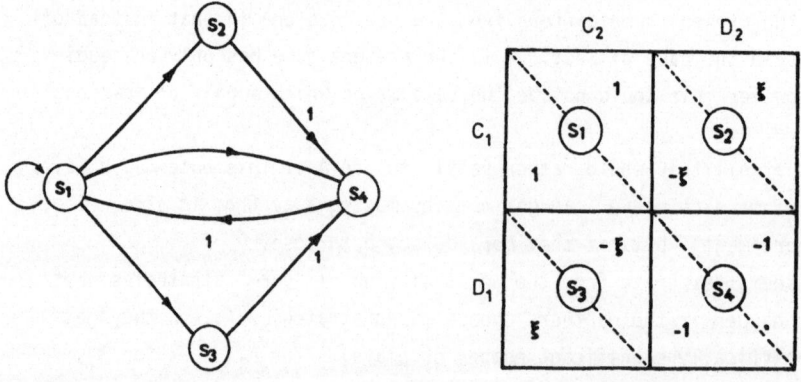

Fig.5.11. The prisoner's dilemma played inductively and the resulting Markov chain (left)

230

probabilties, or "propensities", for cooperation and defection.

For player I

$P(C_1/S_1) = x_1$,
$P(C_1/S_2) = y_1 = 0$,
$P(C_1/S_3) = z_1 = 0$,
$P(C_1/S_4) = \omega_1 = 1$.

For player II

$P(C_2/S_1) = x_2$,
$P(C_2/S_2) = y_2 = 0$,
$P(C_2/S_3) = z_2 = 0$,
$P(C_2/S_4) = \omega_2 = 1$.

This implies an immediate retaliation on the part of the "betrayed" partner and an immediate return to the state S_1 after the paradoxical outcome S_4. The 16 transitional probabilities defining the Markov chain are therefore deduced as follows:

$P_{11} = x_1 x_2$,
$P_{12} = x_1(1 - x_2)$,
$P_{13} = x_2(1 - x_1)$,
$P_{14} = (1 - x_1)(1 - x_2)$,

$P_{21} = y_1 y_2 = 0$,
$P_{22} = y_1(1 - y_2) = 0$,
$P_{23} = y_2(1 - y_1) = 0$,
$P_{24} = (1 - y_1)(1 - y_2) = 1$,

$P_{31} = z_1 z_2 = 0$,
$P_{32} = z_1(1 - z_2) = 0$,
$P_{33} = z_2(1 - z_1) = 0$,
$P_{34} = (1 - z_1)(1 - z_2) = 1$,

$P_{41} = \omega_1 \omega_2 = 1$,
$P_{42} = \omega_1(1 - \omega_2) = 0$,
$P_{43} = \omega_2(1 - \omega_1) = 0$,
$P_{44} = (1 - \omega_1)(1 - \omega_2) = 0$.

This is an aperiodic Markov chain with asymptotic values of the probability of occupancy of the states S_i, u_i, $i \in (1,2,3,4)$, which are calculated from the relations

$$u_i = \sum_{j=1}^{4} u_j P_{ji} \quad \text{and} \quad \sum_{i=1}^{4} u_i = 1 \ . \tag{5.2.12}$$

It turns out that

$$u_1 = \frac{1}{\Sigma} , \qquad u_2 = \frac{x_1(1 - x_2)}{\Sigma} , \tag{5.2.13}$$

$$u_3 = \frac{x_2(1 - x_1)}{\Sigma} , \qquad \text{and} \qquad u_4 = \frac{1 - x_1 x_2}{\Sigma} ,$$

where

$$\Sigma = 2 + x_1 + x_2 - 3x_1 x_2 \ . \tag{5.2.14}$$

The payoffs of the two agents are

$$G_1(x_1, x_2) = u_1 - u_4 + \xi(u_3 - u_2) \qquad \text{and}$$

$$G_2(x_1, x_2) = u_1 - u_4 + \xi(u_2 - u_3) , \tag{5.2.15}$$

respectively.

The Markovian kinetics do not evolve, however, under fixed propensities, i.e. fixed transition probabilities. Since the games are played in iteration, *learning* takes place, which means that the time evolution of the propensities is governed by a system of first-order nonlinear differential equations: the time derivatives of the propensities are proportional to the gradient of the expected payoff with respect to that propensity. Thus the time evolution of the main propensities $x_1(t)$, $x_2(t)$ is determined from the coupled nonlinear differential equations

$$\frac{dx_1}{dt} = \frac{\partial G_1}{\partial x_1} \quad , \qquad \frac{dx_2}{dt} = \frac{\partial G_2}{\partial x_2} \tag{5.2.16}$$

which describe a transitory "learning" process.

The steady-state values of the propensities x_1^*, x_2^* coming out from the system, that is,

$$\frac{dx_1^*}{dt} = \frac{(3\xi + 1)x_2^{*2} + 2(1 - \xi)x_2^* - 2\xi}{(2 + x_1^* + x_2^* - 3x_1^*x_2^*)^2} = 0$$

$$\frac{dx_2^*}{dt} = \frac{(3\xi + 1)x_1^{*2} + 2(1 - \xi)x_1^* - 2\xi}{(2 + x_1^* + x_2^* - 3x_1^*x_2^*)^2} = 0 \quad , \tag{5.2.17}$$

are equal to

$$x_1^* = x_2^* = x^* = \left(\xi - 1 + \sqrt{1 + 7\xi^2}\right)/(3\xi + 1) \quad ;$$

they exist only for $1 < \xi \leqslant 3$.

The value x^* (see Fig.5.12) represents a "threshold" above which "locking-in" at the (CC) regime is ultimately reached; it (x^*) is an unstable steady-state (a saddle point), as can be easily deduced by investigating the conditions of stability, which are

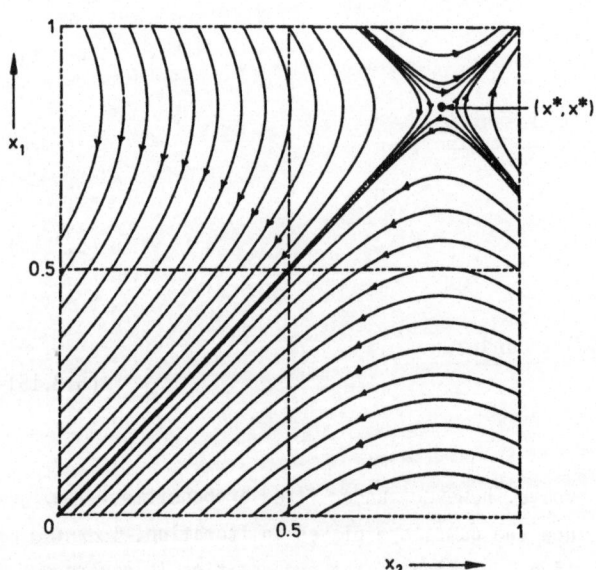

Fig.5.12. Phase-space description of the prisoner's dilemma ($\xi = 1.5$). The point (x^*,x^*) is a saddle point, an unstable steady state

$$\frac{\partial^2 G_1}{\partial x_1^2} + \frac{\partial^2 G_2}{\partial x_2^2} < 0 \qquad\qquad (5.2.18)$$

and

$$\frac{\partial^2 G_1}{\partial x_1^2} \frac{\partial^2 G_1}{\partial x_2^2} > \frac{\partial^2 G_1}{\partial x_1 \partial x_2} \frac{\partial^2 G_2}{\partial x_1 \partial x_2} \quad , \qquad\qquad (5.2.19)$$

for $x_1^* = x_2^* = x^*$.

It turns out that here

$$\left(\frac{\partial^2 G_1}{\partial x_1^2}\right)_{x^*} = 0 \quad\text{and}\quad \left(\frac{\partial^2 G_2}{\partial x_2^2}\right)_{x^*} = 0 \quad ,$$

and also

$$\left(\frac{\partial^2 G_1}{\partial x_1 \partial x_2}\right)_{x^*} = \frac{2\sqrt{1 + 7\xi^2}}{\Sigma^2} > 0 \quad ,$$

$$\left(\frac{\partial^2 G_2}{\partial x_1 \partial x_2}\right)_{x^*} = \frac{2\sqrt{1 + 7\xi^2}}{\Sigma^2} > 0 \quad .$$

We next turn to a modification of the prisoner's dilemma (PD); we present a generalized case where the probability of retaliation of the "betrayed" partner is now a, where $0 \leqslant a < 1$, i.e. $a \neq 1$. We have

$$y_1 = z_1 = y_2 = z_2 = a \qquad\text{and}$$
$$\omega_1 = \omega_2 = 1 \quad .$$

For the new Markov chain we will have:

$$P_{11} = x_1 x_2 \quad , \qquad\qquad P_{21} = a^2 \quad ,$$
$$P_{12} = x_1(1 - x_2) \qquad\qquad P_{22} = a(1 - a) \quad ,$$
$$P_{13} = x_2(1 - x_1) \quad , \qquad\qquad P_{23} = a(1 - a) \quad ,$$
$$P_{14} = (1 - x_1)(1 - x_2) \quad , \qquad\qquad P_{24} = (1 - a)^2 \quad ,$$

$$P_{31} = a^2 \quad , \qquad\qquad P_{41} = 1 \quad ,$$
$$P_{32} = a(1 - a) \quad , \qquad\qquad P_{42} = 0 \quad ,$$
$$P_{33} = a(1 - a) \quad , \qquad\qquad P_{43} = 0 \quad ,$$
$$P_{34} = (1 - a)^2 \quad , \qquad\qquad P_{44} = 0 \quad .$$

$$(5.2.20)$$

The asymptotic values of the probabilities of occupancy of the available states S_i are deduced as

$$u_1 = \frac{2a - 2a^2 - 1}{\Sigma_0} \quad ,$$

$$u_2 = \frac{(a - a^2 - 1)x_1 - a(1 - a)x_2 + x_1 x_2}{\Sigma_0} \quad,$$

(5.2.21)

$$u_3 = \frac{-a(1 - a)x_1 + (a - a^2 - 1)x_2 + x_1 x_2}{\Sigma_0} \quad,$$

$$u_4 = \frac{a^2 x_1 + a^2 x_2 + (1 - 2a)x_1 x_2 + 2a - 2a^2 - 1}{\Sigma_0} \quad,$$

where

$$\Sigma_0 = (3 - 2a)x_1 x_2 + (a^2 - 1)(x_1 + x_2) - 2(2a^2 - 2a + 1) \quad .$$

(5.2.22)

[By putting $a = 0$, we recover the previous expressions for PD (5.2.13).] The steady states of x_1, x_2 follow as

$$x_1^* = x_2^* = x^*$$

$$= \frac{-(a^2 - 1)\xi - (1 - 2a) + \sqrt{(a^4 - 8a^3 + 18a^2 - 16a + 7)\xi^2 + (2a^2 - 4a + 1)}}{(3 - 2a)\xi + 1} \quad .$$

(5.2.23)

This steady state is again unstable since

$$\left(\frac{\partial^2 G_1}{\partial x_1^2}\right)_{x^*} = \left(\frac{\partial^2 G_2}{\partial x_2^2}\right)_{x^*} = 0$$

(5.2.24)

and

$$\left(\frac{\partial^2 G_1}{\partial x_1 \partial x_2}\right)_{x^*} = \left(\frac{\partial^2 G_2}{\partial x_1 \partial x_2}\right)_{x^*}$$

$$= \frac{\sqrt{(a^4 - 8a^3 + 18a^2 - 16a + 7)\xi^2 + (2a^2 - 4a + 1)}}{\Sigma_0^2} > 0 \quad .$$

(5.2.25)

Again,

$$G_1(x_1, x_2) = u_1 - u_4 + \xi(u_3 - u_2) \quad,$$

$$G_2(x_1, x_2) = u_1 - u_4 + \xi(u_2 - u_3) \quad .$$

(5.2.15)

In what follows we will consider the case $a = 1/2$; moreover, we will modify the normalized matrix of the game by putting both payoffs in the state S_4 as -2ξ, see Fig.5.13. Thus the states DC and CD (S_2, S_3) become local equilibria since the S_4 (DD) state is too expensive to afford [the "chicken" game (G)]. After some trivial algebra along the same lines as above concerning the new expression for G_1 and G_2,

$$G_1 = u_1 - 2\xi u_4 + \xi(u_3 - u_2) \quad,$$

$$G_2 = u_1 - 2\xi u_4 + \xi(u_2 - u_3) \quad,$$

(5.2.26)

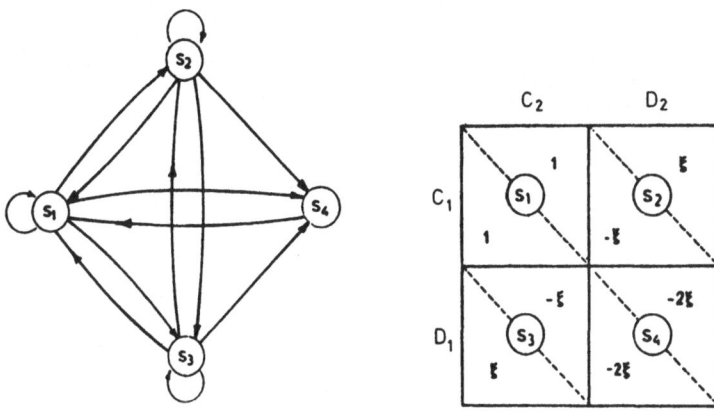

Fig.5.13. The game of chicken and the resulting Markov chain (left)

the system of nonlinear differential equations now takes the form

$$\frac{dx_1}{dt} = \frac{32\xi x_2^2 + 4(4 - 11\xi)x_2 + 12\xi - 6}{[4 - 8x_1x_2 + 3(x_1 + x_2)]^2} \quad ,$$

$$\frac{dx_2}{dt} = \frac{32\xi x_1^2 + 4(4 - 11\xi)x_1 + 12\xi - 6}{[4 - 8x_1x_2 + 3(x_1 + x_2)]^2} \quad .$$

(5.2.27)

We get for the steady state of the propensities x_1, x_2 the expressions

$$x = \frac{(\frac{11}{4}\xi - 1) \pm \sqrt{(1 - \frac{5}{4}\xi)^2}}{4\xi} = \begin{cases} 3/8 \\ \frac{2\xi-1}{2\xi} \end{cases} \quad ,$$

(5.2.28)

which turn out again to be unstable or marginally stable. More specifically, the states $(3/8, 3/8)$ and $(1 - 1/2\xi, 1 - 1/2\xi)$ are saddle points, and the states $(3/8, 1 - 1/2\xi)$, $(1 - 1/2\xi, 3/8)$ are centers as shown in Figs.5.14-16.

In conclusion we note that the continuous nonlinear dynamics governing the time evolution of the propensities $x_1(t)$ and $x_2(t)$ give rise (after a transient time) to discontinuous or *switching* phenomena both for the PD $(1 < \xi \leqslant 3)$ and CG $(1 < \xi < \infty)$. More specifically, in the case of PD, for $\xi > 3$, the temptation to unilateral defection is so great that no "threshold of trust" ($x^* \leqslant 1$ or $u^* \leqslant 1$) —above which stable cooperation takes place— can be established: the system goes towards noncooperation (DD) whatever the initial values of x_1, x_2. The probability of "locking-in" (cooperation) in (CC), u, is related to the threshold probability x^*, beyond which bifurcation takes place, by $u = 2(1 - x^*)^2$ (see Fig.5.12). Its maximum value (for $\xi = 1$) is ~17.2%. In the case of CG, the corresponding probability of cooperation equals $1 - \pi ab$ for $\xi < 2.4$, where a, b are the axes

$$a = \sqrt{\frac{3}{2}} \left(\frac{5}{8} - \frac{1}{2\xi}\right) , \qquad b = \frac{1}{\sqrt{2}} \left(\frac{5}{8} - \frac{1}{2\xi}\right) ,$$

235

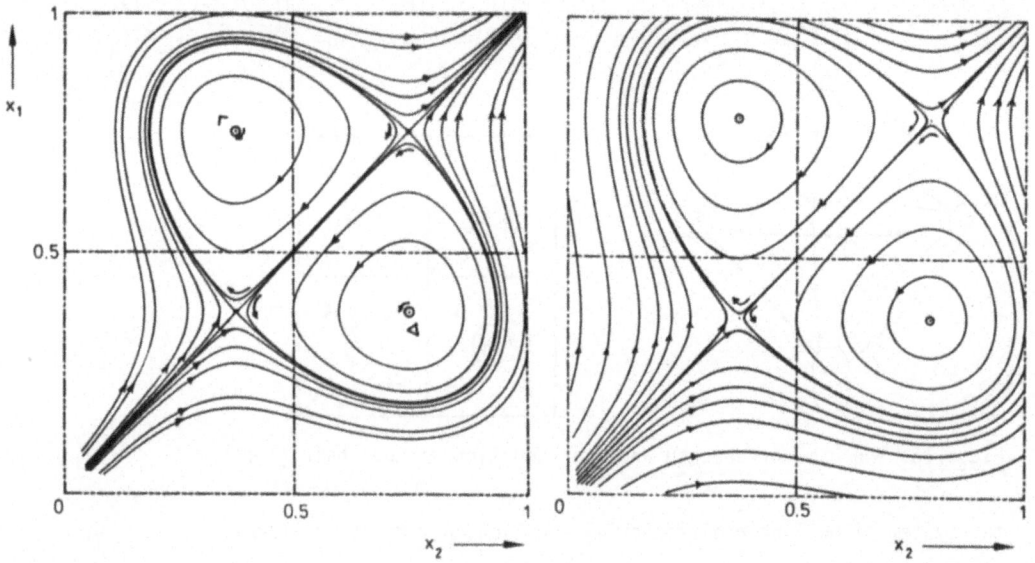

Fig.5.14   Fig.5.15

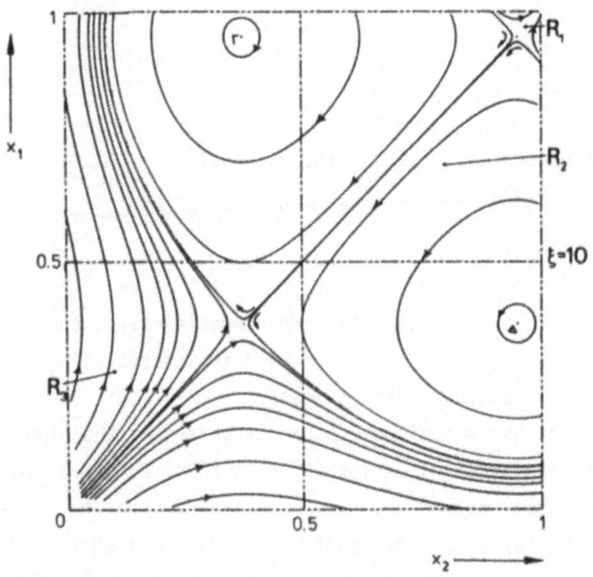

Fig.5.14. State-space description of the game of chicken
($\xi = 2$). Δ and Γ label the two
centers

Fig.5.15. State-space description of the game of chicken
($\xi = 2.4$)

◄ Fig.5.16. State-space description of the game of chicken
($\xi = 10$). Δ and Γ label the two
centers. R_1, R_2, and R_3 show
three distinct areas in the
state space

of the separatrix ellipse in Fig.5.14. For $\xi \geqslant 2.4$ (Figs.5.15,16) the area of the
state space is divided into three regions R_1, R_2, and R_3. The probability for
locking-in on the (CC) state rapidly goes to zero as $\xi \to \infty$ if one assumes "absorbing" boundaries, since now trajectories starting from the "southwest" region may
stop after hitting the boundaries $x_1 = 0$ and $x_2 = 0$, or $x_2 = 0$ and $x_1 = 1$. If, however,
one assumes "gliding" boundaries, so that trajectories starting from the "southwest"
region end up eventually in the (CC) corner (Fig.5.16), then as ξ increases beyond

2.4, the probability of cooperation *increases* and tends to 1 as $\xi \to \infty$. Of course, this outcome agrees more with real cases of high stakes in blackmail conflicts.

For the *asymmetrical* game of chicken the separatrix disappears, and the probability of locking-in at (CC) is always near 1 [5.2].

5.3 Competing Species

Let us start by considering a single population evolving with time, $x(t)$, in such a way that obvious competition is not observed among individual members of this species for the (limited) available resources and that competition does not take place between $x(t)$ and the species which $x(t)$ preys on. Under such circumstances Malthus's law should apply, predicting that the rate at which the population will increase will be proportional to the existing population, that is $dx/dt = \alpha x$ or $x(t) = x_0 e^{\alpha t}$; here $\alpha > 0$ is the so-called birthrate constant.

Accordingly, the population will increase exponentially with time. However, this trend will not go on for long: overcrowding will inevitably initiate dyadic interactions detrimental to the population explosion —either in terms of competition for available space and food, or in terms of an epidemic spreading (Sect.5.8). Last but not least, members of the population die, of course. If β stands for the average death rate (caused by all three above effects) we will end up with the dynamical relation

$$dx/dt = \alpha x - \beta x^2 \quad , \tag{5.3.1}$$

known as the logistic equation.

This equation can be solved analytically; for pedagogical reasons we provide below the solution.

Separating the variables in (5.3.1) we get

$$\int \frac{1}{x(\alpha - \beta x)} \, dx = \int dt \quad . \tag{5.3.2}$$

Since

$$\frac{1}{x(\alpha - \beta x)} = \frac{1}{\alpha x} + \frac{\beta}{\alpha(\alpha - \beta x)} \quad ,$$

$$\frac{1}{\alpha} \int \frac{dx}{x} + \frac{\beta}{\alpha} \int \frac{1}{\alpha - \beta x} \, dx = \int dt$$

and carrying out the integration we obtain

$$\frac{1}{\alpha} [\ln x - \ln(\alpha - \beta x)] = \frac{1}{\alpha} \ln \frac{x}{\alpha - \beta x} = t + c \quad , \tag{5.3.3}$$

where c is a constant of integration.

If the initial population at time $t = 0$ is $x(0)$, we find

$$c = \frac{1}{\alpha} \ln \left[\frac{x(0)}{\alpha - \beta x(0)} \right] ;$$

(5.3.4)

thus the solution becomes

$$\ln \frac{x(t)}{\alpha - \beta x(t)} - \ln \frac{x(0)}{\alpha - \beta x(0)} = \alpha t \qquad \text{or}$$

(5.3.5)

$$\ln \frac{x(t)[\alpha - \beta x(0)]}{x(0)[\alpha - \beta x(t)]} = \alpha t .$$

(5.3.6)

Taking the exponents of both sides we finally get

$$x(t) = \frac{x(0)\alpha e^{\alpha t}}{\alpha - \beta x(0) + \beta x(0)e^{\alpha t}} .$$

(5.3.7)

The equilibrium population will come from (5.3.7) if we take the limit of $x(t)$ as $t \to \infty$, namely (Fig.5.17)

$$\lim_{t \to \infty} x(t) = \frac{\alpha}{\beta} .$$

(5.3.8)

The next logical step in our analysis is to consider the coupling between *two* populations $x_1(t)$ and $x_2(t)$. Competition between the two populations above implies that members of each species exercise an *inhibiting* effect on the reproduction of members of the other. (In the banal case of an *arms race*, the coupling between the constestants is mutually excitatory — leading to unstable equilibria.)

This leads to the following system of equations for the growth rates — which system is clearly a two-dimensional extension of the single-species logistic equation (5.3.1):

$$\frac{dx_1}{dt} = x_1(\alpha_1 - \alpha_{11}x_1 - \alpha_{12}x_2) ,$$

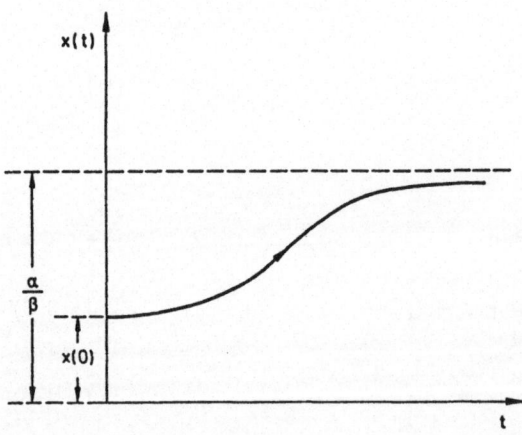

Fig.5.17. The solution of the logistic equation

$$\frac{dx_2}{dt} = x_2(\alpha_2 - \alpha_{21}x_1 - \alpha_{22}x_2) \quad , \tag{5.3.9}$$

where now α_1, α_2 are the individual birth rates; α_{11}, α_{22} are the individual death rates; and α_{12}, α_{21} are the *cross-coupling* coefficients standing for the rates of inhibition of species I by species II, and species II by species I, respectively. We will assume according to the description above that *all* coefficients α_i and α_{ij} are positive.

Let us find the steady states in the system (5.3.9) and investigate the criteria for their stability. These are deduced from the system of equations:

$$x_1(\alpha_1 - \alpha_{11}x_1 - \alpha_{12}x_2) = 0 \quad ,$$
$$x_2(\alpha_2 - \alpha_{21}x_1 - \alpha_{22}x_2) = 0 \quad . \tag{5.3.10}$$

Thus we have two steady states $x_1 = 0$, $x_2 = \alpha_2/\alpha_{22}$ and $x_1 = \alpha_1/\alpha_{11}$, $x_2 = 0$ (which correspond to one species becoming *extinct* and the other species then reaching its equilibrium population when it is alone in the environment) *as well as* a *third* equilibrium point, which is more interesting since it may correspond to a stable coexistence of the two species, namely the point of intersection of the straight lines

$$\alpha_{11}x_1 + \alpha_{12}x_2 = \alpha_1$$
$$\alpha_{21}x_1 + \alpha_{22}x_2 = \alpha_2 \quad , \tag{5.3.11}$$

shown in Fig.5.18.

The coordinates of this point, provided that $\alpha_{11}\alpha_{22} \neq \alpha_{21}\alpha_{12}$, are

$$x_1^* = \frac{\alpha_{22}\alpha_1 - \alpha_{12}\alpha_2}{\alpha_{11}\alpha_{22} - \alpha_{21}\alpha_{12}} \quad \text{and}$$

$$x_2^* = \frac{\alpha_{11}\alpha_2 - \alpha_{21}\alpha_1}{\alpha_{11}\alpha_{22} - \alpha_{21}\alpha_{12}} \quad .$$

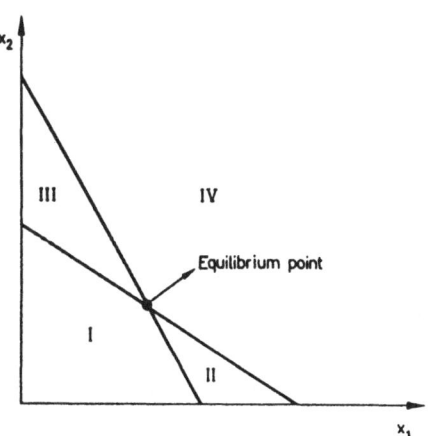

Fig.5.18. Lines of equilibria for two competing species

If these numbers are positive, they represent (if rounded to the nearest integer) populations of the two competing species than can coexist —and remain constant.

In the x_1x_2-plane, the equations (5.3.11) represent straight lines that divide the plane into four distinct regions: specifically, in region (I), x_1 increases and x_2 increases; in region (II), x_1 increases and x_2 decreases; in region (III), x_1 decreases and x_2 increases; and in region (IV), x_1 decreases and x_2 decreases. Under these circumstances the point (x_1^*, x_2^*) *is a stable equilibrium point*, if $\alpha_{11}\alpha_2 > \alpha_{21}\alpha_1$ and $\alpha_{22}\alpha_1 > \alpha_{12}\alpha_2$. If $\alpha_{11}\alpha_2 < \alpha_{21}\alpha_1$ and $\alpha_{22}\alpha_1 < \alpha_{12}\alpha_2$, the equilibrium is unstable. The "winner" in the competition depends on the initial densities of the two species.

5.4 Survival and Extinction

If populations of two species occur together in the same geographical region (the same ecological "niche") and have the same ecological requirements for resources, the hypothesis of natural selection leads us to expect that the "better-adapted" species will completely displace the other one. This is known as the principle of competitive exclusion. Briefly it means that complete competitors cannot coexist.

To model the process of extinction, let us consider the case of two competing species with the same ecological requirements coexisting at some time in an environment that can support exactly N individuals of both species. Suppose that initially there are K individuals of species I and N-K individuals of species II in the environment. Assume that the two species compete by means of a succession of encounters which are such that, on each encounter, the probability that species I increases by one individual is p, and the probability that species II increases by one individual is $q = 1 - p$. The two species are equally well adapted if $p = q = 1/2$, and species I has a selective advantage if $p > q$. We also assume that p is independent of the populations N and N - K of the two species.

Let us define p_K to be the probability that the first species will displace the second species if its initial population is K. (If the initial population is 0 then $p_0 = 0$, i.e., the first species is already extinct. If the initial population is N, then $p_N = 1$ since the first species has already displaced the second one.) After the first encounter, the population of species I is either K +1 or K -1, with probabilities p and q respectively.

The probability p_K is therefore the sum of two terms, namely

$$p_K = p p_{K+1} + q p_{K-1} \ . \tag{5.4.1}$$

Here $p p_{K+1}$ is the probability that the first species has population K +1 after one move, and then displaces the second species. Also $q p_{K-1}$ is the probability that the first species has population K -1 after one move, and then displaces the second species.

Let us look for a solution of (5.4.1) of the form $p_K = \lambda^K$. We have

$$\lambda^K = p\lambda^{K+1} + q\lambda^{K-1} \qquad \text{or} \tag{5.4.2}$$

$$p\lambda^2 - \lambda + q = 0 \qquad \text{or}$$

$$\lambda = \frac{1 \pm \sqrt{1 - 4pq}}{2p} = \frac{1 \pm \sqrt{1 - 4p(1-p)}}{2p} = \frac{1 \pm (1 - 2p)}{2p} \quad ,$$

or

$$\lambda_1 = 1 \qquad \text{and} \qquad \lambda_2 = \frac{1 - p}{p} = \frac{q}{p} \quad . \tag{5.4.3}$$

The two roots above are equal for $p = q = 1/2$. The general solution of (5.4.1) equals therefore

$$p_K = c_1 + c_2 \left(\frac{q}{p}\right)^K \quad , \qquad \text{if} \quad p \neq 1/2 \quad ,$$

and

$$p_K = c_1 + c_2 K \quad , \qquad \text{if} \quad p = 1/2 \quad , \tag{5.4.4}$$

where c_1, c_2 are constants to be determined. We have seen from the definition of p_K that $p_0 = 0$ and $p_N = 1$. Therefore, if $p \neq 1/2$, $c_1 + c_2 = 0$ and $c_1 + c_2(q/p)^N = 1$, which implies

$$c_1 = \frac{1}{1 - (q/p)^N} \quad , \qquad c_2 = -\frac{1}{1 - (q/p)^N} \quad . \tag{5.4.5}$$

If $p = 1/2$, then $c_1 = 0$, and $c_2 N = 1$ or $c_2 = 1/N$. So the solutions of (5.4.1) will be

$$p_K = \frac{1 - (q/p)^K}{1 - (q/p)^N} \quad , \qquad \text{if} \quad p \neq 1/2$$

and

$$p_K = \frac{K}{N} \quad , \qquad \text{if} \quad p = 1/2 \quad . \tag{5.4.6}$$

Thus we have been able to calculate explicitly the probability that one species will displace the other.

Let us consider as examples various possibilities for a total population $N = 1000$. If the initial populations of the two species are both 500 and if $p = q = 1/2$, then the probability that the first species will displace the second is $p_{500} = 500/1000 = 1/2$; neither species has a competitive advantage in this case. However, if $p = 2/3$ and $q = 1/3$, then

$$p_{500} = \frac{1 - (1/2)^{500}}{1 - (1/2)^{1000}} = \frac{1}{1 + (1/2)^{500}} \sim 1 \quad .$$

In this case, therefore, it is overwhelmingly probable that species II will become extinct. Even small competitive advantages (p *slightly* greater than $1/2$) with the above initial populations lead to the same result. Note that if $p = 2/3$ and $q = 1/3$, and the initial populations for species I and II are 1 and 999 respectively, then the probability that the first species displaces the second species is

$$p_1 = \frac{1 - (1/2)^1}{1 - (1/2)^{1000}} \sim \frac{1}{2} .$$

This implies, for example, that if one individual of the better-adapted species invades the territory of the second species, the probability is 1/2 that the invader will completely displace the second species; if $p = q = 1/2$, then $p_1 = 1/1000$. In other words, if the two species are equally well adapted, the probability is very small that the introduction of one individual of one species will lead to the extinction of the other species.

It is also easy to prove that if, say, the first species has a selective advantage (p q), the probability that the *first species* becomes extinct is

$$1 - p_K = \frac{(p/q)^{N-K} - 1}{(p/q)^N - 1} , \tag{5.4.7}$$

where K is the initial population of the first species and N is the total number population. Therefore if N is small, the probability of extinction of the better-adapted species is *not* negligible.

5.5 Some Elementary Knowledge from Genetics: Selection and Fitness

The conflicts between populations studied so far are conflicts between populations of *phenotypes*, i.e., of growing, individual organisms. The concept of *fitness* and *adaptation* has recurred in our treatment on a purely heuristic basis. How can we trace the origins of such an adaptive tendency to the genotype, i.e., to the hereditary "arsenal" of the organism(s) involved? The issue is of some importance since we try, in the last analysis, to understand the whole domain of dynamical interaction between hierarchical systems in terms of mutual simulation processes —which presumably start from the *lowest* biological level —namely the genetic level.

The modern theory of the inheritance of "character", i.e., of capacity for certain modes of behavior at the phenotypic level(s), is couched in terms of the Mendelian principles governing the transmission of genes from parents to offspring. A particular gene may occur in several forms or *alleles*. Let us consider the simplest case of a gene with two alleles: A and α. The genes occur in every cell of an organism and are grouped together on the chromosomes. Except in the reproductive cells, the genes occur in *pairs* and appear on paired chromosomes.

The three possible pairs of a gene AA, Aα, and αα determine the three possible genotypes of the organism relative to this gene. The genotypes AA and αα are called *homozygous* or pure, and the genotype Aα is called *heterozygous* or hybrid.

The reproductive cells (that is the sperm and the egg) have unpaired chromosomes and therefore only one copy of each gene. The genes of the offspring result, therefore, from the pairing of genes of two reproductive cells, one from each parent. If *both* parents are homozygous, the genotype of the offspring is determined. For example, if one parent is genotype AA and the other is genotype αα, the offspring *must* be genotype Aα. On the other hand, if one or both parents are heterozygous, the genotype of the offspring is *not* determined. For example, if both parents are heterozygous, the offspring may be AA, Aα, or αα with probabilities 1/4, 1/2, and 1/4 respectively —assuming equal viability of the offspring. (Many characteristics, such as albinism in human beings, are controlled by a single gene. Other characteristics like height are controlled by the cooperative effects of a very large number of genes and are often strongly influenced by environmental factors.)

One of the two alleles, say A, of a particular gene is said to be *dominant* if the genotypes AA and Aα are indistinguishable from each other. In this case, the allele α is then said to be *recessive* if the genotype αα is observably different from the genotypes AA and Aα. If it were possible to classify the individuals of a population of a given species according to the genotypes AA, Aα, and αα, then we could determine the proportions of the two alleles in the population. (This would not be possible if, for example, AA and Aα were indistinguishable.) Let x, y, z be the proportions of the three genotypes in the population, and let us assume that these proportions can be evaluated. Then the proportions p and q of the two alleles A and α in the population will be $p = x + (y/2)$ and $q = 1 - p = (y/2) + z$. Here we have used the fact that the A allele comprises 100% of the AA genotype (with proportion x) and 50% of the Aα genotype, and similarly for the α allele. (The second equation above comes straight from the first, since $x + y + z = 1$.)

In practice we are often interested in the inverse problem, namely in determining the proportion of the genotypes when the proportions of alleles are known. In general this problem has not a unique solution because the equation $p = x + (y/2)$ has two unknowns x, y. To provide one more independent equation, we can sometimes make the assumption of *random mating* —meaning that the probability of a given individual mating with another (of the same species) is not dependent on the genotype of the other individual. (In numerous cases this assumption is violated.)

Suppose that in a *large* parent population the alleles A and α of a particular gene are present in proportions p and $q = 1 - p$. Assuming that these proportions are the same for males and females, and assuming that the mating is random, let us calculate the proportions according to which the first and subsequent generations of offspring will be composed of the three genotypes AA, Aα, and αα.

As we have seen above, an individual of the first generation is of genotype AA if both parents contribute to the allele A. Since the probability of an allele A coming from either parent is p, the probability of genotype AA in an offspring is

p^2. Similarly, the probability of the genotype $\alpha\alpha$ is q^2. The probability of the geno-type $A\alpha$ is $2pq$, since the $A\alpha$ and αA individuals are identical genotypes. Therefore, the proportions p_1 and q_1 of the alleles A and α in the first generation will be

$$p_1 = p^2 + \frac{1}{2}(2pq) = p(p + q) = p \quad \text{and}$$

$$q_1 = \frac{1}{2}(2pq) + q^2 = q(p + q) = q \quad .$$

Thus under the assumptions of random mating and the equal viability of all three genotypes, the proportions of the two alleles are *invariant* with the generation of the offspring. We conclude therefore that after the initial generation, the propor-tions of the three genotypes AA, Aα, and $\alpha\alpha$ remain constant, i.e., p^2, $2pq$, and q^2.

The discussion in Sect.5.4 about the survival and extinction of *two* competing species did *not* take into account the development of the competing populations in successive generations. Also in that section we did not relate the probabilities p and q of success and failure on each encounter to genetic *or* environmental vari-ables. In some cases, however, it is reasonable to assume that p depends mainly on environmental conditions. For example, it is known from field studies that tempera-ture and humidity are key factors in determining the outcome of competition between two species of beetle. In other cases, the outcome of competition between two popu-lations is determined mainly by genetic factors.

In that respect, it is of interest to examine conditions under which, in a single species population, *competition between alleles* takes place, namely one allele of a particular gene may *displace* another allele (thereby conferring a selective advan-tage). Consider an environment that supports n reproducing individuals in each suc-cessive generation. A particular gene with two alleles A and α has 2n representatives in each generation. Suppose that in the mth generation the allele A occurs λ times and the allele α occurs $2n - \lambda$ times. Let us determine the probability that in the next generation, the allele A occurs μ times and the allele α occurs $2n - \mu$ times, for $\mu = 0,1,2,\ldots,2n$. In the context of Markov chains, the system has $2n + 1$ states corresponding to the population having $0,1,2,\ldots,2n$ copies of the allele A. The evolution of the Markov chain via successive trials corresponds to the successive generations of the population. To determine this Markov chain, we must derive the transition probabilities p_{ij} from the ith state to the jth state (or, in our nota-tion, $p_{\lambda\mu}$). The states u_0 and u_{2n} are *absorbing states* in the sense that the sys-tem may go there but cannot leave (this means $p_{0j} = 0$ and $p_{2nj} = 0$ for *every* $j \neq 0,2n$). These absorbing states u_0, u_{2n} correspond to populations of *all* $\alpha\alpha$ and AA indivi-duals respectively.

In the absence of any indication to the contrary, we may again assume that mat-ing is random and that *none* of the three genotypes AA, Aα, $\alpha\alpha$ has a selective advan-tage which gives an increased probability of successful reproduction. With these assumptions, the population in the $(m + 1)^{th}$ generation is determined by 2n repeated trials of a binomial experiment with probability $1 - (\lambda/2n)$ of allele α occurring in a trial. The probability of μ "successes" (of μ copies of allele A in the next generation) is given by the binomial probability

$$P_{\lambda\mu} = \binom{2n}{\mu}\left(\frac{\lambda}{2n}\right)^{\mu}\left(1 - \frac{\lambda}{2n}\right)^{2n-\mu} . \tag{5.5.1}$$

The set of $P_{\lambda\mu}$ transition probability elements defines the Markov chain. The probabilities that allele A or allele α will disappear in the next generation are

$$P_{\lambda,0} = \left(1 - \frac{\lambda}{2n}\right)^{2n} \qquad \text{and} \tag{5.5.2}$$

$$P_{\lambda,2n} = \left(\frac{\lambda}{2n}\right)^{2n} . \tag{5.5.3}$$

respectively.

So far we have assumed that *neither* allele confers a selective advantage which *increases* the probability that it will be represented in the next generation. It is, however, an observational fact that there is a finite probability that the gene proportions may change from one generation to the next. This effect is known as "sampling error" or "random drift". In rather small populations this random fluctuation of the gene proportions may result in the disappearance of one of the alleles from the genetic pool.

To model a selective advantage for one of the alleles we have to modify the formula for the binomial probability $\lambda/2n$ so that allele A is represented in one of the 2n trials. Suppose again that the population in the $(m+1)^{\text{th}}$ generation is determined by 2n repeated trials of a binomial experiment. Suppose also that the probabilities of alleles A and α occurring in a particular trial are $(\lambda/2n)^{\xi}$ and $1 - (\lambda/2n)^{\xi}$, where ξ is a positive number. Then the probability that there are μ copies of allele A in the next generation is

$$P_{\lambda\mu} = \binom{2n}{\mu}\left(\frac{\lambda}{2n}\right)^{\xi\mu}\left[1 - \left(\frac{\lambda}{2n}\right)^{\xi}\right]^{2n-\mu} , \tag{5.5.4}$$

when there were λ copies of A in the previous generation. Therefore, in this more general model $(\xi \neq 1)$, the probabilities that allele A or allele α will disappear in the next generation are

$$P_{\lambda,0} = \left[1 - \left(\frac{\lambda}{2n}\right)^{\xi}\right]^{2n} \qquad \text{and} \tag{5.5.5}$$

$$P_{\lambda,2n} = \left(\frac{\lambda}{2n}\right)^{2n\xi} , \tag{5.5.6}$$

respectively. If $\xi > 1$, the allele α has a selective advantage, and if $\xi < 1$, the allele A has a selective advantage. The parameter ξ is an example of a "fitness index". In some cases ξ depends on environmental factors such as the temperature.

In summing up Sects.5.4,5, we can say that the selective advantage, either genetically or environmentally dependent, leads to an outcome of a "game" where the partners may *not* necessarily be "personally" involved in a dyadic "strife". Nature automatically, so to speak settles the rule and decides the outcome in absentia of the specific *phenotypes* (individuals). The following example (from [5.3]) illustrates better this assertion, but also casts serious doubt on the classical claim that sexual recombination in the long run eliminates polymorphism, i.e., allele variety. We consider the conflict between a *host* organism and a *parasite* (pathogen).

A species interacts not only with its physical environment and its competitors, but also with pathogens and parasites. This last host-pathogen interaction is different from ordinary predator-prey dynamics studied so far, due to the *disparity* in genome size, generation lifetime, and speed adaptability between the host and the pathogen. Pathogens *would win the contest* (through faster evolution) and leave their hosts defenseless were it not for the counterstrategies of the host. As *Bremermann* proposes [5.3] the basic feature of the counterstrategies of the host is *variety*. It is asserted that sexual recombination, for example, enhances variety by creating ever-novel genotypes through random assortment of alleles. Thus sex *and* immunological polymorphism are "logistic necessities" of the defensive strategies that are open to host species with larger genomes and longer generation time. We would now like to show, following *Bremermann* [5.3], that there is indeed a basic *disparity* between host and micropathogens; the latter, as we said already, can evolve and adapt *faster* to any changes in their hosts. Micropathogens can evolve faster than multicellular hosts by orders of magnitude. As a consequence, micropathogens could break through host defenses if the host species were genetically homogeneous (like a clone). In short, using our "systemic" language, we can say that the micropathogens can in principle "break the code" of the opponent *first*, thereby *simulating* their host's dynamics and *controlling* it.

(We have seen above that typical mathematical genetic models predict that polymorphism should be rare since selection quickly eliminates alleles of inferior selective value. However, in real life, fortunately, this is not always the case — for reasons which are not yet fully understood. For more details, see [5.3].)

If the host ("defender") can indeed, via sexual recombination, evolve faster than the pathogen ("invader"), then it might stay ahead of the invader, so to speak, by mere selective pressure. That is to say, as soon as the invader virus has "camouflaged" itself (by incorporating into its "coat" a portion of the surface membrane of the cell it is attacking), the defender would modify its "self"-defining antigens. If, however, the *invader* evolves faster, then the defender cannot evade it — he is facing, as we are going to see in the next chapter, a "transcomputational" problem.

We will show now, following Bremermann, that micropathogens *can* evolve orders of magnitude faster than their (vertebrate) hosts. Evolution processes do indeed occur on *vastly different time scales*. For example, the selective proliferation of lymphocytes in response to an antigen occurs in a matter of few days, while by contrast, the evolution of mammals is estimated to have taken place during $\sim 10^8$ years. Yet bacteria resistant to antibiotics appeared soon after penicillin was first used (as did flies resistant to the notorious DDT shortly after its introduction).

Consider now the genome of a bacterium or a virus concerned with evolving polypeptides that facilitate its coexistence with the host or its multiplication at the expense of the host. For example, consider the evolution of antigenic polypeptides that would camouflage the pathogen with respect to the host's immune system. (Such imitation on the part of the virus leading to camouflage amounts, in the last analysis, to the ability of the pathogen to compress the code of the host and to simulate its evolution.) It is plausible to consider that the process starts with an amino acid sequence that is *random* with respect to the goal (final) sequence. The amino acid sequence is coded by a nucleotide sequence (Sect.4.5.9). There are *four* nucleotides for each nucleotide locus; each nucleotide locus can be represented by *two bits*. We will assume that transitions (due to mutations) between different binary digits are equiprobable and that the probability is the same for all digits; let it be p.

Similarly the "ideal genotype" of the pathogen is defined by the requirement that its nucleotide sequence be *compatible* with the host's enzymes. Suppose the number of nucleotides in the pathogen's genome is n/2 —which corresponds to n bits. A number, say m, of these nucleotides have to mutate in order to achieve the "ideal genotype". For bacteria, n is of the order of 10^8 and for viruses, of the order of 10^4. In a random polypeptide chain, as many as half of the corresponding digits are expected to be initially "wrong", i.e., $\sim 10^3$.

The probability of at least one improvement among the m improper bits is $1 - (1 - p)^m$. On the other hand, the probability of (nondegenerative) change among the n -m right bits is $(1 - p)^{n-m}$, so the combined probability of at least one favorable mutation and no detrimental mutation will be the product of the above two, namely,

$$F = (1 - p)^{n-m}[1 - (1 - p)^m] = (1 - p)^{n-m} - (1 - p)^n \; . \tag{5.5.7}$$

By differentiating F with respect to p and equating with zero, we obtain the value of p, giving the optimal mutation probability for producing the largest number of improved mutants:

$$\frac{\partial F}{\partial p} = 0 \rightarrow n - m = n(1 - p)^m \quad \text{or}$$

$$p = 1 - \left(1 - \frac{m}{n}\right)^{1/m} , \quad \text{or} \tag{5.5.8}$$

$$p_{opt} \sim \frac{1}{n} \; ,$$

since m $\ll$ n.

This optimum mutation probability is therefore approximately equal to the inverse number of nucleotides in the genome. (Small genotypes have an advantage over large ones.)

Let us now calculate the expected number of generations required for the pathogen to reach the ideal genotype, i.e., the genotype which will be lethal to its host — unless the host meanwhile works out some counterstrategies.

The critical rate of mutation at which mutations act on the already –adapted part of the genome *faster* than they can be screened out by selection is called the "error catastrophe". This critical mutation rate is approximately equal to the optimal mutation rate. Let $p_c = K/n$ (where K is a numerical constant) be the probability associated with the critical mutation rate; then the proportion of error-free genomes is

$$\sim\left(1 - \frac{K}{n}\right)^{n-m} \sim \left(1 - \frac{K}{n}\right)^{n} \sim e^{-K} \quad . \tag{5.5.9}$$

(The fraction of the improved genotypes Km/n is not included.) Thus for K = 1, only about 1/e of the offspring are not worse than their antecedents. For reproduction through *fission*, K = ln 2 would be an upper bound for K, so $p_c = (\ln 2)/n$.

We also assume "perfect selection" in the sense that all mutants that exceed the currently best number of "wrong" bits are screened out immediately, and all improved genotypes of the pathogen are allowed to spread. *In each fission* the probability of an improvement is

$$(1 - p_c)^{n} [(1 - p_c)^{-m} - 1]$$

$$= e^{-\ln 2}[(1 - p_c)^{-m} - 1] \cong \frac{\ln 2}{2} \left(\frac{m}{n}\right) \quad . \tag{5.5.10}$$

Thus the expected number of generations required to reach the ideal genotype is proportional to

$$n \frac{2}{\ln 2} \left(\frac{1}{m} + \frac{1}{m - 1} + \frac{1}{m - 2} + \ldots + \frac{1}{2} + 1\right) \sim \frac{2}{\ln 2} n \ln(m) \quad , \tag{5.5.11}$$

that is, proportional to the number of nucleotides in the pathogen's genome. Therefore small pathogens (viruses) evolve until they reach the "ideal" for them and become lethal for their host genotype, *faster* than larger pathogens (bacteria, fungi, etc.).

5.6 Games Between Animals Adopting Specific Modes of Behavior (Roles). Concepts of Evolutionarily Stable Strategy

A particular category of non-constant-sum games refers to the field of what are known as intraspecific animal conflicts. Restraint used in fighting has been observed in many mammalian species, particularly those possessing lethal weapons.

Usually in such encounters the defeated partner "surrenders" by making essentially a symbolic gesture, for example, exposing his throat to the victor. The victor does *not* kill the defeated opponent but allows him to retreat. It seems, therefore, that nature somehow provides for "limited combat" in intraspecific disputes. The interest lies in imagining and modeling how such limitations might have evolved — in order to favor *individual* selection.

Pioneering work in this field has been initiated by *Maynard-Smith* [5.4]. In what follows we essentially summarize his basic ideas. The field may also be relevant to arms-control philosophy. Consider two strategies employed by each contestant in a dyadic conflict, C (conventional) which when used does not inflict any appreciative damage and D (dangerous) which *may* cause serious damage with a given a priori probability. These tactics are analogous to the cooperative (C) and defective (D) strategies, respectively, in the prisoner's dilemma and more so in the game of chicken, both examined previously in Sect.5.2.2.

It is *assumed* that somehow the genetic structure of the individual determines (for the animal case) the particular string algorithm CCCDDCD... of strategies that the animal employs.

Let us give examples of (more than two) *roles*, i.e., algorithms of essentially C, D alternations that the individual animal may adopt during the conflict:

1) Dove (or "Mouse"). An individual committed to a "dovish" strategy never uses the move D. He begins the game with C and responds thereafter with C to the opponent's C. *Once* the opponent uses D, the "Dove" retreats leaving the field (food, mates) to the opponent.

2) "Hawk". This algorithm uses D always and continues it until either the player becomes incapacitated or the opponent retreats.

3) "Bully". He plays D *if* making the first move. He plays also D in response to C, but plays C in response to D. He retreats if the opponent plays D a second time in a row.

4) "Retaliator". He plays C *if* making the first move. In reply to opponent's C, he uses C, and in reply to D, he uses D.

5) "Prober". He plays C if moving first. Following C, he continues playing C with high probability, occasionally defecting to D (just to probe the opponent). If the opponent retaliates with a D, the prober promptly reverts to C. Otherwise (if the opponent replies with C), the prober takes advantage of it and continues with D.

Here we have five different roles (the reader may invent many more) which may be adopted in a serial manner by two specific opponents.

The question now is, supposing a *large* population is involved, what type(s) of phenotype polymorphism is (are) going to emerge out of a long contest among pairs of individuals? Which form of behavioral polymorphism is *stable* and under what conditions? To answer such questions one has to resort to specific examples, i.e., to specific computing simulations, for given numerical values of some key "control parameters".

We assume, following *Maynard-Smith* and *Price* [5.5], that every contest, unless terminated earlier by either retreat or serious injury of one of the contestants, terminates after a finite (fixed) number of plays.

Control parameters in this specific example have been determined as follows:

- Probability of serious injury from a single D-move by the opponent: 10%.
- Probability that the "Prober" will probe: 5%.
- Payoff for winning (when the opponent either retreats or gets seriously injured): 60.
- Payoff for receiving serious injury: -100.
- Payoff for each D received without series injury: -2.
- Payoff for saving time and energy if not seriously injured varies from 0 (long runs) to +20 (short runs).

One thousand plays have been considered in matching each role above with every other. Table 5.1 gives the average payoffs for each role when matched against every other type.

Table 5.1. The average payoff received by a player adopting different roles or strategies when matched against an opponent following each strategy

Player		D	H	B	R	P
Dove	(D)	29	19.5	19.5	29	17.2
Hawk	(H)	80	-19.5	74.6	-18.1	-18.1
Bully	(B)	80	4.9	41.5	11.9	11.2
Retaliator	(R)	29	-22.3	57.1	29	23.1
Prober	(P)	56.7	-20.1	59.4	26.9	21.9

(Opponent spans columns D, H, B, R, P)

A particular role will be stable if the payoff in the corresponding diagonal entry is the largest of those that are also the largest of their rows in that column, which means, a role will be stable if the corresponding strategy is best against every other. From Table 5.1 we can see that a population composed entirely of *Retaliators* is stable (but not one of Hawks or Bullies, for example). In this way then the predominance of *limited*-combat strategies *can*, in principle, be explained on the basis of *individual selection*. There are, of course, more complex situations, such as the one above where polymorphism gives way to "conformism" —by having every individual adopting the profile of Retaliator.

Suppose a population consists almost entirely of Hawks and Bullies. As seen from Table 5.1, Hawk is a better strategy than Bully (the average payoff for a Hawk is largest if its opponent acts as a Bully, 74.6) and Bully is likewise better against

Hawk (as the average payoff for a Bully against a Hawk opponent is positive). Hence we have a system of frequency-dependent selections leading to a *stable* polymorphism that, in turn, leads to a stable population of Hawks and Bullies. Therefore types other than Hawk and Bully are at a disadvantage and will not spread. This result is not affected by the presence of a few individuals adopting the role of Dove since Hawk and Bully are equally the best strategies (80) if the opponent is a Dove.

All the strategies (roles) considered so far are "pure" strategies. How about mixed strategies? Suppose indeed that the members of a population engage in contests in the above-described five "pure" role game in *random pairs* of roles, and subsequently each individual reproduces its kind (i.e., individuals employing the same strategy) in proportion to the payoff it has accumulated. If there is an "evolutionarily stable strategy" for the game, the population will evolve toward it. An evolutionary stable strategy, more specifically, is a mixed strategy with the property that if most of the members of population adopt it, then no "mutant" strategy can invade the population. In other words, a mixed strategy is evolutionarily stable if there is no mutant strategy that gives higher fitness to the individual adopting it.

Generally speaking, suppose that the pure strategies (modes of behavior) for contests within a species are labeled $1,\dots,n$, and that α_{ij} is the payoff for the player using the pure strategy i when his adversary uses the pure strategy j. Then $\sum_{j=1}^{n} \alpha_{ij}q_j$ is the payoff for the pure strategy i against the mixed strategy given by the probability vector

$$q = (q_1,\dots,q_n) \quad ,$$

and

$$\sum_{\substack{j=1 \\ i=1}}^{n} \sum \alpha_{ij}p_iq_j = pAq \qquad (5.6.1)$$

is the payoff for the mixed strategy $p(p_1,\dots,p_n)$ played against q.

Let A denote the payoff matrix and let $\Omega_n = \{x_1,x_2,\dots,x_n\}$ ($\sum_{j=1}^{n} x_j = 1$) be the simplex of all possible strategies. A strategy $p \in \Omega_n$ is called an evolutionarily stable strategy if, whenever a population using this strategy is perturbed by a mutation introducing a small population with strategy $p \neq q$, then p fares better in the new mixed population than does q. To put it quantitatively, p is an evolutionary stable strategy if for all $q \neq p$ one has $pAq \geqslant qAp$ with $pAq > qAp$ in the case of equality, or,

$$\sum_{i}^{n} \sum_{j}^{n} p_i\alpha_{ij}p_j \geqslant \sum_{i}^{n} \sum_{j}^{n} q_i\alpha_{ij}p_j \quad . \qquad (5.6.2)$$

Therefore p is a best reply against itself, and fares better against any alternative best reply q, than does q against itself.

In a population with mixed strategy $x(x_1,...,x_n)$, the payoff for strategy i is $\Sigma_{j=1}^n \alpha_{ij}x_j$, while the average payoff is $\Sigma_K^n \Sigma_j^n \alpha_{Kj}x_Kx_j$. It is natural to assume that the rate of increase $\dot{x}_i/x_i$ is equal to the difference between these two payoffs. Thus our game "reduces" to a system of ordinary differential equations

$$\frac{dx_i}{dt} = x_i \left(\sum_{j=1}^n \alpha_{ij}x_j - \sum_K^n \sum_j^n x_K\alpha_{Kj}x_j \right) \tag{5.6.3}$$

in the state space Ω_n. This system of equations is much more complicated than the one considered before in the context of constant-sum games and paradoxical, non-constant-sum games. In general, analytical solutions for the stationary regime are not possible even for n = 2. For n > 2, the dynamics involve a three- or higher-dimensional state space, and pertinent discussion must necessarily be postponed until Chap.6. In any case, the solution for the stationary regime of the above system provides the percentages of persistent behaviors in a polymorphic population.

5.7 The Game of Competitive-Cooperative Production and Exchange. The Concept of "Parasite" at a Symbolic Level

Consider two large-scale systems x, y (men, organizations) producing two different substances or items x and y respectively. Each system gives the other a fraction q of what it produces and keeps the remaining fraction p = 1 - q. Following *Rapoport* [5.6] let us analyse the situation.

We are interested in the dynamics of the evolution of the products x and y; thus we have to specify the expected payoffs for both partners. If these payoffs are G_x and G_y, then it is natural to invoke once again the algorithm which essentially says that the rate of increase or decrease of products x and y will depend on the differential payoff each partner appreciates by producing x or y respectively. We adhere to the principle that the two systems are able to adjust, for given "rules of exchange" (i.e., p, q), so as to maximize their respective "utilities" (payoffs) by controlling their productive output. This is presumably what the individual would do if he were alone. The more he works, the more he produces, and the more payoff would accrue to him on that account. However, beyond a certain level of production, "fatigue" of the production system may offset progressively the payoff by adding an ever-increasing amount of negative utility. At some point the net payoff maximizes; this would be the point at which the individual producer would fix his output.

Now, the question arises, to what extent does utility depend on the reward and on the effort? For the effort we make the simplest possible assumption, namely that the (negative) utility due to fatigue is proportional to the effort. For the reward we will assume here the so-called principle of *diminishing returns*: As the reward increases, the utility also increases but at a decreasing rate: a thousand DM rise means much more to the man who is making 3000 DM per month, than to one who is making 20 000 DM per month.

If the increase of utility (u) is to be inversely proportional to the reward (r) already accrued, we will have

$$\frac{du}{dr} = \frac{K}{r} \quad , \qquad (5.7.1)$$

where K is a constant, or $du = Kdr/r$ and $u = K \ln r + A$. We take $A = 0$. Now in our case the total utility will be

$$
\begin{aligned}
G_x &= \ln(r_x) - \beta x \qquad \text{for partner X} \quad , \\
G_y &= \ln(r_y) - \beta y \qquad \text{for partner Y} \quad .
\end{aligned}
\qquad (5.7.2)
$$

(We have assumed the "fatigue factor" β to be the same for both partners, and the same assumptions were made for the individual rations p, q.) What are the expressions for r_x, r_y?

Obviously for partner X the reward r_x will contain the term $px + py$, and for partner Y the reward r_y will correspondingly contain the term $qx + py$. For normalization purposes, however, we put $r_x = 1 + px + qy$ and $r_y = 1 + qx + py$, so that when $x = 0$, $y = 0$, $G_x = 0$, $G_y = 0$ ($\ln 1 = 0$).

Then our dynamical equations are

$$\frac{dx}{dt} = \frac{\partial G_x}{\partial x} = \frac{\partial}{\partial x} [\ln(1 + px + qy) - \beta x] \quad ,$$

$$\frac{dy}{dt} = \frac{\partial G_y}{\partial y} = \frac{\partial}{\partial y} [\ln(1 + qx + py) - \beta x] \quad .$$

$$(5.7.3)$$

Equations (5.7.3) give

$$\frac{dx}{dt} = \frac{p}{1 + px + qy} - \beta$$

$$\frac{dy}{dt} = \frac{p}{1 + qx + py} - \beta \quad .$$

$$(5.7.4)$$

Let us investigate the nature and the stability of the steady states (if any). The steady states are determined by the equation

$$
\begin{aligned}
y_1(x) &= px + qy = (p/\beta) - 1 \\
y_2(x) &= qx + py = (p/\beta) - 1 \quad .
\end{aligned}
\qquad (5.7.5)
$$

We have two straight lines, "the optimal lines", which represent all the points of balance for each of the partners respectively, that is, each has a balance point for each value of the other's output. Therefore each will try to adjust his output so as to bring the common point upon *his* optimal line. We have to keep in mind, however, that each individual controls a separate coordinate: X can move only horizontally on the diagram, Y can move only vertically (Fig.5.19). Equations (5.7.4) tell

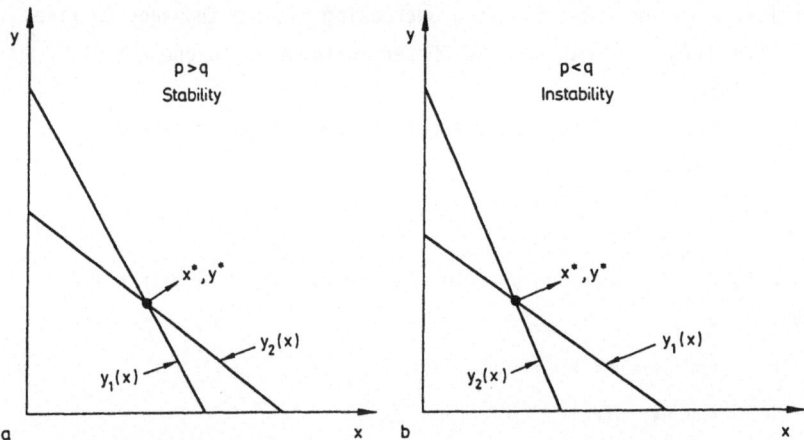

Fig.5.19a,b. Lines of equilibria for the production-exchange game

us that the equilibrium point

$$x^* = y^* = (p/\beta) - 1 \qquad (5.7.6)$$

can be in the positive quadrant only if $p > \beta$. This means that either p is not too small or that β is not too large. In turn, this implies that the fraction kept by each individual should not be too small or that the "fatigue factor" should not be too large.

Let us investigate now the stability of the equilibrium point (5.7.6). This equilibrium will be called stable if in its neighbourhood the point determined by the efforts of the players tends to move *towards* the equilibrium as the players attempt to maximize their *own* respective utilities G_x, G_y. If the point moves away from (x^*, y^*), the equilibrium is unstable.

Let us suppose p q, that is, the optimal line of X has the greater slope. We conceive the motion of the point (x,y) as follows. As one coordinate remains stationary, the player who controls the other brings the point (x,y) into his own optimal line. Then *his* coordinate remains stationary and the other player does the same.

Let player X choose some value $x^{(1)}$ —a perturbation near the equilibrium point (x^*,y^*). According to (5.7.4,5) player Y will then move to a value $y^{(1)}$ such that the point $(x^{(1)},y^{(1)})$ is on Y's curve $y_2(x)$; in other words the first pair of choices results in $(x^{(1)},y_2(x^{(1)}))$. Now player X again chooses a value $x^{(2)}$ so as to bring the point to *his own* optimal line $y_1(x)$. This value is the intersection of the horizontal line $y = y_2(x^{(1)})$ with $y_1(x)$, which is approximately

$$x^{(2)} \sim \frac{y_2(x^{(1)})}{y_1'(0)} \cong \frac{y_2'(0)}{y_1'(0)} x^{(1)} , \qquad (5.7.7)$$

where $y_1'(0)$, $y_2'(0)$ stand for the values of the derivatives of the two straight lines in the neighborhood of the steady state.

254

We shall then have

$$|x^{(2)}| < |x^{(1)}| \quad \text{or}$$

$$|x^{(2)}| > |x^{(1)}|$$

depending on whether

$$|y_2'(0)| < |y_1'(0)| \ , \qquad (p > q) \qquad \text{or}$$

$$|y_2'(0)| > |y_1'(0)| \ , \qquad (p < q) \ .$$

Thus the equilibrium is stable only if $p > q$; when $p = q = 1/2$, we have marginal stability. Stability then occurs when each individual relies on his own efforts *more* than on the partner's, that is, each keeps the *greater* proportion of his own product and exchanges the smaller portion.

What happens now in the case of instability, i.e., when $p < q$? In this case, the "production" point will end up either on the horizontal or on the vertical axis. If the end state is on the horizontal axis, Y does not work; he becomes a "parasite" on X. If the end state is on the vertical axis, X is the parasite. Either result is bound to happen in the unstable case because the slightest disturbance of equilibrium will precipitate it. *Which* one will be the parasite depends on the direction of the initial disturbance. Whoever relaxes his effort *first* will become the parasite. This is so because in the unstable case, any such relaxation on the part of one partner will be immediately compensated by the other and consequently will provoke another lapse of the same sort. (Of course the roles X, Y are interchangeable; the reason that stability is associated with the greater slope of X's optimal line has to do with the way the coordinates have been labeled.)

Suppose now a *stable* balance exists. Under this regime how much utility goes to X and Y? Substituting the values $x^* = y^* = (p/\beta) - 1$ into the expressions giving the payoffs (5.7.2) we find,

$$(G_x)^* = \ln(p/\beta) - p + \beta \quad \text{and}$$

$$(G_y)^* = \ln(p/\beta) - p + \beta \ . \tag{5.7.8}$$

We would next like to check whether the above values are "optima", that is, if they represent (for both partners) the greatest profit they can obtain under the constraints of the process.

In the first place, the maximum production for each partner is *not* given by the steady-state values x^* and y^*, but by the maxima values of those expressions which are equal to $(1/\beta) - 1$. Then the corresponding utilities are $\ln(1/\beta) - 1 + \beta$. Is this more or less the values $(G_x)^* = (G_y)^*$? Let us calculate their difference:

$$\ln\left(\frac{1}{\beta}\right) - 1 - \ln\left(\frac{p}{\beta}\right) + p = \ln\left(\frac{1}{p}\right) - 1 + p = \ln\left(\frac{1}{p}\right) - (1 - p) \ . \tag{5.7.9}$$

This expression is always *non-negative* (being zero for $p = 1$). We thus come to the conclusion that *both* partners might do better if they produce $(1/\beta) - 1$ than if they produce $(p/\beta) - 1$ (except when $p = 1$, in which case they do equally well, but $p = 1$

means that there is no product sharing at all, so this particular case is not to be considered further).

It appears therefore that the balance point, even if it is stable, is *not* the best point from the point of view of either partner, although each does best at the steady-state point provided that the competitor is there too. What is really wrong in this case is the dynamics of purely *selfish* interests. Each producer adjusts only his *own* utility with regard for the utility of his partner.

Before suggesting — and testing — a more "socially oriented" philosophy let us calculate in the case of instability the utility of the "parasite" and its "host". Let X be the parasite. The parasite (X) produces nothing and the host (Y) has to adjust accordingly. The host's optimal line is $qx + py = (p/\beta) - 1$. However, since now $x = 0$, Y must produce so as to satisfy the equation $py = (p/\beta) - 1$ or $y = (1/\beta)$ $- (1/p)$ ($p > \beta$, so this value of y is positive). The corresponding utilities for the parasite (X) and the host (Y) are therefore

$$G_x = \ln\left(1 + \frac{q}{\beta} - \frac{q}{p}\right) \quad \text{and}$$

$$G_y = \ln\left(\frac{p}{\beta}\right) - 1 + \frac{\beta}{p} \, ,$$

(5.7.10)

respectively.

Let us compare these expressions with $\ln(1/\beta) - 1 + \beta$, the "ideal" utility accruing to each if each produces not the "individual optimum" $(p/\beta) - 1$ but rather the "social optimum" $(1/\beta) - 1$. Concerning the host (Y), the difference is $\ln(1/p) + \beta$ $[1 - (1/p)]$. This expression is always positive if $p > \beta$, so, the host (Y) is *always better off* at the "social" optimum than in his role as "host". This should be expected.

Let us see now if the parasite is better off as a "parasite" or as a full cooperator. The difference in X's utilities between what he gets as the social optimum and as a parasite is $\ln p - \ln(\beta p + pq - q\beta) - 1 + \beta$. Whether this quantity is positive or negative depends on β/p, which can vary in the range $0 - 1$. When $\beta = p$, the above expression gives $\ln(1/p) - 1 + p$, which is always positive; when $\beta \ll p$, it gives $\sim - \ln q - 1$. We are dealing with the unstable case where $p < q$, i.e., $q > 1/2$, so $- \ln q - 1$ is negative. But the expression $- \ln q - 1$ is just the asymptotic expression for the above difference in payoffs as β approaches zero.

Therefore the parasite is better off as a "parasite" if β is small (although not so if β is large). This means that it *pays* to be a "parasite" if your "host" is "indefatigable" (small β) and sustains you anyway!

Finally, let us turn to the issue of "social optimum" characterized by the "ideal" payoffs $x_0 = y_0 = (1/\beta) - 1$. How can they be achieved? Suppose that the two "producers" X and Y develop some sort of a "social conscience", that is, the utility of one somehow depends on the utility of the other. Let us take a linear dependence via the matrix

$$G_{x_0} = AG_x + BG_y \quad,$$

$$G_{y_0} = CG_x + DG_y \quad,$$

(5.7.11)

where A, B, C, D are weighting factors from 0 to 1.

We will investigate the extreme case $A = B = C = D$, which simply means that each partner derives from the utility of his competitor as much satisfaction as he derives from his *own* utility. Then our dynamics take the form

$$\frac{dx}{dt} = \frac{\partial}{\partial x} (G_x + G_y) \quad \text{and}$$

$$\frac{dy}{dt} = \frac{\partial}{\partial y} (G_x + G_y) \quad .$$

(5.7.12)

The steady-state regime is then investigated through the conditions

$$\frac{p}{1 + px + qy} + \frac{q}{1 + qx + py} - \beta = 0 \quad \text{and}$$

(5.7.13)

$$\frac{q}{1 + px + qy} + \frac{p}{1 + qx + py} - \beta = 0 \quad ,$$

which lead indeed to the steady state $x_0 = y_0 = (1/\beta) - 1$ with payoffs

$$G_{x_0} = G_{y_0} = \ln \frac{1}{\beta} - 1 + \beta \quad .$$

(5.7.14)

It is fairly likely, therefore, that even a phenomenon as "lofty" as *social conscience* might have arisen in populations via optimizing properties of payoffs through natural selection.

5.8 Epidemiology of Rumors

Rumors constitute an "inferior subset" of information-transfer problems. This mode of "epidemic" thrives in cases where information cannot be "broadcast", so to speak, either because communication networks are at a primitive stage of development or because unqualified diffusion of information is severely restricted by "law", as in the case of totalitarian or "theocratic" organizations (which essentially suppress fluctuations to preserve dogma, i.e., the stability of the status quo). Following *Frauenthal* [5.7], we intend to deal below with the spread of a rumor through a group of loosely connected groups of individuals or "communities". We consider a system of $N + 1$ isolated small communities which can communicate only by means of a primitive telecommunication system. This "phone" system permits one pair of communities to be in contact at any time, one community making the call, the other one receiving the call.

The rumor spreads just by means of these dyadic, *one-way* telephone calls, and the entire community is supposed to know the rumor once the community phone receiver "decodes" it.

The N + 1 communities ("cells") are classified into three categories:

Category S ("Susceptible"): Communities which have not yet received the rumor but which would spread it to other communities if it came.

Category I ("Infective"): Communities which have received the rumor and are active in spreading it to other communities. After spreading the rumor once, "I" types are subsequently "neutralized".

Category R ("Resistive" or "unyielding"): Communities which have received the rumor but either cannot or are not willing to spread it to other communities —although they do not actively "campaign" against it.

The dynamics of the spread of the rumor are assumed to be as follows:

1) A randomly selected community places a call at random to *any* of the other N communities (provided that a telephone line is free).
2) If a community of the I type calls a community of type S, the rumor is spread. Consequently both become type I.
3) If a community of type I calls a community of either type I or R, the community making the call becomes essentially of type R. The community receiving the call stays in the category to which it belonged before.
4) Calls transmitted from communities of either type S or type R have no effect on the spread of the rumor (such calls are meaningless).

Owing to condition 4, we have to deal with calls realized under conditions 2 and 3 above: Indeed *all* calls which alter the state of the system emerge from communities of type I.

Let $s(t)$, $i(t)$, and $r(t)$ be the number of communities of types S, I, and R respectively, at (discrete) time t. We define a state vector

$$P(t) = (s(t), i(t), r(t)) \ . \tag{5.8.1}$$

At $t = 0$ we will have, of course,

$$P(0) = (N, 1, 0) \ . \tag{5.8.2}$$

The first call which changes the state of the system consists of having the one community of type I which knows the rumor call one of the N communities of type S. Thus

$$P(1) = (N-1, 2, 0) \quad . \tag{5.8.3}$$

From now on, however, there are *two* different types of calls which alter the state of the system: (a) one of the two type I communities can call one of the $N-1$ type S communities or (b) one of the two type I communities can call the other type I community. So the precise state of the system becomes uncertain for $t \geq 2$.

Rather than investigate the probability of all the different possible outcomes, we investigate the evolution of an expectation variable, that is, we replace the computation of the behavior of a random variable by the calculation of the sequentially conditioned expected values of that variable P.

This procedure assumes *implicitly* that the expected value of our random variable P is representative of the actual behavior of the system. It is interesting to divert at this point and give a counterexample, i.e., a case where the mean value of a random process is a misleading estimate of the actual behaviour.

Let X be a random variable which can take only two values, as follows:

$$P(X = N^2) = \frac{1}{N}$$

$$P(X = 0) = \frac{N - 1}{N} \quad .$$

The expected value of this process is

$$E(X) = N^2 \frac{1}{N} + 0 \frac{N - 1}{N} = N \quad ,$$

so, as $N \to \infty$, the expected value of X becomes unbounded, yet the probability that $X = 0$ approaches unity.

Let us further calculate the variance $D(X) = E(X^2) - E^2(X)$. Since

$$E(X^2) = N^4 \frac{1}{N} + 0 \frac{N - 1}{N} = N^3 \quad ,$$

$$D(X) = N^3 - N^2 = N^2(N - 1) \quad ,$$

and the standard deviation $N \sqrt{N - 1}$ grows *faster* than the mean value as N increases. When such a thing happens, the mean value does *not* say anything about the system. One has to estimate the variance, but to do that one has to calculate the probabilities for the random variable. Strictly speaking, therefore, it is impossible to assess the accuracy of the results for our model, although the method given here greatly simplifies the mathematical calculation.

Now back to the calculation: For $t = 2$,

$$P_2 = E(P(t = 2)) = (S_2, i_2, r_2) \quad , \tag{5.8.4}$$

and for $t = K \ (= 3, 4, \ldots)$,

$$P_K = E(P(t = K) | P(t = K - 1) = P_{K-1}) = (S_K, i_K, r_K) \quad . \tag{5.8.5}$$

We try, that is, to calculate the vector P_K recursively, assuming that we know the true vector P_K $(t = K - 1)$ —while what we actually know is its expectation value

P_{K-1}. In order to develop the recurrence relationships, we assume that at $t = K - 1$ we know that

$$P_{K-1} = (S_{K-1}, i_{K-1}, r_{K-1}) \quad . \tag{5.8.6}$$

The actual state $P(t = K)$ is, then, either

$$(S_{K-1} - 1, i_{K-1} + 1, r_{K-1}) \tag{5.8.7}$$

or

$$(S_{K-1}, i_{K-1} - 1, r_{K-1} + 1) \quad , \tag{5.8.8}$$

where the first possibility refers to the case where a community of type I calls a community of type S, which occurs with probability S_{K-1}/N, and the second possibility refers to the case where a community of the type I calls either a community of type I or type R, which occurs with probability $(N - S_{K-1})/N$.

Consequently, the first two elements of the vector P_K are

$$S_K = \frac{S_{K-1}}{N} (S_{K-1} - 1) + \frac{N - S_{K-1}}{N} (S_{K-1}) = \frac{N - 1}{N} S_{K-1} \tag{5.8.9}$$

$$i_K = \frac{S_{K-1}}{N} (i_{K-1} + 1) + \frac{N - S_{K-1}}{N} (i_{K-1} - 1) = i_{K-1} + \frac{2}{N} S_{K-1} - 1 \quad . \tag{5.8.10}$$

We try now to solve the system of the above two recurrence relations — which constitute a system of coupled linear difference equations.

Let us start from (5.8.9). It is solved by induction:

$$S_1 = \frac{N - 1}{N} S_0 \quad ,$$

$$S_2 = \frac{N - 1}{N} S_1 = \left(\frac{N - 1}{N}\right)^2 S_0 \quad ,$$

$$\vdots$$

$$S_K = \frac{N - 1}{N} S_{K-1} = \cdots = \left(\frac{N - 1}{N}\right)^K S_0 = N\left(\frac{N - 1}{N}\right)^K \quad , \tag{5.8.11}$$

since $S_0 = N$.

Consider now (5.8.10). We have again

$$i_1 - i_0 = 2\left(\frac{N - 1}{N}\right)^0 - 1 \quad ,$$

$$i_2 - i_1 = 2\left(\frac{N - 1}{N}\right)^1 - 1 \quad , \tag{5.8.12}$$

$$\vdots$$

$$i_K - i_{K-1} = 2\left(\frac{N - 1}{N}\right)^{K-1} - 1 \quad .$$

Adding together the above K equations, we get

$$i_K - i_0 = 2 \sum_{j=0}^{K-1} \left(\frac{N-1}{N}\right)^j - K \quad . \tag{5.8.13}$$

Recalling that $i_0 = 1$ and also that

$$\sum_{j=0}^{K-1} \xi^j = \frac{1 - \xi^K}{1 - \xi} \quad ,$$

we finally obtain

$$i_K = 2N\left[1 - \left(1 - \frac{1}{N}\right)^K\right] + 1 - K \quad , \qquad K = 0, 1, 2, \ldots \tag{5.8.14}$$

From the expressions for S_K and i_K, r_K follows from the formula

$$r_K = N + 1 - S_K - i_K \quad . \tag{5.8.15}$$

For any value N, we can therefore estimate the evolution of the rumor in time.

It remains to investigate the very important question of *how far* the rumor will go. It is clear that the rumor stops propagating once there are no more communities of type I. The answer to our question will therefore be the number or the percentage of communities of type S which *remain* when no type I communities are left. We will therefore calculate the value of K for which $i_K = 0$.

Since the value of i_K required in the above calculation is not ncessarily an integer, we may find $i_\nu > 0$ and $i_{\nu+1} < 0$, where ν is an integer. We therefore replace K by λN in the equation for i_K, and consider λ a continuous variable; we look for values of λ for which $i_{\lambda N} = 0$. We find

$$0 = 2N\left[1 - \left(1 - \frac{1}{N}\right)^{\lambda N}\right] + 1 - \lambda N$$

or

$$\lambda = 2 - 2\left(1 - \frac{1}{N}\right)^{\lambda N} + \frac{1}{N} \quad . \tag{5.8.16}$$

If N is very large ($N \to \infty$), $1/N \sim 0$ and $[1 - (1/N)]^N \sim 1/e$. Thus

$$\lambda_0 = 2\left[1 - e^{-\lambda_0}\right] \tag{5.8.17}$$

and

$$\lambda_0 \sim 1.594 \quad .$$

Therefore it follows that the rumor "dies out" after about $t = \lambda_0 N = 1.594N$ "telephone calls" by type I communities when N is large. Also, since

$$S_{\lambda N} = N\left(\frac{N-1}{N}\right)^{\lambda N} \sim N e^{-\lambda_0} \sim 0.238N \quad ,$$

when the rumor stops circulating, about 23.8% of the communities can claim that they have "never heard of it".

6. Stochasticity Due to Deterministic Dynamics in Three- or Higher-Dimensional Space: Chaos and Strange Attractors

6.1 A Reappraisal of Classical Statistical Mechanics. The Kolmogorov-Arnold-Moser Theorem

In Chap.2 of this book we referred to evidence in support of the observed compatibility among entropy production, progressive differentiation, complexity increase, and emergence of self-organizing properties in large-scale systems. We have already stated that these "self"-organizing properties have to do with the ability of the system to simulate not only its environment but also parts of itself. In turn, simulation was considered as an act of compressing information received from the environment and subsequent use of this compressed form as an input or "program" in a "finite-state automaton" whose output is the time evolution or the "scenario" of the phenomenon involved. It is time now to examine in some detail how the above procedure of self-organization or compressibility might be dynamically implemented, at least in theory. The issue essentially has to do with the properties of attractors in a multidimensional space. To investigate these properties, we must, for the moment, digress somewhat and examine on a very elementary level some rather recent progress in the field of classical (statistical) mechanics —which essentially concerns the way the trajectory of a multicomponent dynamical system is "spread" in state space.

Let us start by reminding ourselves that the condition of perfect ergodicity *and* mixing[1] (or "molecular chaos") holds (for the moment) only for Boltzmann's perfect conservative gas, which then tends to equilibrium from any initial state. Thirty years ago it was still believed that any sufficiently complex conservative system would do the same. Specifically, it was thought that a large system of oscillators affected by *non*linear interactions would probably tend to an equilibrium with equipartition of energy. Furthermore, it was expected that the state-space motion would be ergodic for such complex systems. The assertion of *Kolmogorov* in 1957 [6.1] (see later in this section) that nonlinearly perturbed two-component oscillator systems

---

[1] A system is "mixing" when information about the initial conditions is completely lost. This can happen via either cascading *collisions or* cascading *iterations*. Note here the parallelism between a physical system (such as a perfect gas) and a symbolic system (such as an algorithm).

262

(four degress of freedom) could have invariant tori in state space, effectively ruled out ergodicity in these cases. *Fermi* et al. investigated in 1955 [6.2] a system of coupled oscillators of greater number (up to N = 64), but again found no tendency to equipartition of energy.

Inversely, as we shall see below, the result of applying a small perturbation even to an integrable *two-component* Hamiltonian system is that the nature of the solution, i.e., the (four-dimensional) phase-space trajectory, is utterly changed. Whereas originally the motion is regular, after perturbation we have regions in state space where the motion still behaves in a regular fashion sharply separated from regions of highly irregular ("chaotic") motion.

Let us start by considering again the case of a single nonlinear pendulum. In the language of Hamiltonian dynamics we can say that in this case there is one "generalized coordinate" q (the angle the pendulum makes with the vertical) and one component of momentum p. The Hamiltonian, i.e., the total energy of the system per unit mass, is

$$H = \frac{p^2}{2} - \left(\frac{g}{l}\right)\cos q \quad , \tag{6.1.1}$$

where l is the length of the pendulum. The equations of motion are

$$\dot{q} = p \quad , \quad \dot{p} = - \left(\frac{g}{l}\right)\sin q \quad . \tag{6.1.2}$$

These can be solved in terms of elliptic functions. The phase-space flow is shown in Fig.6.1.

For small deviations near the origin, the behavior looks like simple harmonic motion with an oscillating frequency $\sqrt{g/l}$. For large p, however, the motion is a *rotation* with q increasing without bounds as time increases. The curve that separates the *rotation* region from the *oscillation* region is the *separatrix*, and it includes "X" points (saddle points) that correspond to the *unstable* equilibrium with the pendulum vertical. This is the simplest case in which the period of the oscillatory motion varies from trajectory to trajectory and tends to *infinity* (i.e., the motion becomes *aperiodic*) as the *separatrix* is approached.

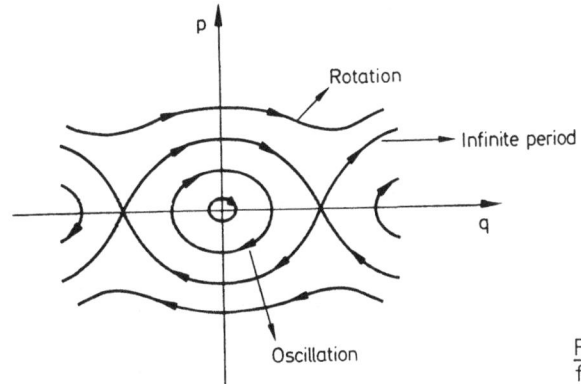

Fig.6.1. The phase-space diagram for the pendulum oscillator

Let us move now to a two-component system, i.e., a two-oscillator, simple-harmonic-motion system. The Hamiltonian is now

$$H = \sum_{i=1}^{2} \frac{1}{2} (p_i^2 + \omega_i^2 q_i^2) \quad . \tag{6.1.3}$$

The amplitudes of the two oscillators A and B are invariants of motion and can be written

$$A = \frac{1}{\omega_2} \sqrt{p_1^2 + \omega_1^2 q_1^2} \quad , \quad B = \frac{1}{\omega_2} \sqrt{p_2^2 + \omega_2^2 q_2^2} \quad . \tag{6.1.4}$$

The orbit of the system lies on the intersection of the two surfaces $A = C_1$ and $B = C_2$ —which intersection is a two-dimensional torus —and thus a knowledge of the values C_1, C_2 restricts the orbit to a limited region of the (four-dimensional) state space. Specifically for varying C_1, C_2, the orbits lie on a family of 2-tori nested one inside the other in phase space. These are the invariant surfaces of our system. If ω_1/ω_2 is rational, the orbit's spirals around the torus are closed (periodic).

If ω_1/ω_2 is irrational, the orbit of the system covers the torus *ergodically*, i.e., it is not only open but also *dense* in the sense that, given enough time, the orbit on the torus passes arbitrarily close to any given point at least once.

What happens if we perturb slightly such a simple system? To answer this question let us write the Hamiltonian of the two-dimensional system in the general form

$$H = \frac{1}{2} (p_1^2 + p_2^2) + V(q_q, q_2) \tag{6.1.5}$$

and assume, following *Henon* and *Heiles*'s [6.3] celebrated numerical example, that our system can be simulated by a particle moving in the following potential:

$$V(q_1, q_2) = \frac{1}{2} (q_1^2 + q_2^2) + q_1^2 q_2 - \frac{1}{3} q_2^3 \quad . \tag{6.1.6}$$

One can see easily that for small values of q_1, q_2, the motion about the equilibrium point (0,0) looks like two-dimensional simple harmonic motion.

As the amplitude of the motion is increased, however, the cubic terms in the potential take over and the motion in the plane q_1, q_2 takes a triangular shape (Fig.6.2). For the Hamiltonian of the form given above, Henon and Heiles have carried out a numerical computation of the orbits as follows:

The difficulty of representing pictorially the calculations of orbits in four-dimensional state space can be overcome

a) if attention is confined to orbits having a particular total energy H_0 so that $H(q_1, q_2, p_1, p_2) = H_0$. This eliminates one variable, say p_1, and the system has now three degrees of freedom q_1, q_2, p_2; and
b) if one considers, instead of the three-dimensional trajectory, the intersection of this orbit with a plane, say $q_1 = 0$.

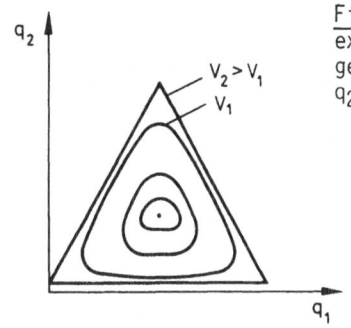

Fig.6.2. Equipotential lines for the Henon-Heiles examples of a perturbed two-oscillator system. The generalized coordinates of the two oscillators are q_1, q_2 and V_1, V_2 are two values of the potential function

If the motion is bounded in state space and the orbits confined to a closed volume, then any particular orbit of the system will pass around and around the closed region, crossing the plane $q_1 = 0$ repeatedly. This "cutting" of the state-space trajectory with a (plane) surface is called a "Poincaré mapping" and clearly amounts to an "analogue-to-digital conversion". Figure 6.3 is taken from the Henon-Heiles calculations, and shows the successive points at which an orbit of the system above crosses the plane $q_1 = 0$.

As time progresses, these points trace out a closed integral curve. The orbits then map out a one-parameter family of closed curves on the plane $q_1 = 0$ — the cross sections of the invariant tori. Therefore, by calculating orbits for a given value of the energy of our simple system and plotting the intersection of the orbits with the plane $q_1 = 0$, it is possible to represent in *two* dimensions the behavior of a dynamical system with a *four*-dimensional phase-space portrait. Henon and Heiles

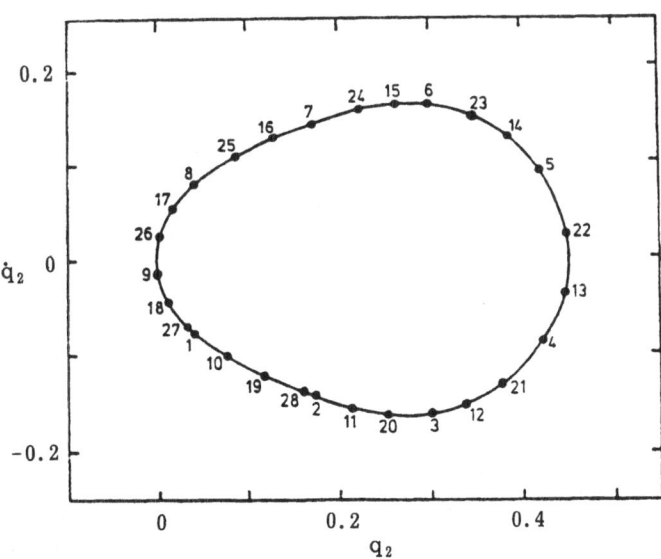

Fig.6.3. The successive points at which an orbit of the Henon-Heiles system inter-sects the plane q_2, $\dot{q}_2$. After [6.4]

plotted the Poincaré mapping of the system for a range of energy values: the extremely surprising and unsuspected results they obtained are reported in this section.

Figure 6.4 shows again the Poincaré mapping for $H_0 = 0.083$, for the totality of the integral curves. The whole area of phase space that is available to the system with this energy appears to be filled with invariant curves of rather complicated shape. Remember that in the absence of cubic terms in the potential function, the diagram above would consist of concentric circles — the cross sections of nested tori. The various singular points — "O" points (centers) and "X" saddle points — correspond to the simple periodic orbits of the motion.

Now when the Hamiltonian is raised to the value $H_0 = 0.125$, we see the extraordinary outcome illustrated in Fig.6.5. There are *still* some invariant curves centered on a periodic orbit, but the curves no longer fill the whole of the state space available to "particles" with $H_0 = 0.125$. In the intervening region, the orbits wander about in an irregular fashion. The scattered dots in the Fig.6.5 represent the crossings of a single *aperiodic* orbit. As the energy is increased further, the area filled by invariant curves is reduced still further. One is left with the impression that invariant curves are predominant for small amplitude motion and small perturbations, but that for higher energies the invariants are progressively swept away, so to speak, and give way to completely irregular or chaotic behavior which one would not "expect" *from a simple deterministic system with few degrees of freedom.*

How could one justify such really surprising behavior? The answer comes essentially from the celebrated Kolmogorov-Arnold-Moser (KAM) theorem — perhaps the greatest breakthrough in classical mechanics since the time of Poincaré. We do not intend to give a proof here, which would be irrelevant to our main theme, we will simply close the introductory section by making a number of qualitative comments.

The crucial thing is to understand what happens to the invariant tori in our four-dimensional state space — or to the family of the corresponding concentric circles on the Poincaré plane $(q_2, \dot{q}_2)$ — when we apply small perturbations to our simple two-oscillator system. Under increasing perturbation most of the "rational" surfaces (tori for which $\omega_1/\omega_2 = n/m$ is rational) break up progressively, leaving an *even* number of alternate elliptic (O) and saddle (X) (hyperbolic) fixed points on the Poincaré map (Fig.6.6), that is, an even number of periodic trajectories.

[The "irrational" tori (for which ω_1/ω_2 is irrational), although modified in shape by a (small) perturbation, are *not* destroyed: their traces on the Poincaré map persist as invariant continuous closed curves.]

Now consider what happens near the hyperbolic fixed points. They should be connected by integral curves — the *separatrices* which we have already seen in the simple pendulum's two-dimensional phase portrait. Note that these invariant curves entering an X point can be thought of as ingoing or outgoing, depending on whether points on these curves move towards or away from the fixed point.

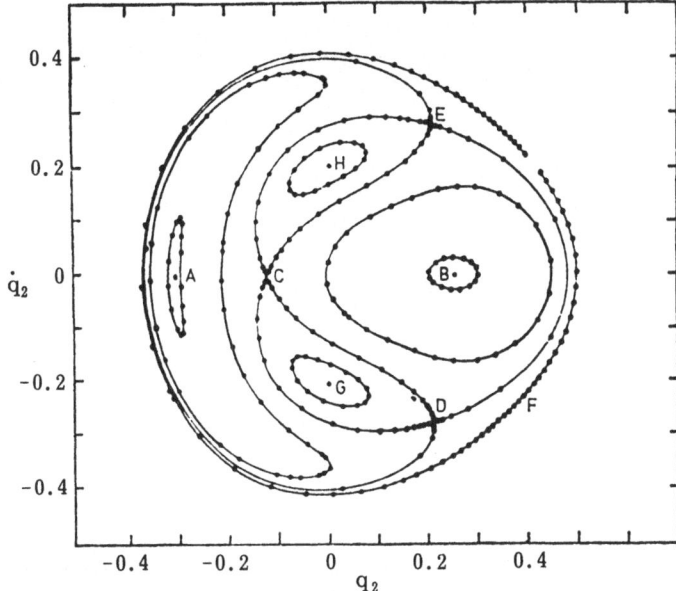

Fig.6.4. Invariant curves of a Poincaré mapping of the Henon-Heiles dynamical system by the plane q_2, $\dot{q}_2$. Different contours are labeled by A to H. After [6.4]

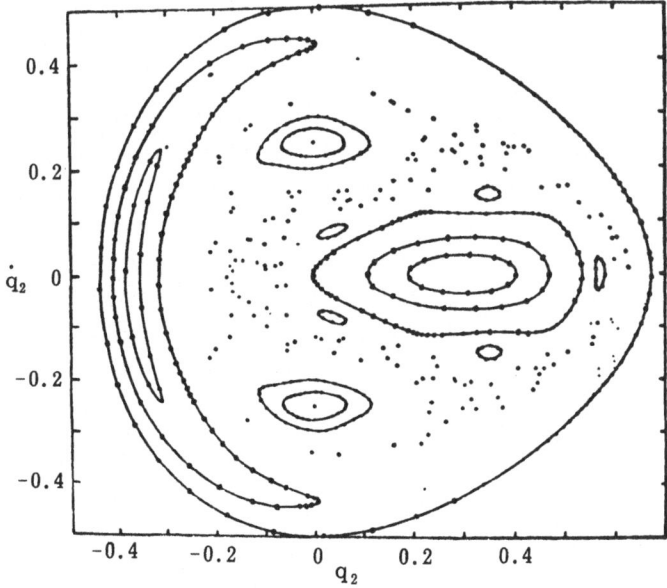

Fig.6.5. Invariant curves of a Poincaré mapping of the Henon-Heiles dynamical system by the plane q_2, $\dot{q}_2$, showing regions of regular motion (orbits lying on invariant curves) surrounded by a region of chaotic behavior —indicated by the traces of a single aperiodic orbit. After [6.4]

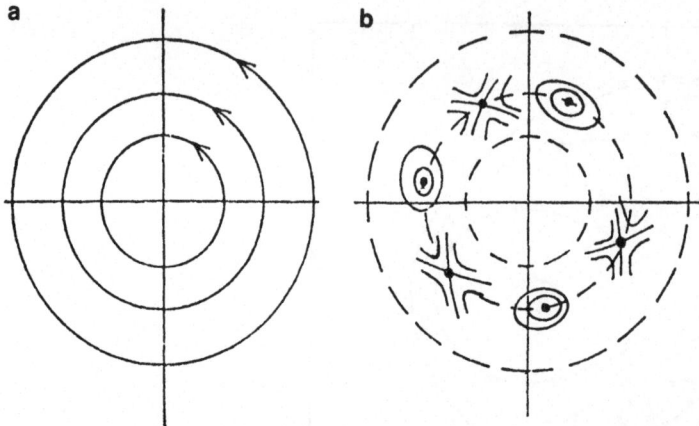

Fig.6.6a,b. Under small perturbations a "rational" torus (**a**) breaks and (**b**) in its place appear alternate elliptic (O) and saddle (X) (hyperbolic) fixed points. After [6.4]

Under special circumstances the outgoing curve leaving one X point may become an ingoing curve of a neighboring X point. However in general this does *not* happen; it has been shown that usually an outgoing invariant curve from one X point will intersect an invariant curve through a neighboring X point. *Smale* [6.5] proved that if it does this once, it will do it an *infinite* number of times. The points where the invariant curves from neighboring X points intersect are called *homoclinic* points (Fig.6.7).

Therefore the final picture on the Poincaré map has the following characteristics. Between the remaining integral curves of the irrational tori are regions of irregular behavior due to the chaotic orbits whose traces on the map are the set of homoclinic points. These regions of chaotic behavior are very small if the perturbation is very small, but increase with the perturbation parameter. Contrastingly, the behavior close to the O points produced by the breakup of the rational surfaces looks pretty much like the original situation, where a periodic orbit is surrounded by

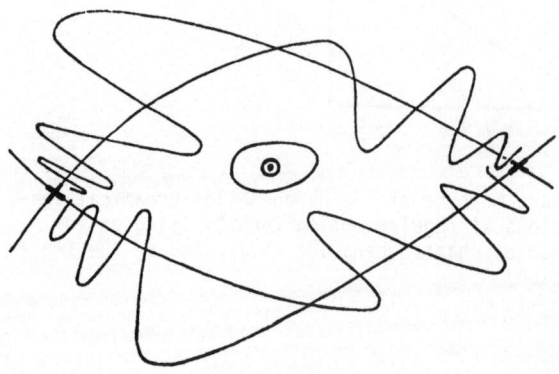

Fig.6.7. An invariant curve from one (X) point intersects an invariant curve from another (X) point at an *infinite* number of points on a Poincaré map. These are the "homoclinic points"

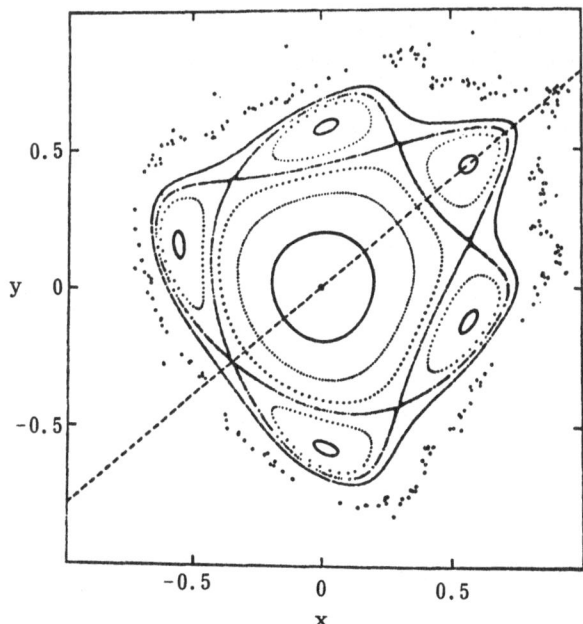

Fig.6.8. Invariant curves surrounded by a region of chaotic
behavior. After [6.4]

invariant (rational) tori. Again, but on a smaller scale, as the perturbation increases, each rational surface is destroyed, producing O and X points. Looking at
the map of Fig.6.8 with a sufficiently powerful microscope, one would see this picture reproduced on a smaller and smaller scale.

Let us sum up: modern classical mechanics, greatly aided by extensive computing
simulation, produces some new profound results bearing on classical *statistical*
mechanics. On one hand, it offers evidence that the Boltzmann H theorem —and hence
ergodicity and mixing —should not perhaps be taken for granted for systems composed
of many interacting components of *any* kind except *hard spheres*. It is true that
simulations like the one performed by *Fermi* et al. [6.2] do not extend to thousands
or millions of interacting oscillators, but just to about a hundred. On those
grounds we should perhaps argue that even if we consider just 100 hard interacting
spheres, we should not necessarily expect monotonic entropy increase. The issue
thus remains inconclusive —but fortunately this aspect is irrelevant to our theme.

On the other hand, work associated with the KAM theorem shows that even a very
small number of interacting oscillators, even a system with *three* degrees of freedom (i.e., *one* nonlinear oscillator under harmonic external excitation [6.6-8])
is, under given values of its control parameters, capable of displaying a completely chaotic —that is an aperiodic —trajectory indistinguishable from random noise.
Moreover, recently it has become increasingly apparent that the Hamiltonian "chaos"
is not fundamentally different from a similar behavior found in low-dimensional dissipative systems —except, of course, that Hamiltonian systems conserve the volume
in phase space, whereas in dissipative systems the phase-space volume *contracts*
onto an *attractor* of lower dimensionality than the original space.

6.2 Dynamics in Three-Dimensional State Space (Three Degrees of Freedom). Steady States, Limit Cycles, Attracting Tori

In the previous section we saw that one of the most interesting results of the KAM theorem is the *persistence* of "irrational" tori under small perturbations inflicted on the total energy of a Hamiltonian system. This simply means that the irrational tori display a peculiar "structural stability" for small values of the perturbing parameter ε in the expression of the Hamiltonian

$$H = H_0 + \varepsilon H_1 \quad , \tag{6.2.1}$$

where H_0 is the energy of the integrable, unperturbed system. Of course, this sort of stability has nothing to do with the *asymptotic* stability of attractors we have met so far in nonlinear dissipative systems with two degrees of freedom, namely the limit cycles. In the first place, the irrational tori (like the rational ones) form sets whose members are "nested" one within another, so they are *not* attractors. Second, in the case of asymptotic stability displayed by attractors (which characterize only dissipative systems), the attractor is structurally stable as far as perturbations of the *initial conditions* are concerned, but may be extremely sensitive to small variations of the control parameters in the vicinity of critical values.

Therefore the persistence of irrational Hamiltonian tori over small perturbations of the *control parameter*(s) of the system, although impressive in itself, has nothing to do with asymptotic stability characterizing the (stable) attractors of dissipative systems. Nevertheless, this discussion makes us wonder what type of attractors we meet in higher than two-state space in dissipative systems. More generally, what type of singularities do we meet in higher than two-state space?

Let us remind ourselves that in two-dimensional state space the only types of singularities we meet are fixed *points* (steady states), *separatrices*, and *limit cycles*. Of these, the only types of attractors (for dissipative systems) are stable steady states and stable limit cycles.

Let us start our study of the dynamics of state space in *three dimensions* by extending first what we already know to hold in two dimensions, that is, by studying a system of *three* coupled nonlinear differential equations of the autonomous form

$$\frac{dX_1}{dt} = f_1(X_1, X_2, X_3; \mu) \quad ,$$

$$\frac{dX_2}{dt} = f_2(X_1, X_2, X_3; \mu) \quad , \tag{6.2.2}$$

$$\frac{dX_3}{dt} = f_3(X_1, X_2, X_3; \mu) \quad ,$$

exactly as we did in Sect.2.2 for a two-dimensional system.

The first step in our analysis has to do with a linearized stability analysis of the steady states X_i^*, that is, with the investigation of the eigenvalues of the interaction matrix

$$A \equiv \begin{bmatrix} \alpha_{11} & \alpha_{12} & \alpha_{13} \\ \alpha_{21} & \alpha_{22} & \alpha_{23} \\ \alpha_{31} & \alpha_{32} & \alpha_{33} \end{bmatrix} ,$$

where $\alpha_{ij} = (\partial f_i / \partial X_j)_*$.

The characteristic equation will now be a cubic one, so the eigenvalues of A will be either all three *real* or one *real* and two *complex* conjugates. The eigenvector, i.e., the solution of the differential system, will then be derived from the linear superposition

$$\mathbf{x} = \sum_{i=1}^{3} c_i x_i e^{\lambda_i t} , \tag{6.2.3}$$

where λ_i are the eigenvalues of A, i.e., the solutions of the characteristic equation, c_i are integration constants, i.e., numbers depending on the initial conditions, and x_i are the solutions of the linearized system:

$$(\alpha_{11} - \lambda)x_1 + \alpha_{12}x_2 + \alpha_{13}x_3 = 0 ,$$

$$\alpha_{21}x_1 + (\alpha_{22} - \lambda)x_2 + \alpha_{23}x_3 = 0 , \tag{6.2.4}$$

$$\alpha_{31}x_1 + \alpha_{32}x_2 + (\alpha_{33} - \lambda)x_3 = 0 .$$

Now let us try to classify the possible categories of singularities.

Case A: All the Eigenvalues are Real and Negative. In this case, we get a *stable steady state that is an attractor* in three-dimensional state space. (If all eigenvalues are real and positive, we simply get an unstable steady state, i.e., a "repeller".)

Case B: All the Eigenvalues are Real, Two Negative (Say λ_1, λ_2) and the Third λ_3 Positive. Then we have a steady state which attracts all trajectories in the plane x_1, x_2 and repels them along the x_3 dimension. This is then a (three-dimensional) *saddle point*.

Case C: One Eigenvalue Real and Negative (e.g., $\lambda_3 < 0$) and λ_1, λ_2 Complex Conjugates with Negative Real Parts. In this case, on the subspace characterized by the plane x_1, x_2, the singular point behaves as a stable focus. Along the x_3 direction, the trajectory spirals *toward* the focus at the level x_1, x_2 on the surface of either a paraboloid with the x_3 axis as the axis of symmetry (the case where $0 > \mathrm{Re}\{\lambda_1,\lambda_2\} > \lambda_3$), or a "concave" cone with the x_3 axis as the axis of symmetry (the case where $\mathrm{Re}\{\lambda_1,\lambda_2\} < \lambda_3 < 0$). When the real parts of the eigenvalues are all *positive*, then one gets an unstable focus on the plane x_1, x_2.

Case D: One Eigenvalue Real and Positive (e.g., λ_3 0) and λ_1, λ_2 Complex Conjugates with Negative Real Parts. This is a more interesting case. Here, as in Case C, the singular point behaves like a stable focus on the plane x_1, x_2, but *unlike* the previous case, since now $\lambda_3 > 0$, the trajectory in three dimensions does not spiral "down" converging on the focus, but instead goes away, as shown in Fig.6.9. We call this type a *focus-saddle* singularity.

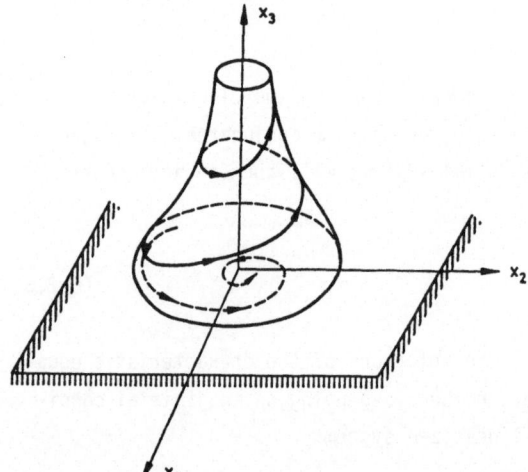

Fig.6.9. A focus-saddle singularity

An even more interesting case arises when we invert the signs of the eigenvalues, namely the case of $\lambda_3 < 0$ and $\text{Re}\{\lambda_1,\lambda_2\} > 0$. In this fashion we meet again the conditions leading (in two-dimensional state space x_1, x_2) to the Hopf bifurcation: the passage from an unstable focus to a stable limit cycle.

What we have said up to now simply (and straightforwardly) generalizes in three-dimensional state space what we already know from the two-dimensional analysis of Chap.2. Let us try next to examine the possible existence of attractors intrinsically associated with three-dimensional state-space dynamics without any counterpart on the plane. The first thing that comes to mind is to search for the possible existence of two-dimensional, *attracting* tori (inevitably, "irrational" ones, since the probability of selecting by random two frequencies ω_1, ω_2 with irrational ratios is far greater than coming up with $\omega_1/\omega_2 = n/m$, where n, m are integers). Let us remind ourselves that in such a case the trajectory on the surface of the torus (Hamiltonian or not) is quasi periodic, that is open and dense, and hence *ergodic*. A dissipative nonlinear oscillator possessing a fundamental frequency ω_1 and driven by an external periodic frequency ω_2 is a system with three degrees of freedom — displaying a trajectory in three-dimensional state space. If the oscillator in free running conditions has a limit cycle with fundamental frequency ω_1, we can imagine that the corresponding 2-torus (Fig.6.10) will have no other torus closely nested within or outside it, and that it will constitute an *attractor* in three-dimensional

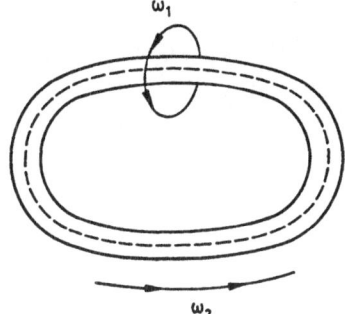

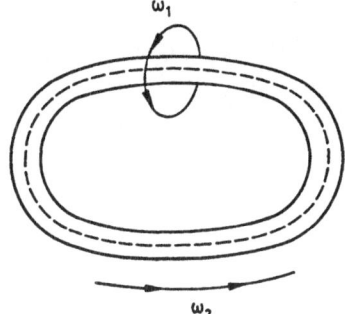

ω_1

Fig.6.10. A 2-torus

ω_2

state space. A Poincaré mapping of this torus will then give on the plane just this limit cycle.

But how can we understand the *emergence* of such a *three-dimensional attractor* via bifurcation? It is instructive to proceed in stages.

Let us then write again our three-dimensional differential system in such a form that the whole thing comes essentially into being through a nonlinear coupling of a linearized two-dimensional system x_1, x_2 with the third variable x_3.

Let x_1, x_2, x_3 stand for the perturbations around some singularity X_1^*, X_2^*, X_3^*. Suppose further that the singularity X_1^*, X_2^* is a focus. The two-dimensional system x_1, x_2 alone would then read (in canonical form)

$$\frac{dx_1}{dt} = \alpha x_1 + \beta x_2 \quad , \qquad \frac{dx_2}{dt} = -\beta x_1 + \alpha x_2 \quad , \tag{6.2.5}$$

and when coupled with the third variable x_3, this gives

$$\frac{dx_1}{dt} = \alpha x_1 + \beta x_2 + x_1 x_3 \quad , \qquad \frac{dx_2}{dt} = -\beta x_1 + \alpha x_2 + x_2 x_3 \quad ,$$

$$\frac{dx_3}{dt} = f(x_1, x_2, x_3) \quad . \tag{6.2.6}$$

Let us write $x_1 = r\cos\varphi$, $x_2 = r\sin\varphi$ so that

$$\frac{d\varphi}{dt} = -\beta \quad , \qquad \frac{dr}{dt} = \alpha r + r x_3 \quad ,$$

$$\frac{dx_3}{dt} = f(r, \varphi, x_3) \quad . \tag{6.2.7}$$

The change of variables shows explicitly that the angular velocity of rotation (say $\omega_2 = \beta$) at the plane x_1, x_2 does *not* depend on the way x_3 is coupled to x_1 and x_2. Consequently, if the parameters of the system of equations with respect to r and x_3 lead to the emergence of a limit cycle with fundamental angular frequency, say, ω_1, the most probable case will be ω_1/ω_2 is irrational. Thus the system above will display an ergodic quasi-periodic behavior.

In order then to provoke the emergence of a 2-torus attractor, it is enough to couple nonlinearly a nonlinear oscillator with a *third* variable x_3. Let us give a specific example, that is, let us specify our function $f(x_1,x_2,x_3)$. Take

$$f(x_1,x_2,x_3) = \sigma x_3 - (x_1^2 + x_2^2 + x_3^2) = \sigma x_3 - (r^2 + x_3^2) \quad . \tag{6.2.8}$$

We concentrate our analysis on the system (r,x_3), namely,

$$\frac{dr}{dt} = \alpha r + r x_3 \quad , \qquad \frac{dx_3}{dt} = \sigma x_3 - r^2 - x_3^2 \quad , \tag{6.2.9}$$

and determine under what conditions this system can give rise to a limit cycle. The steady states are

a) $r^* = 0$, $x_3^* = 0$,

b) $r^* = 0$, $x_3 = \sigma$, and

c) $x_3^* = -\alpha$, $r^* = \sqrt{-\alpha^2 - \sigma\alpha}$, $-\alpha(\alpha + \sigma) > 0$.

Let us investigate the stability properties of the above steady states one by one for small perturbations, by putting $r = r^* + \rho$ and $x_3 = x_3^* + \xi$ where $\rho \ll r^*$ and $\xi \ll x_3^*$.

a) We get by substituting in the original system:

$$\frac{d\rho}{dt} = \alpha\rho \quad \text{and} \quad \frac{d\xi}{dt} = \sigma\xi \quad . \tag{6.2.10}$$

If λ and σ are both negative, the steady state $(0,0)$ is *stable*. It becomes a *saddle* point if $\sigma > 0$, $\alpha < 0$ or $\sigma < 0$, $\alpha > 0$; it finally becomes a *repeller* if $\alpha > 0$, $\sigma > 0$.

b) In this case we get

$$\frac{d\rho}{dt} = \rho(\alpha + \sigma) \quad \text{and} \quad \frac{d\xi}{dt} = -\sigma\xi \quad . \tag{6.2.11}$$

If $(\alpha + \sigma) < 0$ and $\sigma > 0$, the steady state $(0,\sigma)$ is *stable*. It behaves as a *saddle* point for $\sigma < 0$, $(\sigma + \alpha) < 0$ and as a *repeller* for $\sigma < 0$, $(\sigma + \alpha) > 0$.

c) For this case the linearized system reads:

$$\frac{d\rho}{dt} = \alpha(r^* + \rho) + (r^* + \rho)(x_3^* + \xi)$$

$$= \underbrace{\alpha r^* + r^* x_3^*}_{0} + \alpha\rho + r^*\xi + \rho x_3^* + \rho\xi \cong r^*\xi \quad , \qquad (\rho\xi \ll 1) \quad , \tag{6.2.12a}$$

$$\frac{d\xi}{dt} = \sigma(x_3^* + \xi) - (r^* + \rho)^2 - (x_3^* + \xi)^2$$

$$= \underbrace{\sigma x_3^* - r^{*2} - x_3^{*2}}_{0} + \sigma\xi - 2r^*\rho - \rho^2 - 2x_3^*\xi - \xi^2$$

$$\cong - 2r^*\rho + (\sigma + 2\alpha)\xi \quad , \qquad (\rho^2 + \xi^2 \ll 1) \quad . \tag{6.2.12b}$$

The characteristic equation of the above system is

$$\lambda^2 - \lambda(\sigma + 2\alpha) + 2r^{*2} = 0 \qquad \text{or} \tag{6.2.13}$$

$$\lambda^2 - \lambda(\sigma + 2\alpha) - 2\alpha(\alpha + \sigma) = 0 \quad . \tag{6.2.14}$$

The state r^*, x_3^* is a *focus* if the roots of the characteristic equation are complex conjugates, that is, if $(\sigma + 2\alpha)^2 < -8\alpha(\sigma + \alpha)$. Obviously for $\sigma < -2\alpha$ the real parts of the eigenvalues are negative so the focus is stable.

Recalling our discussion in Sect.2.2.9 concerning the destabilization of a focus and the emergence of a limit cycle, we find easily in the present case that for $\sigma \geq -2\alpha$, the focus is destabilized and a limit cycle emerges with angular frequency

$$\omega_1 = \sqrt{2r^*} = \sqrt{-2\alpha(\alpha + \sigma)} \tag{6.2.15}$$

— which in general will be incommensurable with ω_2. Therefore for $\sigma_c = -2\alpha$ the periodic solution ω_2 becomes unstable, and a quasi-periodic *attracting* 2-torus appears, provided that the constraints $\sigma + 2\alpha \geq 0$, $\alpha(\alpha + \sigma) < 0$ are compatible.

Let us sum up: in three-dimensional state space a dissipative system has been found so far to possess stable steady states, limit cycles [generally of period K, where K = 2,4,6,...,2ν, which means that the trace of the trajectory on a Poincaré map is not necessarily a couple of points (K = 2), but rather a set of K ≈ 2ν discrete points visited by the trajectory on a rotating basis], and 2-tori, on which the motion is quasi periodic.

Is there anything left? The answer is yes. We are going to reveal below the existence of another most important category of attractors emerging in three-dimensional state space — called "strange" attractors.

Nonperiodic Attractors. The question which arises then is whether in three-dimensional space there exist attractors more "exotic" than steady states, limit cycles, and 2-dimensional tori. If such attractors do exist, they should first of all possess *zero* volume in three-dimensional state space. We can imagine a one-dimensional Poincaré map of such an attractor as an infinite ensemble of non-denumerable, not-interconnected segments of points along a straight line, which points the attractor visits every time it goes through a "zero-crossing" in a *nonperiodic* "chaotic" way.

Do such structures exist *in principle* along say the normalized interval [0,1]? The answer is yes, and a well-known example is the *Cantor set*, which is something between a finite ensemble of points and a continuous curve (Fig.6.11). Consider the

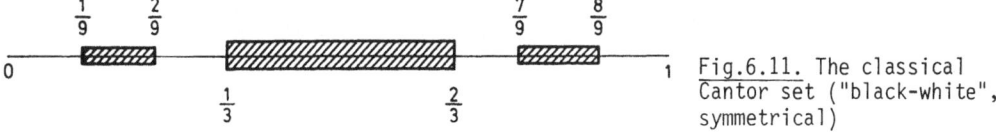

Fig.6.11. The classical Cantor set ("black-white", symmetrical)

closed segment [0,1], divide it into, say, three intervals, and take away the middle interval, leaving the limiting points 1/3 and 2/3. Repeat this procedure an infinite number of times (e.g., next take away the segment 1/9-2/9 excluding the limiting points, and the segment 7/9-8/9 likewise). We are left with an irregular, infinite, non-denumerable, and not-connected set of points of *measure null*. We can immediately see the last property of the ensemble by calculating the total length of its complement, that is, the part of the interval [0,1] taken away.

The length of this taken-away part is

$$\frac{1}{3} + \frac{2}{9} + \ldots + \frac{2^n}{3^{n+1}} + \ldots = \sum_{n=0}^{\infty} \frac{2^n}{3^{n+1}} = 1 \quad . \tag{6.2.16}$$

The corresponding (still hypothetical) attractor in three-dimensional space is neither a closed curve nor a surface with well-defined Euclidean dimensionality. It is rather a non-closing, nonintersecting curve imprisoned within a finite (subset of) volume in state space, forming within such a volume an infinite ensemble of two-dimensional sheets. It visits these sheets irregularly and has along a given direction on at least one of them a Poincaré map which constitutes a Cantor or a Cantor-like set.

6.3 Strange Attractors

6.3.1 One-Dimensional Maps on the Interval. The "Logistic" Model

Let us start this time not with a system of three coupled nonlinear ordinary differential equations but with one nonlinear *difference* equation of the "logistic" type, say

$$X_{t+1} = aX_t(1 - X_t) \quad , \qquad 0 \leqslant X_t \leqslant 1 \quad . \tag{6.3.1}$$

We shall see later that equations of this type describe fairly well a one-dimensional, single-hump Poincaré map of a number of very interesting three-dimensional "strange" attractors. For the time being, however, we will suppose that our dynamical system is merely described by the above equation. Examples of physical systems modeled in such a way include population models with non-overlapping generations. The basic question in such problems is the estimation of the nth generation from the population of the first one. Since this is our first encounter with *discrete* dynamical systems, it is natural to start our investigation (of the above difference equation) by inquiring about the stability criteria of the steady states or, in general, stationary solutions of such systems.

In the logistic equation

$$F(X_n) = X_{n+1} = aX_n(1 - X_n) \quad , \tag{6.3.2}$$

$F(X_n)$ becomes maximum for $X_n = 1/2$ and

$$F_{max} = \frac{a}{4} \quad , \quad 1 \leqslant a \leqslant 4 \quad .$$

The steady states of the above system (that is the values at which future generation populations converge) will be determined as the point or points where the
straight line passing through the origin with a slope of $45°$ meets the parabola
$F(X_n)$. We get two such points, one at the origin itself

$$X_n^* = 0 \quad , \tag{6.3.3}$$

and another for which

$$X_{n+1} = X_n = aX_n(1 - X_n) \quad \text{or}$$

$$X_n = 1 - \frac{1}{a} \quad . \tag{6.3.4}$$

In general, then, the steady states are the solution of the equation $X_n = F(X_n)$,
in the interval $[0,1]$.

Let us inquire next about the stability of the steady states. So far such criteria are established for systems observed in continuous time. For systems observed
"stroboscopically", i.e., in discrete time intervals, the situation changes: in
order to see exactly how, let us write the equations of a linearized continuous
system with respect to the (small) perturbations $x_i(t)$ around a given steady state,
namely

$$\frac{dx_i}{dt} = \sum_{j=1}^{N} a_{ij}x_j \quad , \tag{6.3.5}$$

$(i \in 1,\ldots,N)$ as

$$x_i(t + \tau) - x_i(t) = \tau \sum_{j=1}^{N} a_{ij}x_j \quad , \tag{6.3.6}$$

where N is the number of degrees of freedom of the system, $a_{ij} = (\partial f_i/\partial X_j)_*$ and τ
is the period of time between successive observations. The interaction coefficients
above are exactly the same as in the continuous system, i.e, they are the elements
of the interaction matrix A of the continuous system.

The system of (N) equation (6.3.6) can be written in a more compact form as

$$x(t + \tau) = Bx(t) \quad , \tag{6.3.7}$$

where now the elements b_{ij} of the $N \times N$ interaction matrix B are given by

$$b_{ij} = \tau a_{ij} + \delta_{ij} \quad \text{or}$$

$$B = \tau A + I \tag{6.3.8}$$

where I is the unit matrix,

$$I \equiv \begin{pmatrix} 1 & 0 & 0 & \dots & 0 \\ 0 & 1 & 0 & \dots & 0 \\ 0 & 0 & 1 & \dots & 0 \\ \vdots & & & & \\ 0 & 0 & 0 & \dots & 1 \end{pmatrix} \quad .$$

Putting $x_i(t) \sim (\lambda_i)^{t/\tau}$ and taking into account that $x(t+\tau) = \lambda x(t)$, we get instead of (6.3.7) the system

$$\lambda x(t) = B x(t) \quad . \tag{6.3.9}$$

The eigenvalues λ will be deduced now from the solutions of the characteristic equation

$$\det(B - \lambda I) = 0 \quad . \tag{6.3.10}$$

This equation *differs* from the characteristic equation of the continuous system. In the case of the discrete system, for a steady state to be stable it is necessary and sufficient that the real parts of *all* eigenvalues are within the interval $(-1,1)$, $|Re\{\lambda\}| < 1$. In our example, this means that the steady state(s) is (are) stable if the slope of the tangent to the curve $F(X_n)$ at the corresponding point(s) is between $-45°$ and $45°$ (Fig.6.12).

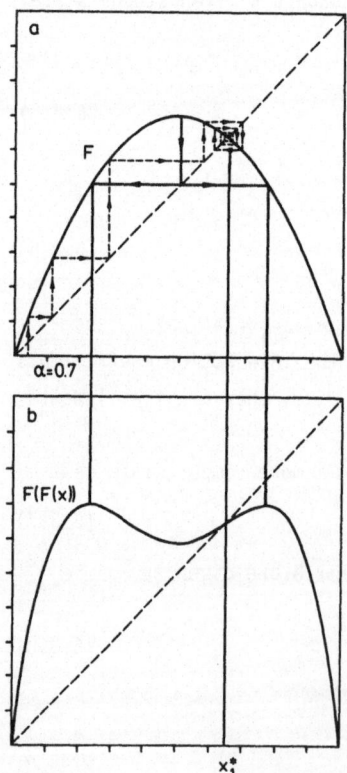

Fig.6.12a,b. The logistic map. One stable steady state. (a) The iterative process (indicated by the *dotted lines*) on the $F(x)$ map. (b) The $F(F(x))$ function. The *dashed line* has a slope of 45° and passes through the origin. After [6.9]

Let us calculate these slopes. At a point X^* the tangent to the curve
$F(X_n) = X_{n+1} = aX_n(1 - X_n)$ equals

$$\lambda^{(1)}(X^*) = \left(\frac{dF}{dX}\right)_{X=X^*} = \frac{d}{dX} [aX(1 - X)]_{X=X^*} = a[(1 - 2X)]_{X=X^*} \quad . \tag{6.3.11}$$

For the steady state $X^* = 0$, $\lambda^{(1)}(0) = a$, but since for nontrivial cases a has to be greater than one, the origin is *always* an unstable steady state. For the steady state $X_1^* = 1 - (1/a)$, we get

$$\lambda^{(1)}(X_1^*) = a - 2a\left(1 - \frac{1}{a}\right) = 2 - a \quad . \tag{6.3.12}$$

This steady state is stable if $\lambda^{(1)}(X_1^*) \leqslant 1$ or for $1 < a \leqslant 3$.

Indeed as the hump of the curve becomes steeper and $F(X)_{max}$ approaches one, a critical value of the control parameter is reached ($a_{c_1} = 3$) beyond which the steady state X_1^* becomes unstable. To find out what happens beyond the critical values $a_{c_1} = 3$, we examine the second iteration of the function $F(X)$, namely

$$X_{n+2} = F(F(X_n)) = F^{(2)}(X_n) \quad . \tag{6.3.13}$$

The steady states are now the values X_2^*, which appear as real solutions of the equation $X_2^* = F^{(2)}(X_2^*)$ in the interval $[0,1]$. As we see from Fig.6.13 we now have four

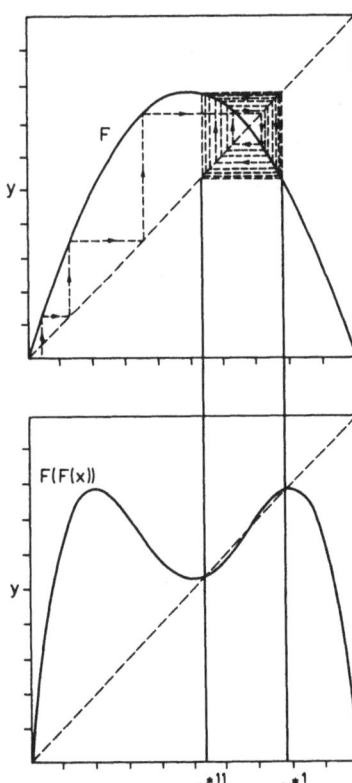

Fig.6.13. The logistic map. The previous steady state becomes unstable and a stable limit cycle of period 2 appears. After [6.9]

steady states, namely the origin, the previous point $X_1^* = 1 - (1/a)$, and two new "satellites" $X_2^{*\prime}$ and $X_2^{*\prime\prime}$ around X_1^* —solutions of the equation

$$a^2 X^* (1 - X^*)[1 - aX^*(1 - X^*)] = X^* \quad , \qquad \text{or} \tag{6.3.14}$$

$$a^3 X^3 - 2a^3 X^2 + a^2(a + 1)X + 1 - a^2 = 0 \quad , \tag{6.3.15}$$

which are all real and positive between 0 and 1.

More precisely, one root is the old state $X_1^* = 1 - (1/a)$. The two new satellite states are given by the other two roots:

$$X_2^{*\prime} = \frac{1 + a + \sqrt{a^2 - 2a - 3}}{2a} \qquad \text{and} \tag{6.3.16}$$

$$X_2^{*\prime\prime} = \frac{1 + a - \sqrt{a^2 - 2a - 3}}{2a} \quad . \tag{6.3.17}$$

Let us investigate the tangent $\lambda(x)$ to the point X_1^* on the new iterative curve $F^{(2)}(X)$. We get

$$\frac{\partial F^{(2)}(X)}{\partial X} = \lambda^{(2)}(X) = a^2 - 2a^2(a + 1)X + 6a^3 X^2 - 4a^3 X^3 \tag{6.3.18}$$

and

$$\lambda^{(2)}(X^*) = a^2 - 4a + 4 = (2 - a)^2 = \left[\lambda^{(1)}(X^*)\right]^2 \quad . \tag{6.3.19}$$

This result, relating the tangents at a given point of successive iterations, can be generalized:

$$\lambda^{(K)}(X) = \left[\lambda^{(1)}(X)\right]^K \quad . \tag{6.3.20}$$

(This implies that when the "parent" point X^* becomes unstable the emerging twin satellites are stable.)

Thus for $a > 3$ we get, via a bifurcation (at $a_{c_1} = 3$), the destabilization of the previous steady state and the appearance of two new steady states around the former. The system oscillates between these two states (Fig.6.13) on a stable asymptotic trajectory —a limit cycle of "period 2". The oscillation is asymptotically stable in the sense that it attracts any initial value X_n during the process of mapping $F(X)$ onto itself.

When a is further increased, there comes a critical value $a_{c_2} = 3.414$ beyond which the tangents at the points $X_2^{*\prime}$, $X_2^{*\prime\prime}$ (which are equal) acquire slopes above $45°$. This means that the cycle of period 2 becomes unstable, whereupon *each* of the satellites $X_2^{*\prime}$, $X_2^{*\prime\prime}$ becomes unstable and acquires now two lateral satellites —which gives rise to the emergence of an asymptotically stable oscillation of period 4. This process of cascading bifurcations of the points X^*, leading to progressive "period doubling" of the emerging limit cycles, continues as a increases further: beyond the critical value $a_{c_{n-1}}$, the cycle of period 2^{n-1} becomes unstable and the ensuing bifurcation gives rise to a stable cycle of period 2^n.

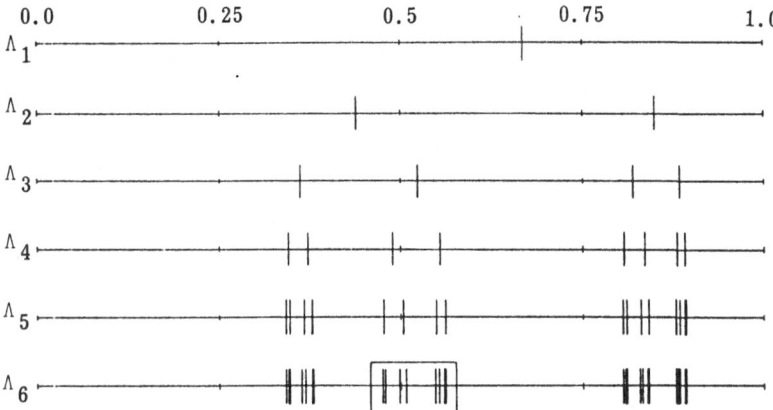

Fig.6.14. Sequential scaling of the subintervals within which the orbits are nested for successive critical values of the control parameter a, leading to period-doubling cascades before the accumulation point $a_{C\infty} = 3.57$. These intervals form in the limit an asymmetrical "black-white" Cantor-like set, whose character is determined by the Feigenbaum scaling constant μ. The successive critical values a_{ci} are related through Feigenbaum's constant δ. After[6.10]

The individual points X^* on the axis X, as it appears (see Fig.6.14), do not give rise to symmetrical ensembles: the sequence of the points X^* in the interval [0,1] is indeed very irregular. The interval [0,1] is thus partitioned into regions (subsets or bands) within which orbits exist, separated by "blank" subsets where orbits do not enter.

Exactly at the critical value $a_{C\infty} \cong 3.5700$ (known as the accumulation point) *no* periodic orbit is practically stable and the successive scheme gives rise also to an *aperiodic* trajectory of infinite period. This regime is usually called "chaos", and in spite of the fact that *no* orbit is stable, it consists of *dense, attracting* subsets on the interval [0,1] rather than discrete points. These intervals of traces of the iterating "polygon" on the X-axis form, as we say, a Cantor-like set —an asymmetric one, see Sect.6.4. Before we proceed further, however, we have to examine the way chaos *is* achieved via the process of period-doubling bifurcations.

A few years ago *Feigenbaum* [6.9] discovered, essentially with the aid of a pocket calculator, that the road to chaos is *recursive* in the following sense. The succession of the critical values a_c between 3 and ~3.57 is subjected to two "universal" constants: (1) a *scaling* coefficient $\mu = 2.50290787\ldots$, which gives the "density-packing index" of the critical points X^* between two successive generations (this means that this distance in the [0,1] interval on the X-axis between the more distant bifurcating steady state and the point 0.5 is μ times *smaller* than the distance between its "parent" and the point 0.5); and (2) a *convergence* coefficient $\delta = 4.66920\ldots$, which gives the convergence rate of the critical values a_c between three generations by the recursive formula

$$\frac{a_{c_n} - a_{c_{n-1}}}{a_{c_{n+1}} - a_{c_n}} = \delta \ .$$

(6.3.21)

Beyond the accumulation value of $a_{c_\infty} \cong 3.57$, the periodic trajectories give way to the appearance of essentially a Markov chain of stochastically interdependent strings of 0's and 1's (if we label say the left half of the interval as 0 and the right half as 1), the interdependence which distinguishes them from fair-coin-tossing random strings being due to the logistic curve. (All possible strings 01001101111010... can appear only at the limit $a = 4$, where the whole interval becomes involved.)

For $a > a_{c_\infty}$, there are an infinite number of fixed points with different periodicities and an infinite number of different periodic cycles. There are also an uncountable number of initial points X_0 that give totally aperiodic trajectories.

As a increases above 3.57, at first all these cycles have *even* periods. There comes, however, a value of the control parameter at $a = 3.6786$ at which the first *odd* period cycle appears. At first these odd cycles have very long periods, but as the control parameter continues to increase, cycles with shorter and shorter odd periods emerge, until at last the three-point cycle appears at $a = 3.8284$. Beyond this point, there are cycles with periods of *every* integer as well as an uncountable number of asymptotically aperiodic trajectories. "Period 3 implies chaos". The term "chaos" means here the existence of dynamical trajectories which are indistinguishable from some stochastic processes. [The truth is that for the logistic equation involved here, for any specific parameter value there is *one* unique cycle that is stable and that attracts essentially *all* initial points; that is, there is *one* cycle that "owns", so to speak, almost all initial points. The remaining infinite number of other cycles, together with the asymptotically aperiodic trajectories, own a set of points which, although uncountable, have measure zero (we will come back to this point in Sect.6.4.1, a propos of the discussion of the dimension of the Cantor set). Any one particular stable cycle happens for an *extraordinarily* narrow window of the control parameter. This fact, together with the long transient times associated with the initial conditions, make it *impossible to reveal* the unique cycle in practice, and a stochastic description of the dynamics is likely to be appropriate, in spite of the underlying deterministic character of the process.]

So far we have explained how the original steady state X^* bifurcates in a cascading fashion to give harmonics of period 2^n (where $n \to \infty$ at the accumulation point $a_{c_\infty} = 3.57$). The problem is how new cycles of period $K \times 2^n$ emerge. The process is illustrated in Fig.6.15, which shows how period-3 cycles arise on the curve $X_{t+3} = F^{(3)}(X_t)$.

If the hump in $F(X)$ is sufficiently steep (i.e., a sufficiently large), the threefold iteration will produce a function $F^{(3)}(X)$ with four humps, as shown in

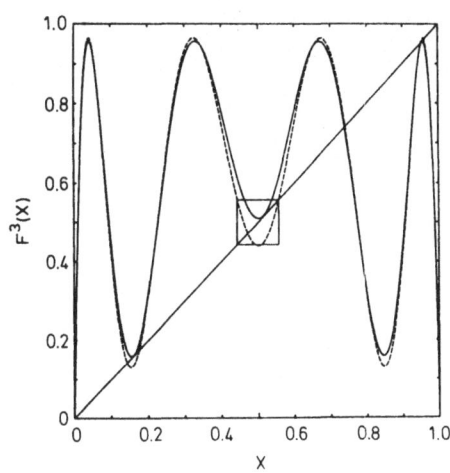

Fig.6.15. The initiation of the tangent bifurcation process (see text)

Fig.6.15. At first (for a < 3.8284), the 45° line intersects this curve only at a single point X^{*} (and at the origin, of course). As the hump steepens, however, the "hills" and "valleys" in $F^{(3)}(X)$ become more and more pronounced, until *simultaneously* the first two valleys on the left and the final hill on the right rise just to touch the diagonal line and subsequently to intercept it at six *new* points, as shown, for a value of a > 3.8284. These six points divide into two distinct three-point cycles.

Indeed, the stability-determining slope of $F^{(3)}(X)$ at three of these points has a common value, which is $\lambda^{(3)} = +1$ at their "birth" and thereafter steepens beyond +1, so this period-3 cycle is never stable. The slope of $F^{(3)}(X)$ at the other three points begins at $\lambda^{(3)} = -1$ and then levels towards zero, resulting in a stable cycle of period 3. As F(X) continues to steepen further, the slope $\lambda^{(3)}$ of this initially stable three-point cycle decreases beyond -1, so the cycle becomes unstable and gives rise, by the bifurcation process known thus far, to stable cycles of period 6,12,24,..., in general, 3×2^{n}. This birth of a stable *and* unstable pair of period-3 cycles and the subsequent harmonics, which emerge as the initially stable cycle becomes unstable, has also an "accumulation point" at a = 3.8495. Between a ≅ 3.57 and a ≅ 3.8284, no cycle is stable, and also beyond a = 3.8495, no cycle is stable.

Thus we witness two different kinds of bifurcation processes:

1) Truly *new* cycles of period K arise in pairs (one stable the other unstable) as the hills and valleys of higher iterations of F(X) move respectively up and down to intercept the 45° line. Such cycles are born at the moment when the hills and valleys become tangent to the 45° line, and the initial slope of the curve $F^{(K)}(X)$ at these points is $\lambda^{(K)} = +1$. This type of bifurcation is called *tangent bifurcation.*

2) Alternatively, an originally stable cycle of period K may become unstable as F(X) steepens. This happens when the slope of $F^{(K)}(X)$ at these period-K points

steepens beyond $\lambda^{(K)} = -1$, whereupon a new and initially stable cycle of period 2K emerges. This type of bifurcation is called *pitchfork bifurcation*.

We conclude, therefore, that as the control parameter in F(X) is varied, the *fundamental* stable dynamical behaviors are cycles of basic period K, arising by tangent bifurcations, along with their associated cascade of harmonics of period $K \times 2^n$, which arise by pitchfork bifurcations. The entire range of the control parameter values $1 < a < 4$ may thus be regarded as made up of infinitely many "windows" of parameter values — some large, some very small — each corresponding to a single value of K, that is, to one group of the above basic dynamical units.

These windows are divided from each other by the accumulation points of the harmonics of period $K \times 2^n$, *at which* the system is really chaotic with no attractive cycle. Although there are infinitely many such parameter values, they have measure zero on the interval [1,4]. At $a = 4$, the whole interval is mapped onto itself. The most important result, which has so far arisen from the above analysis of the logistic equation, is that even in the simplest, one-dimensional nonlinear discrete system — in modeling or simulating, for example, the evolution of a population of non-overlapping generations — forecasting is virtually impossible. Consequently, when one starts with a fixed population in the present generation, for sufficiently large values of the control parameter one comes to the conclusion that due to extreme sensitivity of the dynamical behavior to the slightest fluctuations in the initial conditions or in the control parameter a, the population of even the next generation may display an essentially stochastic behavior, that is, it may oscillate in a non-periodic fashion without converging toward any definite stable value. To put it another way, two neighboring points X_1 and X_2 along the interval [0,1] — used as initial conditions — may within a very small number of iterations give rise to utterly different orbits; they essentially diverge exponentially with time.

6.3.2 Fractal Dimensionality. The Cantor Set

Feigenbaum's "contraction" (or scaling) ratio μ imposes on the distribution of points/states on the interval [0,1] a *self-similar character*.[2] Before proceeding further in investigating other salient parameters, it is fundamental to examine what we call the "fractal" properties and the "fractal" dimensionality of a wide category of curves or processes.

Suppose we are given a very irregular segment of a curve (non-differentiable at an infinite number of points — such as the trajectory described by the Brownian motion of a particle in a liquid) and we are asked to calculate its length between two specific points A and B as well as its dimensionality. [The question is meaning-

---

2 By self-similarity we imply the property of an object whose structure as observed on one length scale is repeated on successively smaller scales.

ful in our context because we are interested in knowing *more formally* something
about the structure of the set of states left on the interval [0,1] beyond the
first accumulation point $a_c \cong 3.57$, and especially the "total length" the states
occupy and the dimensionality of that length. One is perhaps tempted to conclude
that the dimensionality of the structure is 1, but we will see soon that, in general,
this is not the case.

We are given, then, a very irregular curve and are asked to calculate its length.
To proceed, we first choose the length of a measuring stick, say G. Then we try
essentially to count the number of sides, all of length G, of an open polygon whose
corners lie on the curve. If G is small enough, it does not matter whether we start
from the end A or the end B of the curve. Thus we get an estimate of the length
L(G). The crucial point is that unlike the case of a regular, perfectly differen-
tiable curve with radius of curvature $R_i \geqslant G$, in our case L(G) depends greatly on G.
Consequently, it is necessary to know L(G) for several values of G; it would be
preferable to know an analytic formula for the function L(G), say of the form
$L(G) \sim AG^{1-D}$, where A is a (positive) constant and D is a constant equal to or
greater than unity; This D is called the "fractal" dimensionality of the curve under
investigation, and may be a *non-integer* number.

More generally, in order to understand the nature of fractal dimensionality, let
us start with the straight line in Euclidian space which has dimension one. Hence
for every positive integer N, the segment $0 \leqslant x < X$ can be exactly decomposed into N
non-overlapping segments of the form

$$\frac{(n-1)X}{N} \leqslant x < \frac{nX}{N} \quad , \tag{6.3.22}$$

where n runs from 1 to N. Each of these parts is deducible from the whole by a similarity with ratio $r(N) = 1/N$.

Similarly a plane has dimension two. Hence for every perfect square number N,
the rectangle $0 \leqslant x < X$, $0 \leqslant y < Y$ can be decomposed exactly into N non-overlapping rec-
tangles of the form

$$\frac{(K-1)X}{\sqrt{N}} \leqslant x < \frac{KX}{\sqrt{N}} \quad ,$$

$$\frac{(\xi-1)Y}{\sqrt{N}} \leqslant y < \frac{\xi Y}{\sqrt{N}} \quad , \tag{6.3.23}$$

where K and ξ run from 1 to $\sqrt{N}$. Each of these parts is deduced from the whole by a
similarity with ratio $r(N) = 1/\sqrt{N}$. In general, whenever $N^{1/D}$ is a positive integer,
a D-dimensional rectangular parallelepiped can be decomposed into N parallelepipeds
deducible from the whole by a similarity ratio $r(N) = 1/N^{1/D}$. Thus the dimension D
is characterized by the relation

$$D = -\frac{\ln N}{\ln r(N)} = \frac{\ln N}{\ln \frac{1}{r(N)}} \quad . \tag{6.3.24}$$

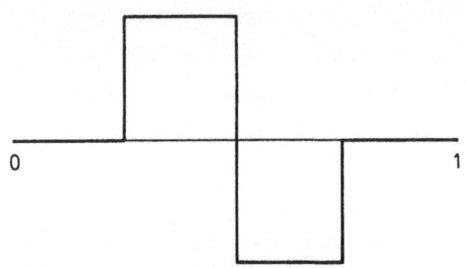

Fig.6.16. The first step in the formation of a fractal curve through self-similar scaling (see text)

0   1

Example

Consider the curve of Fig.6.16 in the interval (0,1). It is composed of $N = 8$ "legs". Replace each of its N parts by a curve deduced from the whole drawing through a similarity ratio $r(N) = 1/4$. One is left with a curve made up of N^2 legs of length $(1/4)^2$. Next replace each leg of the new curve by a curve obtained from the whole drawing through a similarity with ratio $(1/4)^2$. The desired self-similar curve is obtained, or rather approached, by an *infinite* sequence of such steps. Clearly the total length of such a curve in the unit interval tends to *infinity*, and its fractal dimensionality is

$$D = \frac{\ln 8}{\ln 4} = 1.5 \quad .$$

Let us now try to calculate the fractal dimensionality of a *Cantor set*. It will be defined (for any bounded set of points) via the number N of small "balls" of size ε needed to cover the set. If this number increases for $\varepsilon \to 0$ as $N(\varepsilon) = \text{const.}\varepsilon^{-D}$, then D is the "fractal" or Hausdorff dimension of the set.

We will have again

$$D = \lim_{\varepsilon \to 0} \frac{\ln N(\varepsilon)}{\ln(1/\varepsilon)} \quad . \tag{6.3.25}$$

If we are content to know where the set lies within an accuracy ε, then to specify the location of the set we need only specify the positions of $N(\varepsilon)$ "spheres" covering the set. Hence the Hausdorff dimension may be interpreted as telling us *how much information is necessary to specify the location of the set to within a given accuracy*. For the Cantor set $\varepsilon = 1/3$ and $N = 2$ for the first division, $\varepsilon = 1/9$ and $N = 4$ for the second division, ..., $\varepsilon = (1/3)^\rho$, $N = 2^\rho$ for the ρth division, so

$$D = \frac{\ln 2}{\ln 3} = 0.630 \quad .$$

Therefore the fractal dimensionality of the Cantor set is *less* than its geometrical dimensionality —a necessary (but not sufficient) condition to specify an attractor that is a *compressor* of information. We note finally that the Cantor set as constructed above has the property of *scale invariance*, that is, the set between 0 and 1 looks precisely the same as, for example, that part between 0 and 1/3, if the latter is examined under a magnification of factor 3.

It remains now to see whether or not the set of states left in the unit interval at the accumulation point $a_c = 3.57$ of the logistic curve *does*, in fact, constitute a Cantor-like set.

The scaling coefficient μ of Feigenbaum shows that in the interval [0,1] and
at the critical accumulation point $a_c = 3.57$, we have a set of disjoint intervals
with self-similar characteristics. In that respect our set is a Cantor-like set,
and its fractal dimensionality is calculated to be ~0.538, see Sect.6.4. However,
the fact that the fractal dimensionality of such a set of points is below one does
not necessarily prove that we have an attractor. For that we need to investigate
yet another parameter which provides some measure of the degree of randomness generated by the deterministic equation —a parameter related to *orbital stability*.

Orbital stability for a given cycle depends on the behavior of its neighboring
orbits. If points near an orbit converge toward it, then it is stable to small perturbations and is said to be locally stable. An orbit will be attracting, that is,
it will be *asymptotically* stable, if on average the orbit is locally stable. In the
case of one-dimensional maps, these stability criteria are estimated directly from
the slope of the map at points visited by an orbit. In particular, if the slope at
a point is less than one, nearby points will be brought closer to it at the next
iteration of the map. Similarly, if the slope at a point is greater than one, nearby points will be taken further away under iteration of the map. An asymptotically
stable orbit then requires *the average* of these slopes along the orbit to be *less*
than one. When the average is greater than one, the orbit is unstable; consequently, initial small deviations from the orbit will increase under iteration of the
map. To put it in a different way, an average slope above one is an indication that
the flow generates *variety* (potential information), while an average slope below
one is an indication that *compressibility* and dissipation of the generated variety
take place; hence relevation of information.

6.3.3 The Concept of the Lyapounov Exponents for the Period-Doubling and Chaotic Regimes

Sensitivity to small fluctuations, i.e., to some initial uncertainty in the initial
conditions (in specifying or measuring a state), is of decisive importance in characterizing the behavior of a dynamical system. In such cases, the initial uncertainty
grows exponentially with time until one can no longer predict the next state of the
system. The information about the initial state is lost in a finite amount of time
and so beyond this time (to be calculated in specific examples later on) the system
is effectively unpredictable. This sensitivity can be considered a basic feature of
chaos. (Remember our "precursor" example in Sect.2.3.2.)

To measure the average local stability or the rate at which new information is
generated by the dynamical flow or, alternatively, the rate at which information
about an initial state is lost, the pertinent parameter to use is the Lyapounov

characteristic exponent λ. The information change per iteration on the map is computed from the slope dF/dx of the map as

$$I = \log_2 \left| \frac{dF}{dx} \right| \quad [bits] \quad . \tag{6.3.26}$$

If $\left| dF/dx \right| < 1$, the map at the specific point acts as an information sink, and if $\left| dF/dx \right| > 1$, as an information source.

If $P(x)$ is the asymptotic probability distribution of an orbit at a given parameter value a, then the *average* amount of information change $<I>$ is the Lyapounov characteristic exponent for that particular value of the control parameter, namely

$$<I> = \lambda(a) = \int_0^1 P(x) \log_2 \left| \frac{dF(x,a)}{dx} \right| dx \quad . \tag{6.3.27}$$

If one assumes ergodicity of the orbit within the attractor, there is an alternative form for $\lambda(a)$, namely

$$\lambda(a) = \lim_{N \to \infty} \frac{1}{N} \sum_{i=1}^N \log_2 \left| \frac{dF(x,a)}{dx} \right|_{x_i} \quad , \tag{6.3.28}$$

where N is the number of iterations that the particular orbit undergoes in the map.

The calculation of the asymptotic probability distribution of the specific orbit goes as follows: We arbitrarily assume some initial probability density $P_1(x)$ (say equal to one everywhere) along the interval. The resulting value of its first iteration $P_2(x)$ corresponds to the original $P_1(x)$, transformed by the action of the map. The number density of trajectories impinging on some small interval in $F(x)$ is equal to the densities at the inverse points of the mapping weighted by the slope at those points. This expresses the law of the conservation of probability or the conservation of trajectories. If $x_i = F^{-1}(x)$ is the inverse function of $F(x)$, then

$$P_2(x) = \sum_{i=1}^2 \frac{P_1(x_i)}{\left| dF/dx \right|_{x_i}} \quad . \tag{6.3.29}$$

In general, if $x_i^{(n)} = [F^{-1}(x)]^{(n)}$ is the inverse function of $F^{(n)}(x)$, then

$$P_n(x) = \sum_{i=1}^{2^{n-1}} \frac{P_1(x_i^{(n-1)})}{\left| \frac{dF^{(n-1)}(x)}{dx} \right|_{x_i^{(n-1)}}} \quad . \tag{6.3.30}$$

Successive iterations of this procedure will produce closer and closer approximations to the correct equilibrium value of $P(x)$. [If the map has a stable periodic solution, $P(x)$ converges to sharp spikes on the periodic orbits, i.e., in this case $P(x)$ will be composed of a series of delta functions. In the chaotic regime, where only part of the interval is covered, $P(x)$ may also be discontinuous but nonzero over a finite range of x.]

Thus, with the above definition of the Lyapounov exponent, we see that

for $\lambda(a) < 0$, the orbit is stable and periodic (the iterative process acts as an information sink),

for $\lambda(a) = 0$, the orbit is marginally stable (information is neither produced nor dissipated); and

for $\lambda(a) > 0$, the orbit is locally unstable and chaotic (the iterative process acts as an information source).

A *steady state* in three-dimensional space has three Lyapounov exponents, all negative $(-, -, -)$; a *limit cycle* in three-dimensional space has three Lyapounov exponents, two of them negative and one (along the flow) zero $(-, -, 0)$; and a *strange attractor* is characterized by $(+, 0, -)$, that is, one positive Lyapounov exponent λ_+, one zero λ_0, and one negative λ_-. In other words, a *strange* attractor creates information along a given direction and compresses information along another direction. Along the flow, the Lyapounov exponent is zero.

However, to be an *attractor* the system must have $\Sigma_{i=1}^3 \lambda_i < 0$, i.e., the absolute value of the negative Lyapounov exponent must be greater than the value of the positive Lyapounov exponent. The above sum gives the average volume contraction rate in state space.

Now let us investigate for the logistic map the behavior of the Lyapounov exponent as a function of the control parameter a. The slope of the logistic curve at a given point x_i is

$$\left[\frac{dF(x,a)}{dx} \right]_{x=x_i} = a(1 - 2x_i) \quad . \tag{6.3.31}$$

For a period K orbit $\{x_i\}$ the probability density $P(x)$ is a set of K delta functions $\delta(x - x_i)$; so

$$\lambda(a) = \sum_{i=1}^K \ln \left| \frac{dF(x,a)}{dx} \right|_{x_i} \quad . \tag{6.3.32}$$

(The choice of the logarithmic base does not matter; when the result is quoted in "bits", the base is 2.)

For the period-1 orbit $(1 < a < 3)$ the characteristic exponent is then

$$\lambda(a) = \ln |2 - a| \quad .$$

It is obvious that for this periodic orbit $\lambda(a) < 0$. However, for $a = 1$ and $a = 3$, $\lambda(1) = 0$ and $\lambda(3) = 0$, while for $a = 2$, $\lambda(a) \to \infty$. What is this supposed to mean? The period-doubling "tangent" bifurcations occur where $\lambda(a)$ vanishes. That is to say, an orbit must first pass through a neutrally stable attractor before it can take on qualitatively different structures. Between these bifurcations, $\lambda \to -\infty$ as the point $x_c = 0.5$, where $F(x_c)$ is maximum, becomes a point on the periodic orbit, since

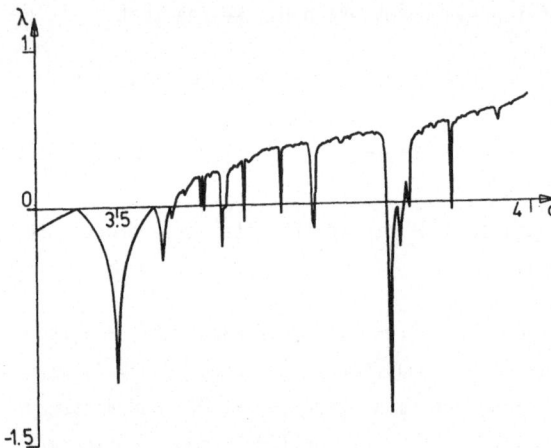

Fig.6.17. Variation of the Lyapounov exponent $\lambda(a)$ as a function of the control parameter a
for the logistic map

at this point the tangent becomes infinite. For the period-1 orbit, this happens
for a = 2. For the period-2 orbit (a > 3), we have

$$\lambda(a) = \ln|a^2 - 2a - 4| \quad .$$

Again, $\lambda(a)$ is everywhere ≤ 0. It becomes zero for a = 1 +$\sqrt{6}$, and goes again to $-\infty$
when the critical point x_c = 0.5 becomes part of the period-2 orbit; this happens
for a = 1 +$\sqrt{5}$ = 3.236068. For higher-order orbits up to the accumulation point
a_c = 3.57, this story repeats itself (see Fig.6.17), that is, the Lyapounov characteristic exponent remains *non*-positive (the system behaves as an information sink)
and diverges at orbits for which the running point x_i approaches the critical point
x_c = 0.5 —where the slope vanishes.

Finally, let us see what happens as the accumulation point of the control parameter $a_c \cong 3.57$ is reached and exceeded. When a = a_c exactly, $\lambda(a_c)$ = 0, and for
a > a_c, λ > 0, i.e., the system acts as an information source. It can be shown that
the envelope of $\lambda(a)$ near a_c displays universal behavior (reminiscent of an order
parameter near the critical point of a phase transition). That is to say, one can
write

$$\lambda(a) = \lambda_0(a - a_c)^\tau \quad , \tag{6.3.33}$$

where τ =~0.4498 and λ_0 is a constant. In general, the curve $\lambda(a)$ beyond a_c smoothly rises as the parameter a is increased, reaching finally one bit per iteration as
the map becomes strictly two-to-one, that is, as a →4. Cascades of higher periodicities in the chaotic regime are indicated by windows of negative λ.

It is now time, before attempting to investigate possible relationships between
the Lyapounov characteristic exponent and the fractal dimensionality of the map,
to examine the "real thing", that is, a three-dimensional strange attractor, and to
try to find additional appropriate parameters for the characterization of its dynamical behavior, especially concerning its *compressibility* properties. We should
not forget that in the present context we are interested in such dynamical systems

only as far as their use as *information processors* goes, so we will try to understand how, via devices such as strange attractors, information is generated and
dissipated, that is, *processed*. We intend to propose that information in general
is produced either via cascading *bifurcations*, giving rise to *broken symmetries*,
or via cascading *iterations*, giving rise to *increases in resolution*. As the resolution increases, types of intrinsic microscopic noises, washed out when the
"window" of integration in space and time was wide enough, appear now as fluctuating
entities, and they result in a passage (of these microfluctuations) from microscopic to macroscopic hierarchical levels. It is this process which ultimately
produces *new* variety, i.e., information and entropy along the flow of a dynamical
system possessing a strange attractor.

6.3.4 A Typical Three-Dimensional Strange Attractor. The Lorenz Model

The combination of sensitivity to initial conditions with only approximate knowledge about these initial conditions leads to an inability to make accurate long-
term predictions about the evolution of systems composed by even *three* nonlinearly
coupled variables. From the practical point of view, one of the most important
sensitive systems is the atmosphere. Lorenz suggested in 1963 [6.11] that its dynamics is severely sensitive to initial conditions. The implications of this for
weather prediction are quite serious —even if massive improvements in modeling and
data collection were made.

Yet Lorenz found that by severely truncating the Navier-Stokes equations (describing a system with an infinite number of degrees of freedom) one could end up with
a nonlinear system with just *three* variables preserving many of the characteristics
of the initial system. This "compressed" model we intend to study now in some detail, because it happens to be very representative of three-dimensional systems possessing strange attractors and capable of an unbelievably rich behavior repertoire.

Let us consider the system:

$$\frac{dX}{dt} = \sigma Y - \sigma X \quad , \qquad \frac{dY}{dt} = -Y + rX - XZ \quad ,$$

$$\frac{dZ}{dt} = -bZ + XY \quad ,$$

$$(6.3.34)$$

where σ, r, and b are control parameters; what X, Y, Z stand for is of no importance in the present context. Let us keep b fixed (as happens in the case of a
number of hydrodynamics applications) and investigate the dynamics of the above
model by changing the (real and positive) values of the parameters σ and r.

As usual, we start with the study of the simplest singularities, that is, of
the steady states, and their (linearized) stability analysis. The steady states are

$$\text{(a)} \quad X_1^* = Y_1^* = Z_1^* = 0 \quad ,$$

and

(b) $X_2^* = Y_2^* = \pm\sqrt{b(r-1)}$, $Z_2^* = r - 1$.

Let us now linearize the Lorenz equations around each of the above two steady states, i.e., we put $X = X^* + x(t)$, $Y = Y^* + y(t)$, $Z = Z^* + z(t)$.

For the steady state (a) we find

$$\frac{dx}{dt} = \sigma(y - x) \quad , \qquad \frac{dx}{dt} = rx - y \quad ,$$

$$\frac{dz}{dt} = -bz \quad .$$

(6.3.35)

The characteristic equation of this system is

$$\begin{vmatrix} -\sigma-\lambda & \sigma & 0 \\ r & -\lambda-1 & 0 \\ 0 & 0 & -b-\lambda \end{vmatrix} = 0 \quad , \qquad \text{or}$$

$$(b + \lambda)[\lambda^2 + (\sigma + 1)\lambda + \sigma(1 - r)] = 0 \quad ,$$

(6.3.36)

where λ are the eigenvalues of the interaction matrix.

For $r > 0$, the characteristic equation has three real values. For $r < 1$, all three eigenvalues are negative so the steady state (a) is stable. For $r > r_c = 1$, one eigenvalue becomes positive, a bifurcation takes place, and the state (a) becomes a saddle point in three dimensions; nevertheless, this state behaves as a stable steady state in two dimensions.

For the steady state (b), the linearized system becomes

$$\frac{dx}{dt} = \sigma(y - x) \quad , \qquad \frac{dy}{dt} = x - y - \sqrt{b(r-1)}z \quad ,$$

$$\frac{dz}{dt} = \sqrt{b(r-1)}(x + y) - bz \quad .$$

(6.3.37)

The corresponding characteristic equation is now

$$\begin{vmatrix} -\sigma-\lambda & \sigma & 0 \\ 1 & -\lambda-1 & -\sqrt{b(r-1)} \\ \sqrt{b(r-1)} & \sqrt{b(r-1)} & -b-\lambda \end{vmatrix} = 0 \qquad \text{or}$$

$$\lambda^3 + (\sigma + b + 1)\lambda^2 + b(\sigma + r)\lambda + 2b\sigma(r - 1) = 0 \quad .$$

(6.3.38)

When $r > 1$, the product of the roots of (6.3.38) is a *negative* real number, which means that at least one root, say λ_1, is *real and negative*, the others (λ_2, λ_3) being either real and of the same sign or complex conjugates. In the vicinity of $r \geqslant r_c \sim 1$, the real parts of these eigenvalues are negative.

The problem is now to estimate what happens to the above steady state when the parameter r increases further. A necessary condition for an instability is that the roots λ_2 and λ_3 are complex conjugates, since $2\sigma b(r-1) > 0$. Let us try to find the critical value r_c', for which $\lambda_{2,3} = \pm j\Lambda$. Under such conditions, the characteristic equation becomes

$$\lambda^3 - \lambda_1 \lambda^2 + \Lambda^2 \lambda - \lambda_1 \Lambda^2 = 0 \quad . \tag{6.3.39}$$

By comparing (6.3.39 and 38), we deduce

$$r'_c = \frac{\sigma(\sigma + b + 3)}{\sigma - b - 1} \quad ,$$

$$\tag{6.3.40}$$

$$\Lambda = \pm\sqrt{b(r'_c + \sigma)} \quad , \quad \text{and}$$

$$\lambda_1 = -(\sigma + b + 1) \quad .$$

If $\sigma < b + 1$, no positive value exists for r'_c.

Now the crucial point is what happens for $r \gg r'_c$? This behavior of the system was first investigated numerically by *Lorenz* in 1963 [6.11], for $r = 28$, $\sigma = 10$, and $b = 8/3$. Lorenz found that the system starts with a rotation around one of the (unstable) focuses with an amplitude increasing with time, thereby forming a divergent spiral. After a number of such oscillations, the system *suddenly* leaves this regime and goes monotonically towards the second available focus, around which it again begins an oscillatory motion along a divergent spiral. Again, after a certain number of oscillations around this focus, the system jumps towards the vicinity of the previous focus, from which it again begins a divergent oscillatory trajectory, and so on. The interesting thing is that the time intervals the system spends in the vicinity of each focus before jumping into the vicinity of the other are stochastically distributed, and there is no regularity whatsoever in the process, which nevertheless is created by the unfolding of a deterministic (nonlinear) dynamics. Also, the number of divergent oscillations the system undergoes around each of the two available foci are apparently irregular and therefore totally unpredictable.

Figure 6.18 gives the projections of the trajectories of the Lorenz system for the above values of the parameters in state space, on the XY- and XZ-planes.

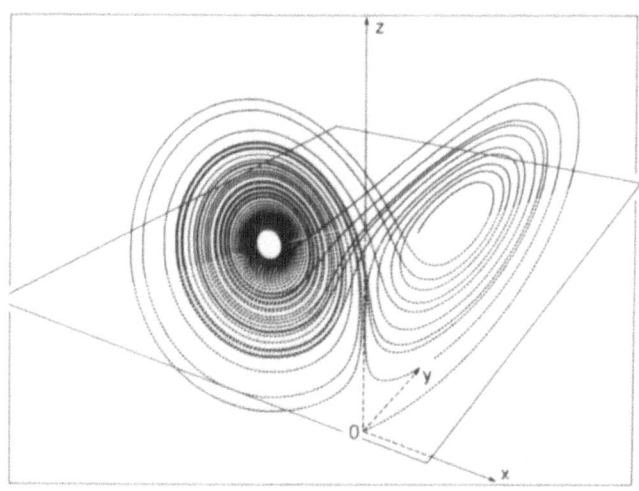

Fig.6.18. The trajectory of the Lorenz attractor on the XY- and XZ-planes for $r = 28$, $\sigma = 10$, $b = 8/3$ in both cases

The "chaotic" behavior shown in this figure is just one of many in a rich repertoire of different modes of chaos that the simple Lorenz model is capable of. Indeed, for higher values of the control parameter r, a succession of new dynamical regimes appear —always investigated via computing simulation. We end the discussion on the Lorenz system proper by giving a brief description of two most interesting phenomena, namely "intermittency" and "metastable" chaos.

Let us review more precisely the behavior repertoire of the Lorenz system (we still do not know if it is an "attractor"!) as investigated by computer simulation for a rather wide range of the control parameter r. For r <1 the origin (0,0,0) is the only stable steady state.

At r = 1 one observes a continuous supercritical bifurcation towards a new regime represented by the two *new* stable steady states $(\pm\sqrt{b(r-1)}, \pm\sqrt{b(r-1)}, (r-1))$. These two states remain stable until the value

$$r'_c = \frac{\sigma(\sigma + b + 3)}{(\sigma - b - 1)}$$

is reached; for b = 8/3, σ = 10, r'_c = 24.74.

At r = r'_c = 24.74, as the complex eigenvalues cross the imaginary axis, there is a Hopf bifurcation in which the hitherto stable states lose their stability. (The bifurcation is "supercritical" if each state loses its stability by "expelling" a stable periodic orbit. It is "subcritical" if each loses its stability by "absorbing" an unstable periodic orbit. In this case it has been shown that the bifurcation is subcritical.) For r > 24.74 both steady states become unstable. The flow, linearized around each one of them, has one *negative real eigenvalue* and a complex conjugate pair of eigenvalues with *positive* real parts. Therefore at r = 24.74 an inverted or "subcritical" bifurcation takes place, separating the stationary state(s) from a time-dependent non-periodic state. Below 24.74 the situation is complicated: there exist several particular values of r where new features appear, such as unstable periodic orbits as well as aperiodic ones, but there is no attracting limit cycle corresponding to a stable periodic motion. This regime reveals itself through a "metastable" chaos and a hysteretic behavior.

To see this more clearly, consider a sketch of the one-dimensional Poincaré map performed on the YZ-plane of the attractor (Fig.6.19). A portion of this map has a slope less than one, and it has a stable fixed point at the origin. Any trajectory falling at points with x less than x_0 will be attracted to the origin. However, a trajectory initiated *above* x_0 may move chaotically for quite a while before becoming trapped in the region between zero and x_0. The parameter r can be adjusted to make the trapping region $0 - x_0$ as small as desired, resulting in arbitrarily long "half-lives".

Metastable chaos can, then, be described as a strange attractor with a finite probability of decaying into a fixed point or to a limit cycle; it implies a transitory regime whose decay times have an exponential distribution. Almost every

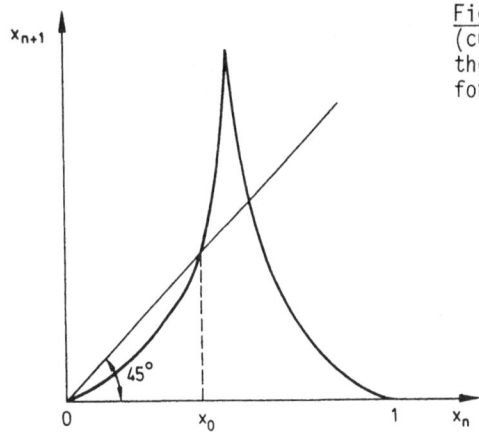

Fig.6.19. Sketch of the one-dimensional (cusplike) map of the Lorenz attractor, on the YZ-plane, illustrating the conditions for the existence of metastable chaos

trajectory eventually settles down to a periodic orbit after some initial period of irregular oscillation. The mean duration of the chaotic regime generally is not long, typically about 50 iterations, but by trial and error initial conditions on the map can be found which remain chaotic much longer. In other situations a metastable chaotic regime might decay to a steady state or to one which is also chaotic, but which has significantly different behavior, i.e., which has a different probability density function and occupies a different subinterval of the whole attractor. On the other hand, the new regime might also be metastable, eventually decaying to a third regime or possibly back to the original one.

The strange aperiodic attractor proper appears for $r > 24.74$ and then slowly evolves up to $r \sim 145$. From this point on, a complicated sequence of changes occurs in the topology of the attractor until $r = 148.4$, where a stable limit cycle sets in [6.12].

Beyond $r_c = 166.07$, the limit cycle to which the strange attractor degraded becomes unstable, and the system displays a new dynamical behavior called "intermittency". Loosely speaking, intermittency (realized for some windows of the control parameter r in the Lorenz system) refers essentially to the physical inability to achieve perfect entrainment or stable, strict locking between the phase of a dissipative nonlinear oscillator and the phase of an external periodic excitation. At the time the limit cycle in state space is about to close, a "scrambling" process, as it were, intervenes and destroys the orderly trajectory. It creates for a brief time interval a chaotic motion amounting to broadband noise or to a broadening of peaks in the power spectrum, and then the building up of a limit cycle (a different one) starts all over again. After a while, a new bursting transition to "turbulent" chaos occurs, followed by another limit cycle appearance, and so on, the time sequence of intermittent points being essentially random.

More formally, we can understand the principle of intermittency by resorting to the Poincaré map of the Lorenz attractor around the value $r \sim 167$ of the control parameter, in the YZ-plane, as before (Fig.6.20).

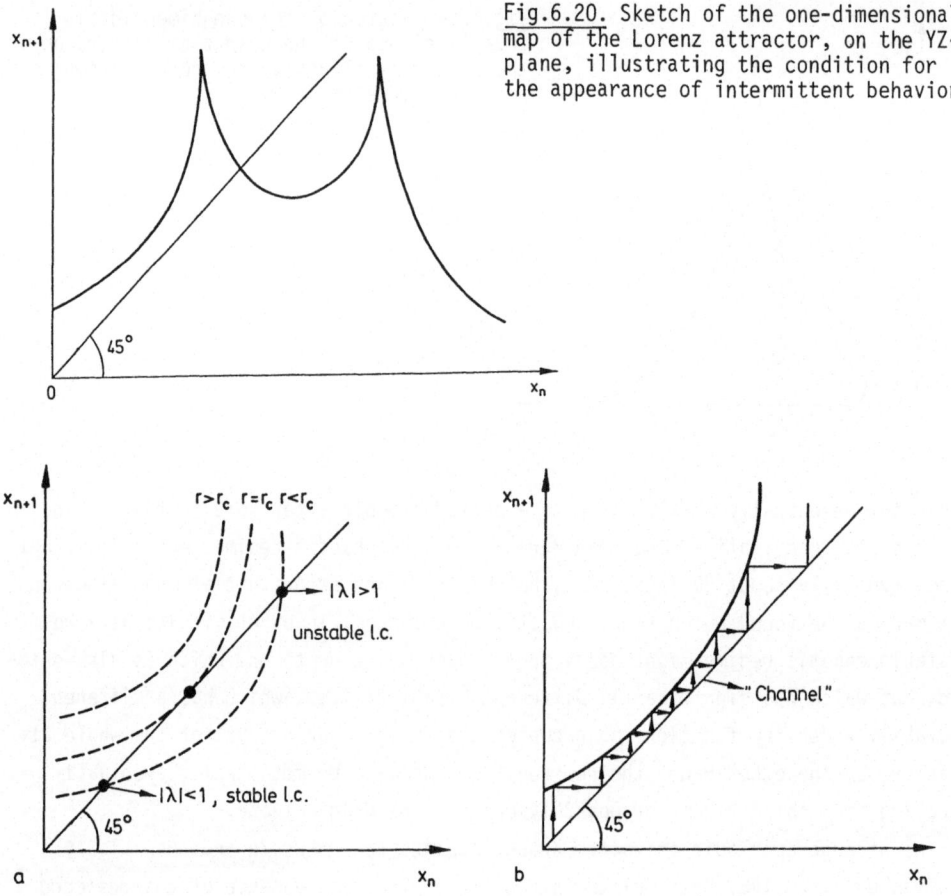

Fig.6.20. Sketch of the one-dimensional map of the Lorenz attractor, on the YZ-plane, illustrating the condition for the appearance of intermittent behavior

Fig.6.21a,b. Detailed conditions for the onset of intermittency regime. In **a**, l.c. stands for limit cycle

As sketched in the detailed Fig.6.21a, as long as the curve of the map inter-sects the first bisectrix, we have a *stable* fixed point in the map which corres-ponds to a stable periodic motion, but as the curve is moved up, the *two* fixed points (one stable, one unstable) first collapse and then disappear.

Just after this disappearance, and as long as the curve stays nearly tangential to the first bisectrix, the system retains a memory of the formerly stable fixed point, and we observe an almost stable limit cycle except at the very beginning and at the very end. Then, far enough from the old fixed point, as the iteration proceeds starting from the left, the representing point enters a kind of "channel", slows down, goes through the channel, and then speeds up and leaves the channel on the right (Fig.6.21b). Even farther from the old fixed point, the system enters a "turbulent" regime, during which correlations concerning phase locking are broken and the system explores randomly a large portion of the state space. This bursting phase ends, sooner or later, when the moving point, as it explores the state space, comes close to the formerly stable limit cycle and so "reinjects" the trajectory in the vicinity of the old fixed point.

Due to a certain randomness in the reinjection process, in the vicinity of the old fixed point the channel is not always covered in its entire length, and there may be a certain dispersion in the duration of the "laminar" intervals. In conclusion, then, the intermittent behavior is merely a consequence of the translation of the parabolic part of the Poincaré map (as the control parameter increases) so that it first intersects, then is tangent to, and finally loses contact with the bisectrix.

6.3.5 The Rate of Information Production by the Lorenz Attractor

In order to calculate the information generated by the Lorenz attractor with $\sigma = 10$, $b = 8/3$, $r = 28$, we first construct a Poincaré map on the ZX-projection of the attractor by making a cut transverse to the flow and then recording successive passes through a maximum along, say, the z-direction. As the attractor is symmetrical with respect to the transformation $x \to -x$, $y \to -y$, $z \to z$, the cut on one "leaf" will yield the same map as a symmetrically placed cut on the other.

The time $t(x)$ between successive passes through the Poincaré map is also recorded [6.13]. Having the return map $F(x)$ and the function $t(x)$, we next calculate the probability density $P(x)$ using the iteration method developed during the treatment of the logistic map (Sect.6.3.1). The information production is then equal to

$$<I> = \int_0^1 P(x) \log \left| \frac{dF(x)}{dx} \right| dx \sim 0.98 \quad \text{bits per iteration} \quad . \tag{6.3.41}$$

This result may be compared with the corresponding value of $<I>$ for the logistic map and a = 4, which can be deduced as follows: Introducing the transformation $x' = (2/\pi)\sin^{-1}\sqrt{x}$ on the logistic map $F(x) = 4x(1-x)$, we obtain the symmetrical "tent" map:

$$F'(x') = \begin{cases} 2x' , & 0 < x' < \frac{1}{2} \\ 2(1 - x') , & \frac{1}{2} < x' < 1 \end{cases} .$$

Obviously, for this map $P'(x') = 1$. From the conservation of probability $P'(x')dx' = P(x)dx$, we find

$$P(x) = \frac{dx'}{dx} = \frac{1}{\pi} \frac{1}{\sqrt{x(1-x)}} \quad ,$$

and so

$$<I> = \frac{1}{\pi} \int_0^1 \frac{\log_2[4(1-2x)]}{\sqrt{x(1-x)}} dx = 1 \quad \text{bit per iteration} \tag{6.3.42}$$

— something we should expect since for $r = 4$ the logistic map is exactly two onto one. For the values $r = 28$, $b = 8/3$, $\sigma = 10$ of the control parameters the cusp-shaped Lorenz attractor Poincaré map is nearly symmetrical and almost two onto one, but not quite.

In order to describe the information-production characteristics of the Lorenz attractor, we calculate

$$\left\langle \frac{dI}{dt} \right\rangle = \int_0^1 \frac{P(x)}{t(x)} \log_2 \left| \frac{dF(x)}{dt} \right| dx \quad \text{[bits/s]} \quad , \tag{6.3.43}$$

or, when $P(x)$ is not known,

$$\left\langle \frac{dI}{dt} \right\rangle = \lim_{n \to \infty} \frac{1}{n} \sum_{i=1}^{n} \frac{\log_2 |dF(x)/dx|}{t_n} \quad , \tag{6.3.44}$$

where t_n is the length of time taken by the nth pass. The information production rate for the Lorenz attractor was calculated [6.13] as ~ 1.19 bits/s.

Suppose finally that the initial point in the interval [0,1] in the map, from which we start iterating in any practical application, is not known *exactly* but with a given uncertainty, expressed by a distribution $P_0(x)$ over the interval. The a priori uncertainty we have about the initial point simply equals the asymptotic probability density function (p.d.f.) $P(x)$ for the specific value of the control parameter, so the "information value" of the initial condition is

$$S = \int_0^1 P_0(x) \log_2 \left(\frac{P_0(x)}{P(x)} \right) dx \quad \text{[bits]} \quad . \tag{6.3.45}$$

Consequently, the "memory" of the system, measured as the time elapsing until the attractor gets causally disconnected from the initial conditions, is given by

$$T = \frac{S}{\langle dI/dt \rangle} = \frac{\int_0^1 P_0(x) \log_2 \left(\frac{P_0(x)}{P(x)} \right) dx}{\int_0^1 \frac{P(x)}{t(x)} \log_2 \left(\left| \frac{dF(x)}{dx} \right| \right) dx} \quad \text{[s]} \quad . \tag{6.3.46}$$

After T, the dynamical evolution of the system no longer reveals "hidden" information associated with the initial condition(s), but generates *new* information implemented by the flow itself (see also Sect.6.5).

It is time now to direct our attention to the study of the parameters which may qualify a strange attractor as an information processor. Obviously, we expect that for values of the control parameters below the initiation of chaos (for which only steady states and periodic orbits exist), the trajectories in state space will only attract, and the corresponding iteration, in the Poincaré map in the interval, will monotonically converge, giving probability densities which are just a set of delta-like functions. Under such subcritical circumstances $\langle I \rangle < 0$, i.e., the system will behave asymptotically as an information sink, that is, as an agent which constrains variety, thereby revealing information — essentially, then, a source of "negentropy".

Beyond the critical value(s) of the control parameter(s) for which continuous chaos is established (almost all orbits are unstable), $\langle I \rangle > 0$ along some direction(s) in state space, thereby creating variety or increasing entropy [source(s)]. Also, $\langle I \rangle < 0$ along some other directions in state space, thereby dissipating variety or revealing information [sink(s)]. It is out of this information-production-dissipation interplay that one expects to derive the salient features which characterize the dynamical system as a more or less efficient information processor.

Apart from the Lorenz model developed above in some detail, there are some very familiar systems, such as the nonlinear "Duffing" oscillator or the van der Pol oscillator, which under harmonic external coupling (e.g., $\ddot{x} + \nu x + \omega_0^2 x^3 = A \cos\omega t$, and $\ddot{x} - \varepsilon(1 - x^2)\dot{x} + \omega_0^2 x = A \cos\omega t$, respectively), except for the limit cycle regime due to entrainment and phase locking, also display all the characteristics of chaotic behavior mentioned so far, for different windows of the control parameters. The interested reader may refer, for example, to [6.6-8] and also to Appendix C.

6.4 Parameters Characterizing the Average Behavior of Strange Attractors: Dimensions, Entropies, and Lyapounov Exponents

Three basic concepts are of crucial importance (and need therefore a quantitative measure) in characterizing the dynamical properties of strange attractors:

a) The degree of "compressibility" displayed by the system (which acts as an efficient "vacuum cleaner", so to speak, in functional space).
b) The exact way that entropy is generated through the passage (and amplification) of the intrinsic fluctuations of the system from microscopic to macroscopic hierarchival levels as the resolution produced by cascading iterations increases.
c) The stability properties of the attractor or the rates of *variety generation* along given directions, and *information dissipation* along other directions in state space.

In this section we intend to search for pertinent parameters quantifying these average characteristics of chaotic attractors and to look for possible relations among them.

We are concerned in this chapter with dynamical systems which alter in discrete time stages n such as "maps" $x_{n+1} = F(x_n)$, or with continuous flows, i.e., ordinary first-order nonlinear differential equations $dx(t)/dt = f(x(t);\mu)$, where in both cases x is generally a vector in a multidimensional state space. Given an initial condition (a value of x at $n = 0$ for the map, or $t = 0$ for the continuous system), *an orbit* $x_1, x_2, \ldots, x_n$ is generated for the map, or *a trajectory* $x(t)$ for the differential system.

Here we are interested in attractors manifested in such systems. An attractor is a compact subset C of the state space, with the property that it attracts initial conditions from a region around it, once transients have subsided. "Attraction" here means that for every initial condition, the limit set of the orbit, as n or t goes to ∞, is the compact set C. Every trajectory in this neighborhood of C passes, then, arbitrarily close to every point of the set C. The dimension of an attractor is the first parameter necessary to characterize its compressibility properties. The *dimension* gives the amount of information necessary to specify the position of a point on the attractor to within a given accuracy.[3] It is suggested that perhaps a suitably defined dimension of an attractor is generally *less* than that of the state space containing it. For trajectories on the attractor, this reduction of dimensionality brings about an accompanying reduction in the information needed to specify an initial condition. Knowledge of the information dimension of an attractor allows an observer to estimate the information gained when a single, isolated measurement is made at a given level of precision.

When dealing with a *dissipative* dynamical system, we may begin with a Euclidean state space of initial conditions of large or even infinite dimensionality. After some time, however, the transients die out, and the point in state space that describes the state of the system approaches the attractor. The number of degrees of freedom is reduced in the sense that the number of independent variables inherent to the motion is much less than the number of independent variables needed to specify an arbitrary initial condition. The difference between the Euclidean dimensionality of the state space where the dynamical system is embedded and the information dimensionality of the attractor gives, loosely speaking, the average degree of "compressibility" manifested by the attraction.

For simple attractors, defining and investigating the information dimension is trivial. For example, a stationary, time-independent equilibrium (steady state) has dimension *zero*, a stable periodic oscillation (limit cycle) has dimension *one*, and a 2-torus has dimension *two*. In these simple cases the dimension takes values that are integers. This, however, is not the case for strange attractors.

In order to appreciate the properties of a chaotic attractor, we must take into account not only the set itself but also the *distribution* or *density* of the points on the attractor. (The reason is that at any fixed level of precision, most information about initial conditions is lost in a finite time T, which we calculated in the Sect.6.3.5. For times greater than T, knowledge of the future is limited to the information contained in the probability distribution of points on the attractor.)

---

3 Alternatively the information dimensionality can be considered as the number of bits one can dynamically *store* on the attractor.

6.4.1 The Concept of Information Dimension

The notion of information dimension is intimately linked to the measuring process.
Consider a measuring instrument with a uniform scale of resolution ε, so that the
measurement of any of the N variables yields one of $1/\varepsilon$ possible numbers. If such
a measuring instrument is assigned to each of the N variables of our dynamical
system, the quantization of the scales of those instruments induces a *partition* of
the state space, that is a set of non-empty nonintersecting measurable "hypercubes"
(see Fig.6.22 for a two-dimensional projection). The region of the state space containing the attractor is therefore divided into ε^{-N} boxes, each of dimension N. The
collection of these boxes is the partition, and each measurable state corresponds
to a box. Let $n(\varepsilon)$ be the number of boxes covered by the attractor with nonzero
probability (that is, containing at least one point of the attractor). Some boxes
will contain more points than others since certain regions in state space are visited by the flow far more frequently than others. The "natural measure" of a given
region is proportional to the frequency with which it is visited. Following Farmer
[6.14] let $P_i(\varepsilon)$ be the "probability density"[4] of occurrence of the attractor some-
where in the ith box. The collection $\{P_i(\varepsilon)\}$ is referred to as a "coarse-grained"
probability distribution of resolution ε. The (average) information contained in a
single isolated measurement made at resolution ε is, then,

$$I(\varepsilon) = - \sum_{i=1}^{n(\varepsilon)} P_i(\varepsilon)\log_2 P_i(\varepsilon) \quad \text{[bits]} \quad . \tag{6.4.1}$$

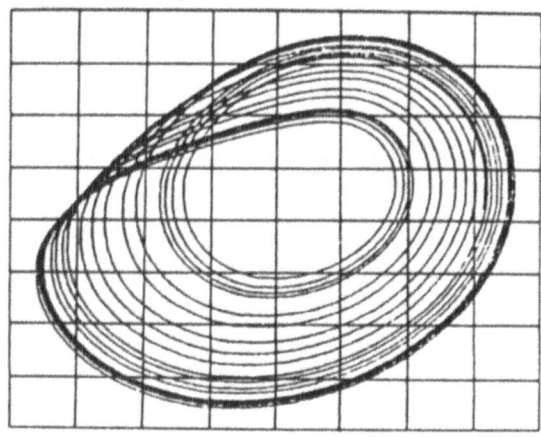

Fig.6.22. A sketch of a two-dimensional partition of an attractor
into state space "boxes" (squares)

---

4 For deterministic systems, use of the word "probability" requires some qualifica-
tion [6.14]. For each cube C on the attractor and initial condition x, define $\mu(x,C)$
as the fraction of time that the trajectory originating from x stays in C. If every
x gives the same value of $\mu(x,C)$, we call this common value $\mu(C)$ the natural "measure" of the attractor. It gives the relative probability of different regions of
the attractor as obtained from time averages. The measure of the set C is the integral of the density on C, i.e., $\mu(C) = \int_C P(x)dx$.

As the scale of resolution ε decreases, the number of boxes increases and we get a sequence of successively more refined, less coarsely grained probability distributions. The information dimension D_I is the rate at which the information scales as the precision of measurement is increased, namely

$$D_I = \lim_{\varepsilon \to 0} \left[\frac{I(\varepsilon)}{|\log_2 \varepsilon|} \right] . \qquad (6.4.2)$$

D_I is given, then, by the asymptotic value of the slope of the graph of $I(\varepsilon)$ versus $\log_2(1/\varepsilon)$.

Thus if the information dimension of an attractor is known, the information $I(\varepsilon)$ contained in a measurement of a state of resolution ε is estimated as

$$I(\varepsilon) = D_I |\log_2 \varepsilon| . \qquad (6.4.3)$$

If the probability of all the boxes in the partition is equal, then

$$I(\varepsilon) = \log_2 n(\varepsilon) , \qquad (6.4.4)$$

and the information dimension acquires its maximum value, which coincides with the *fractal* dimension

$$D_F = \lim_{\varepsilon \to 0} \frac{\log_2[n(\varepsilon)]}{\log_2(1/\varepsilon)} . \qquad (6.4.5)$$

The fractal dimension is then an *upper bound* of the information dimension.

What limits ε in practice is, of course, the level of intrinsic (internal) fluctuations in the system *and* the measuring instrument —collectively labeled as "observational noise". This noise is manifested only during the measuring process and does not affect the phenomenological dynamics of the system. In the above definition of D_I or D_F, external noise is excluded.

Now before proceeding to a number of applications let us pause and ask the following question: why should we stretch the resolution ε to zero? In practice we never do that. Furthermore, who assures us that the maximum of dynamical storage capacity of the attractor, D_I, is achieved for $\varepsilon \to 0$? After all, in the long struggle for survival, biological organisms have been trying to maximize their dynamical memory storage capacity —without jeopardizing compressibility. To survive, an organism had better be perhaps a "jack-of-all-trades", namely obtain "fuzzy" information about a large sector of environment, rather than concentrate on a thin sector of environmental stimuli with perfect resolution. Shouldn't we then rather try to calculate the "optimum nonzero resolution" ε^* for which $D_I(\varepsilon^*)$ becomes maximum, that is $\partial D_I/\partial \varepsilon = 0$?

To do this we proceed as follows [6.15]: Let us consider the expression $C_I = N - D_I$, representing the degree of *average* compression realized by the particular attractor and try to calculate a critical resolution ε^* for which D_I becomes maximum or $\partial C_I/\partial \varepsilon = 0$. To take into account the best case let us perform the above calculation for the maximum value of

$$D_I = \frac{- \sum\limits_{i=1}^{n(\varepsilon)} P_i(\varepsilon) \log_2 P_i(\varepsilon)}{\log_2(1/\varepsilon)} \quad , \qquad (\varepsilon \text{ finite}) \quad . \tag{6.4.6}$$

The maximum value of the information dimension under given finite resolution ε is the "fractal dimension" of the attractor and is realized for $P_i(\varepsilon) = \text{const} = 1/n(\varepsilon)$. Therefore,

$$D_I = - \frac{\ln[n(\varepsilon)]}{\ln \varepsilon} \quad \text{and} \tag{6.4.7}$$

$$C_I = N + \frac{\ln[n(\varepsilon)]}{\ln \varepsilon} = \frac{N \ln \varepsilon + \ln[n(\varepsilon)]}{\ln \varepsilon} = \frac{\ln[\varepsilon^N n(\varepsilon)]}{\ln \varepsilon} \quad . \tag{6.4.8}$$

We intend now to express $n(\varepsilon)$ in terms of the Lyapounov exponents of the attractor and the resolution length ε. Let the number of points on a (one-dimensional) Poincaré return map of the attractor concerned be M; this means that we determine the orbit with strings M digits long. The number of cells representing the attractor, with degree of coarseness ε, is just $n(\varepsilon)$. Now, for the degree of resolution ε to be meaningful, it is necessary that each of the possible outcomes of an orbit of length M digits can be determined with precision. Since we deal here with systems whose dynamics can be adequately represented in each step by two possible states, say 0 and 1, the number of outcomes is clearly 2^M. Hence ε should be chosen such that

$$\varepsilon \sim \frac{1}{2^M} \quad . \tag{6.4.9}$$

Let t_c be the sampling time or, alternatively, with sampling time period $\Delta t = 1$, the number of samplings (iterations) probing our attractor. For an attractor represented by M points, it is therefore meaningful to choose

$$\frac{n(\varepsilon)}{M} \sim t_c \quad . \tag{6.4.10}$$

Owing to the chaotic character of the dynamics, after t_c iterations in the map the system is completely disengaged from the initial condition. This means that t_c can be estimated from the relation

$$\varepsilon \exp(\lambda_+ t_c) \sim 1 \tag{6.4.11}$$

where λ_+ is the positive Lyapounov exponent. Therefore

$$t_c \sim \frac{\ln(1/\varepsilon)}{\lambda_+} \tag{6.4.12}$$

and finally

$$n(\varepsilon) = \frac{(\ln \frac{1}{\varepsilon})^2}{\lambda_+ \ln 2} \quad . \tag{6.4.13}$$

Substituting into the expression for C_I (6.4.8) we get

$$C_I = N + \frac{2\ln(\ln\frac{1}{\varepsilon}) + \ln(\frac{1}{\lambda_+\ln 2})}{\ln\varepsilon} \quad \text{or}$$

(6.4.14)

$$C_I = N + \frac{\gamma + 2\ln(\ln\frac{1}{\varepsilon})}{\ln\varepsilon}$$

(6.4.15)

where $\gamma = \ln[1/(\lambda_+\ln 2)]$.

The requirement for maximum fractal dimension, $\partial C_I/\partial\varepsilon = 0$, gives for the optimum resolution

$$\varepsilon^* = \exp\left[-e^{(2-\gamma)/2}\right]$$

(6.4.16)

or the optimum code length

$$M^* = \frac{\ln(1/\varepsilon^*)}{\ln 2}$$

(6.4.17)

from which

$$C_I(\varepsilon^*) = N - 2e^{-(2-\gamma)/2} \quad \text{[bits]} \quad .$$

(6.4.18)

These are our final formulas for the minimum average compressibility, i.e., the maximum fractal dimensionality of the attractor concerned and the corresponding optimum code length M^*.

It remains to ascertain that the extremum of the function

$$f(\varepsilon) = -D_I(\varepsilon) = \frac{\gamma + 2\ln(\ln\frac{1}{\varepsilon})}{\ln\varepsilon} \quad , \quad \varepsilon \neq 0$$

(6.4.19)

is indeed a minimum or $D_I(\varepsilon)$ is maximum.

In fact we have

$$\frac{\partial f(\varepsilon)}{\partial\varepsilon} = \frac{2 - [\gamma + 2\ln(\ln\frac{1}{\varepsilon})]}{\varepsilon(\ln\varepsilon)^2}$$

(6.4.20)

and

$$\frac{\partial^2 f(\varepsilon)}{\partial\varepsilon^2} = \frac{-6 + 2\gamma + 4\ln(\ln\frac{1}{\varepsilon}) + (\ln\varepsilon)[\gamma - 2 + 2\ln(\ln\frac{1}{\varepsilon})]}{\varepsilon^2(\ln\varepsilon)^3} \quad ;$$

(6.4.21)

for $\varepsilon = \varepsilon^* = \exp\{-\exp[(2-\gamma)/2]\}$ we get

$$\left(\frac{\partial^2 f(\varepsilon)}{\partial\varepsilon^2}\right)_{\varepsilon=\varepsilon^*} = \frac{2}{\exp[-2\exp(\frac{2-\gamma}{2})]\exp[\frac{3}{2}(2-\gamma)]}$$

(6.4.22)

which is always positive. Therefore $f(\varepsilon)$ possesses a minimum at ε^* and consequently $D_I(\varepsilon) = -f(\varepsilon)$ becomes maximum.

$D_I(\varepsilon)$ becomes zero for $\varepsilon_0 = \exp[-\exp(-\gamma/2)]$. Negative values of D_I do not, of course, possess any physical meaning. Notice that the value of $D_I(\varepsilon^*)$ is much less than the fractal dimensionality D_I calculated in the limit $\varepsilon \to 0$. As a matter of fact $D_I(\varepsilon^*)$ no longer refers to a geometrical object. Rather, it provides us with a convenient measure of the efficiency of an information-producing device.

Let us now apply formulas (6.4.17 and 18) to a number of simple cases.

a) *The Lorenz Attractor*: $\dot{x} = \sigma(y-x)$, $\dot{y} = -y + rx - xz$, $\dot{z} = xy - bz$ (for $b = 4$, $\sigma = 16$ and $r \approx 45.92$). We calculate with the computer $\lambda_+ \sim 1.5$, $\gamma \sim -0.03895$, $C_I(\varepsilon^*) \sim 3 - 0.72157 \sim 2.28$, $\varepsilon^* \sim 0.062$ and $M^* \sim 4$. That is, the optimum code which maximizes the storage capability of the Lorenz attractor is ~ 4 bits long. Nevertheless the resulting compressibility is impressive too —much larger than what one calculates from the information dimensionality with $\varepsilon \to 0$. [$C(0) = 3 - 2.06 = 0.94$.]

b) *The Bernoulli Map*: $x' = \mathrm{mod}_1(2x)$. In this case of a one-dimensional map —assumed here to be uniform —the number of covering segments of resolution ε is

$$n(\varepsilon) \sim \ln(\tfrac{1}{\varepsilon})/\lambda_+ \tag{6.4.23}$$

from which we obtain

$$C_I \sim \frac{N + [\ln(\ln \tfrac{1}{\varepsilon}) - \ln\lambda_+]}{\ln\varepsilon} . \tag{6.4.24}$$

The requirement $\partial C_I/\partial\varepsilon = 0$ gives for the optimum resolution

$$\varepsilon^* \sim \exp[-\exp(1 + \ln\lambda_+)] \tag{6.4.25}$$

from which

$$C_I(\varepsilon^*) = N - \exp[-(1 + \ln\lambda_+)] . \tag{6.4.26}$$

The numerical calculations for $\lambda_+ = 2$ in this case, give $C_I(\varepsilon^*) = 0.81606$, $\varepsilon^* = 0.004354$, and $M^* \sim 7$ to 8 bits.

After the above diversion let us give as a first example the information dimension of a symmetrical Cantor set "in shades of grey". This means that rather than successively deleting the middle third of each piece —as we did in Sect.6.3.2 where we calculated the fractal dimensionality of a Cantor set apropos of the logistic map —the middle thirds are only partially erased or, to put it another way, the middle thirds are made *more* or *less* probable than the surrounding thirds. In other words, we intend to treat the Cantor set as a probability density. [In fact, this view allows us to see now more clearly *why* in the case of the logistic map, the interval [0,1] becomes a Cantor-like set around the first accumulation point $a = 3.57$, signaling the end of period-doubling pitchfork cascade bifurcations.]

When we move from the first to the second iteration of the map $F(x)$, the orbits (that is, the first limit cycle of period two) occupy with probability one the interval between the two first satellites $x_1^{*'}$ and $x_2^{*'}$, and the two surrounding portions $0 - x_1^{*'}$, $x_2^* - 1$ are empty (or orbits). However, at each successive pitchfork bifurcation leading to period doubling, that is, to the next iteration, the central *third* of the *newly created interval(s)* is *not* visited by stable orbits (it stays

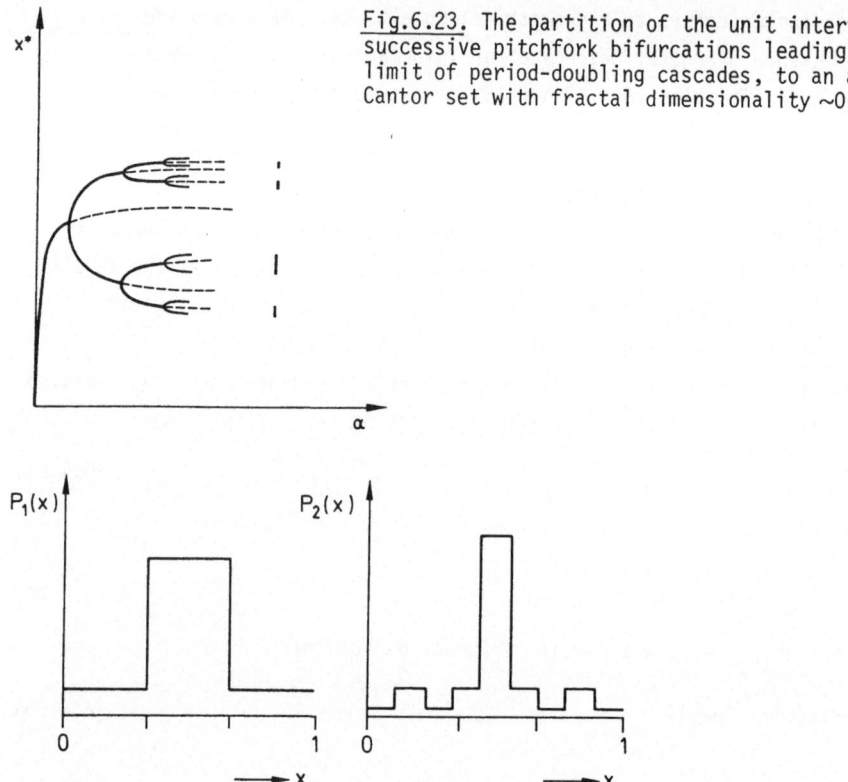

Fig.6.23. The partition of the unit interval through successive pitchfork bifurcations leading, in the limit of period-doubling cascades, to an asymmetrical Cantor set with fractal dimensionality ~0.530

Fig.6.24. The first two steps of a cascade process of rescaling the p.d.f. in order to construct a symmetrical "grey" Cantor set

empty) while the two surrounding subintervals remain occupied, and so on (Fig. 6.23). At the accumulation point a = 3.57 we have a Cantor-like set (black-white) which is, however, asymmetrical (see later in this section). Nevertheless, we can envisage other values of the control parameter (a = 4 for example), for which the orbits occupy the *whole* interval with a *continuous* probability density $P(x)$. In this case the fractal and information dimensions of the logistic map are both one.

We begin by partially erasing the outside parts and leaving the middle thirds intact as in Fig.6.24. Let the probability of each of the two outer pieces be P_0, and the probability of the piece in the middle be P_m. The first approximation is

$$P_1(x) = \{3P_0, \; 3P_m, 3P_0\} \quad , \tag{6.4.27}$$

$$\int_0^1 P_1(x)dx = 1 \quad , \tag{6.4.28}$$

$$P_0 = \frac{1 - P_m}{2} \quad . \tag{6.4.29}$$

We form the second approximation $P_2(x)$ by dividing again each piece into thirds and *redistributing* the probability within each of these nine pieces, so that the ratios within each third are the same as those in $P_1(x)$ (in other words, we seek self-similarity of the p.d.f., or we scale the probability density with increasing resolution). We continue this way ad infinitum. The informational dimension of this "grey" Cantor set is

$$D_I = \frac{-P_m \log_2 P_m - 2P_0 \log_2 P_0}{\log_2 3} \quad , \qquad (6.4.30)$$

but its fractal dimension is $D_F = 1$ bit, since the p.d.f. is nonzero along the entire interval.

In the case of the *black-white* symmetrical Cantor set, both D_I and D_F were equal to $\log_2 2/\log_2 3$. If now we move to the case of an asymmetrical Cantor set (in order to approach more closely "real" cases such as the case of the logistic map) the calculation becomes more involved [6.14].

The idea is to appreciate how asymmetry in the spacing of the elements of a set can modify the information (and fractal) dimension. Beginning, as usual, with the interval [0,1] we delete this time the segment from 1/2 to 3/4, that is, the third fourth (see Fig.6.25). Then we delete the third fourth of each remaining piece, and so on. The limit is an asymmetrical Cantor set with a fractal dimension

$$D_F = \frac{\log_2 3}{\log_2 4} \sim 0.792 \quad . \qquad (6.4.31)$$

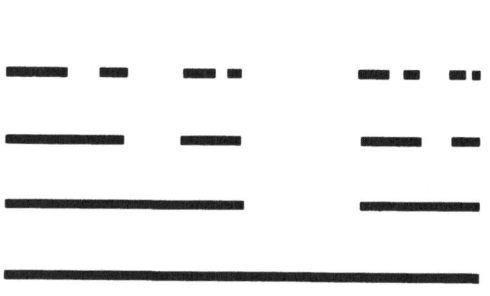

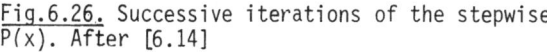

Fig.6.25. Successive approximations to an asymmetrical Cantor set formed by deleting the third fourth of each piece. After [6.14]

Fig.6.26. Successive iterations of the stepwise P(x). After [6.14]

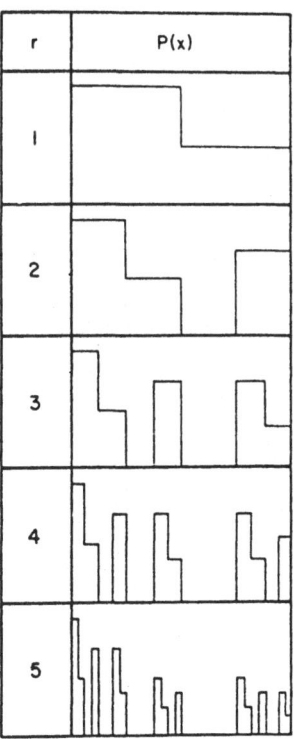

In order to estimate the information dimension, however, the metric properties of the set used above for the calculation of D_F are not enough: we need also to assign a certain arbitrary probability density distribution to the elements of the set. Let us assign a stepwise uniform probability density to the points of this set (Fig.6.26), thereby making it not only asymmetrical but also grey.

To compute the information dimension we partition the interval [0,1] successively into 2^n portions. Each time the resolution is doubled, the asymmetry of the set causes neighbouring portions to acquire different probabilities. The detailed calculation is rather involved due to the difficulty of finding out the scaling law that the p.d.f. follows with increasing resolution, and the reader is referred to the original investigation performed by *Farmer* [6.14]. The result is

$$D_I = -3 \left[\frac{\frac{1}{3} \log_2(\frac{1}{3}) + \frac{2}{3} \log_2(\frac{2}{3})}{4 \log_2 2} \right] \sim 0.6887 \text{ [bits]} \quad . \tag{6.4.32}$$

The calculation of the information dimension of the logistic map is even more involved if we take into account the intricate dynamics (period-doubling pitchfork bifurcation cascades) via which the partition of the set takes place. The calculation gives, of course, a different result for each value of the control parameter a and makes use of Feigenbaum's scaling factor .

For a ≅ 3.57 (the first accumulation point) the *fractal dimensionality* has been calculated by *Grassberger* [6.16] to be $D_F \sim 0.530$ bits. No calculation has been performed for the information dimension.

For a = 3.7, the *information dimension* as calculated by *Farmer* [6.14] is $D_I = 1$ bit. Again, the detailed calculations are very involved since now the probability density function has to be calculated for each iteration from the rather involved recursive relationship in [6.14]. Detailed calculations of that sort are outside our main theme.

6.4.2 The Concept of Characteristic Lyapounov Exponents and Their Relation to Information Dimension

We have appreciated already that the main difficulty in calculating the information dimension comes from the p.d.f. and particularly the way it scales as the resolution increases to infinity. For the logistic map, for example, this probability density is calculated from the recursive relation

$$P_n(x) = \sum_{i=1}^{2^{n-1}} \frac{P_1(x_i)}{\left| \frac{dF^{(n-1)}}{dx} \right|_{x_i}} \quad , \tag{6.4.33}$$

where $x_i = [F^{(n-1)}]^{-1}$.

In essence, therefore, the difficulty springs from the fact that the Jacobian matrix or determinant of the map (or the attractor in general) is not constant, that is, it changes with the order of the iteration. In fact, this is also the main reason for the difference between the information dimension and the fractal dimension.

In turn, the Jacobian matrix is related to the *stability properties* of the attractor, so we are tempted to conjecture that the dimensions of chaotic attractors can be determined directly from the dynamics of the system, or in terms of the "spectrum of Lyapounov exponents" of the attractor. (We have already given two examples in Sect.6.4.1).

The spectrum of Lyapounov characteristic exponents provides a qualitative estimation of the local stability properties of an attractor. The local stability properties of a system are determined by the response the system manifests under small perturbations. A dynamical system can be stable to perturbations along certain directions and unstable to perturbations in others. All possible perturbations can be examined simultaneously by following the evolution of an ensemble of points that are initially contained in a small N-dimensional sphere, where N is the state-space dimension in which the attractor forms a subset.

Imagine, then, an initially infinitesimal ball that has a radius $\delta(0)$ at time $t = 0$. As the ball evolves under the action of a nonuniform flow, it will eventually distort. Since we have assumed an infinitesimal state-space element, the change in shape, we presume, will be determined only by the linear part of the flow, and it remains an ellipsoid as it evolves.

Let $\delta_i(t)$ be the ith member of the set of the principal axes of the ellipsoid at time t. The spectrum of Lyapounov exponents λ_i for a given starting position is determined as

$$\lambda_i = \lim_{t \to \infty} \lim_{\delta(0) \to 0} \frac{1}{t} \log_2\left(\frac{\delta_i(t)}{\delta(0)}\right) \ . \tag{6.4.34}$$

There are N Lyapounov exponents in the spectrum of an attractor of an N-dimensional dynamical system. Positive Lyapounov exponents measure *average* exponential spreading of nearby trajectories and negative exponents measure average exponential convergence of trajectories onto the attractor. The sum of the Lyapounov exponents is the average divergence, which for a dissipative system (possessing an *attractor*) must always be *negative*. (For a Hamiltonian system this divergence is zero.) For a number of dissipative systems, there are examples where computing simulation indicates that the values of the Lyapounov exponents are invariant for all tried initial conditions. If this is so, then the spectrum of Lyapounov exponents may be taken to be a property of the attractor.

It is customary to arrange the Lyapounov exponents in decreasing order. Thus, for instance, the symbolism +, 0, - indicates an attractor in a three-dimensional state space with (on average) exponential expansion along, say, the $\hat{x}$-direction, neutral stability along the flow (say the $\hat{y}$-direction) and exponential contraction of trajectories along the $\hat{z}$-direction. It is important to realize that attractors other than stable steady states *always* have at least one exponent equal to zero since, on average, points along a trajectory *confined to a compact set* can neither separate nor merge. Also, for a dissipative system $|\lambda(-)| > \lambda(+)$.

Consider now a map $x_{n+1} = F(x_n)$ of the attractor, where x is a p-dimensional vector. The definition elaborated above suggests that the Lyapounov exponents of the discrete system are related to the eigenvalues of the Jacobian matrix of the map (in the case of a stable steady state, the Lyapounov exponents are the real parts of eigenvalues of the interaction matrix).

Specifically, let

$$J_n = J(x_n) \cdot J(x_{n-1}) \cdot \ldots \cdot J(x_1) \quad , \qquad \text{where} \qquad (6.4.35)$$

$$J(x_1) = \left(\frac{\partial F}{\partial x}\right)_{x_1} \quad ,$$

$$J(x_{n-1}) = \left(\frac{\partial F}{\partial x}\right)_{x_{n-1}} \quad ,$$

$$J(x_n) = \left(\frac{\partial F}{\partial x}\right)_{x_n} \quad .$$

Then

$$J_n = \left(\frac{\partial F}{\partial x}\right)_{x_1} \cdot \left(\frac{\partial F}{\partial x}\right)_{x_2} \cdot \ldots \cdot \left(\frac{\partial F}{\partial x}\right)_{x_n} = \left(\frac{\partial F^{(n)}}{\partial x}\right)_{x=x_0} \quad , \qquad (6.4.36)$$

where $x_i = F^{(i)}(x_0)$. Here we made use of the known fact that the slope of the nth iterated map at some point x_0 equals the products of the slopes of the original map at the points $x_1, x_2, \ldots, x_n$.

Let $\sigma_1(n) \geqslant \sigma_2(n) \geqslant \ldots \geqslant \sigma_p(n)$ be the magnitudes of the eigenvalues of J_n. The Lyapounov numbers are

$$\lambda_i = \lim_{n \to \infty} [\sigma_i(n)]^{1/n} \quad , \qquad i = 1, 2, \ldots, p \quad , \qquad (6.4.37)$$

where the positive real n^{th} root is taken. The log of λ_i above gives the "Lyapounov exponents". For the one-dimensional case, the above definition coincides with the result we found in Sect. 6.3.3 if we take the algorithm above, namely

$$\lambda = \lim_{n \to \infty} \frac{1}{n} \sum_{i=1}^{n} \left| \frac{\partial F}{\partial x} \right|_{x_i} \qquad (6.4.38)$$

for the Lyapounov exponent.

Let us give an example in the case of a *two-dimensional* map and try from this example to derive a heuristic relationship between the spectrum of Lyapounov ex-

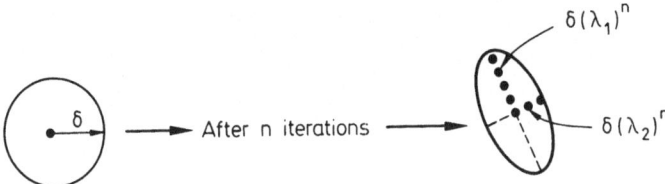

Fig.6.27. Understanding the action of Lyapounov exponents on a two-dimensional system

ponents and the information dimension. For a two-dimensional map, let λ_1, λ_2 be the average distorting coefficients of an infinitesimal circular disk of radius δ (Fig.6.27).

Since the system is discrete, the Lyapounov exponent for which divergence takes place corresponds to say $\lambda_1 > 1$, and the Lyapounov exponent for which shrinking takes place corresponds to $\lambda_2 < 1$. After n iterations of the two-dimensional map, the initial small circle of radius δ will be transformed into an ellipse with major and minor half-axes $\delta(\lambda_1)^n$ and $\delta(\lambda_2)^n$ respectively. Now let us try, following [6.17], to relate this spectrum of Lyapounov exponents to the information dimension of the map. We start, as usual, by covering the two-dimensional attractor with $n(\epsilon)$ squares of side ϵ. Now let us iterate the map ξ times. For ϵ small enough (high resolution), the action of the mapping is practically linear over each square, so each element of the partition will be stretched into a parallelogram. The average dimensions of these parallelograms will be $\lambda_1^\xi \epsilon$ and $\lambda_2^\xi \epsilon$ respectively ($\lambda_1 > \lambda_2$). Suppose now we increase the resolution (we make the partitioning finer) by using squares of side $\lambda_2^\xi \epsilon$. To cover each parallelogram, we used $(\lambda_1/\lambda_2)^\xi$ smaller squares, that is, the number of elements of the new partition will be

$$n(\lambda_2^\xi \epsilon) = \left(\frac{\lambda_1}{\lambda_2}\right)^\xi \cdot n(\epsilon) \quad . \tag{6.4.39}$$

Let us put now $n(\epsilon) \cong (1/\epsilon)^{d_L}$ and substitute into (6.4.39). We get

$$\left(\frac{1}{\lambda_2^\xi \epsilon}\right)^{d_L} = \left(\frac{\lambda_1}{\lambda_2}\right)^\xi \cdot \left(\frac{1}{\epsilon}\right)^{d_L} \tag{6.4.40}$$

or, taking logarithms of both sides,

$$d_L = 1 + \frac{\log_2 \lambda_1}{\log_2 \frac{1}{\lambda_2}} \quad . \tag{6.4.41}$$

This formula holds for two-dimensional maps. It may be generalized, however, for higher dimensionalities. We call d the "Lyapounov" dimension and ask next how it may be related to the information dimension.

Clearly the Lyapounov dimension coincides with the fractal dimension. In the derivation above all squares in the partition have been considered as equiprobable. Let us note that the Lyapounov numbers are *average* quantities, and to compute an average, each element of the partition must be weighted according to its probability, that is, its measurements. To proceed further, we need a specific example that is a particular (but hopefully representative) two-dimensional map. Following again *Farmer* et al. [6.17] we choose as such an example Baker's transformation[5], which is defined as

$$x_{n+1} = \begin{cases} \lambda_a x_n & \text{for} \quad y_n < a \\ \frac{1}{2} + \lambda_b x_n & \text{for} \quad y_n > a \end{cases},$$

and

$$y_{n+1} = \begin{cases} \frac{1}{a} y_n & \text{for} \quad y_n < a \\ \frac{1}{1-a} (y_n - a) & \text{for} \quad y_n > a \end{cases},$$

(6.4.42)

where $0 \leqslant x_n \leqslant 1$, $0 \leqslant y_n \leqslant 1$. In deriving from the Lyapounov exponents of the above map the information dimension of the map, we have to be aware of possible self-similarity manifested as the number of iterations of the map increases. So let us examine what algorithm the above equations implicitly contain.

First of all, we consider the unit square. Let us choose $\lambda_b > \lambda_a$ and a, λ_a, $\lambda_b \leqslant 1/2$. The mapping divides the unit squares into two horizontal strips of heights a and $1-a$ (Fig.6.28). Then it compresses the two strips horizontally until the lower one acquires a width λ_a and the upper one acquires a width λ_b. Next it stretches both strips along the vertical direction until each acquires a height

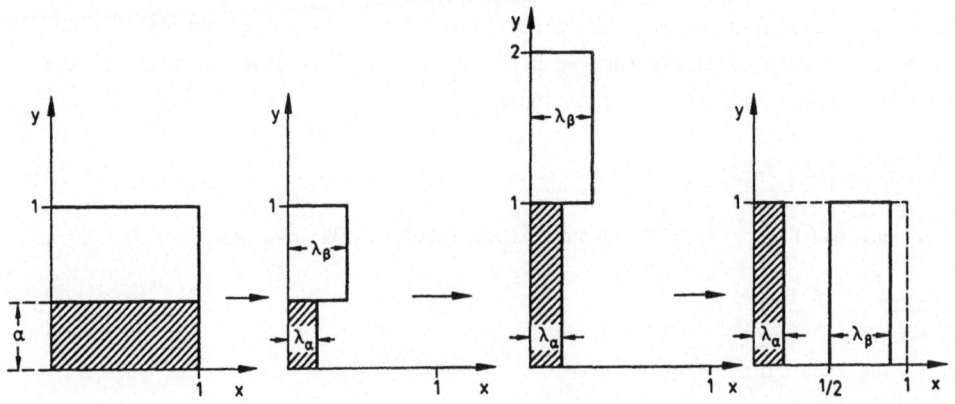

Fig.6.28. An example of the first iteration of the asymmetrical Baker's transformation. After [6.17]

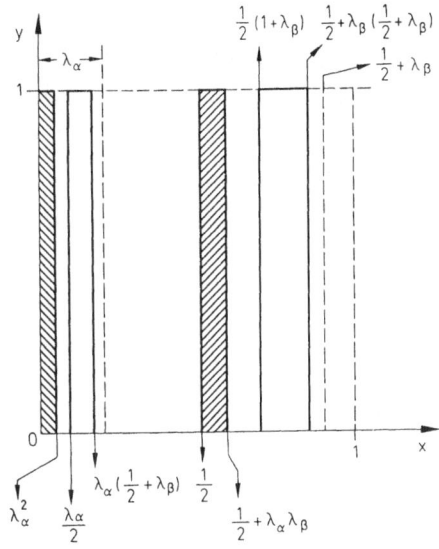

Fig.6.29. The second iteration of the Baker's transformation example of Fig.6.28

In the figure, the labels shown are:

$\frac{1}{2}(1+\lambda_\beta)$ $\frac{1}{2}+\lambda_\beta(\frac{1}{2}+\lambda_\beta)$

$\frac{1}{2}+\lambda_\beta$

λ_α

$\lambda_\alpha(\frac{1}{2}+\lambda_\beta)$ $\frac{1}{2}$

λ_α^2 $\frac{\lambda_\alpha}{2}$ $\frac{1}{2}+\lambda_\alpha\lambda_\beta$

equal to one. Finally, leaving the lower strip intact, it places the higher one on the x-axis between 1/2 and $(1/2)+\lambda_b$.

Applying the map *twice* to the unit square, one creates the arrangement shown in Fig.6.29.

Clearly we see that if the x-interval $[0,\lambda_a]$ is magnified by $1/\lambda_a$, it yields a replica of the previous first iteration. Likewise, if the x-interval $[1/2, (1/2)+\lambda_b]$ is magnified by $1/\lambda_b$, it gives again a replica of the previous iteration. Therefore the map displays self-similarity.

From the structure of the equations of Baker's map we observe that the cover along the y-direction is *ergodic* in the interval [0,1]. Then the Jacobian of the system is diagonal and only y dependent, i.e.,

$$J = \begin{pmatrix} I_2(y) & 0 \\ 0 & I_1(y) \end{pmatrix}, \quad \text{where} \tag{6.4.43}$$

$I_2 = \lambda_a$,   if $y < a$;

$I_2 = \lambda_b$,   if $y > a$;   and

$I_1 = \dfrac{1}{a}$,   if $y < a$;

$I_1 = \dfrac{1}{1-a}$, if $y > a$.

The Lyapounov numbers will thus be given by

$$\lambda_1 = \lim_{n \to \infty} [I_1(y_n)I_1(y_{n-1}) \cdots I_1(y_1)]^{1/n} \tag{6.4.44}$$

and

$$\lambda_2 = \lim_{n \to \infty} [I_2(y_n) I_2(y_{n-1}) \cdots I_2(y_1)]^{1/n} \ . \tag{6.4.45}$$

The elements I_1 take on only two values: $1/a$ for $y < a$, $1/(1-a)$ for $y > a$; likewise the elements I_2 take on only two values: λ_a for $y < a$, λ_b for $y > a$.

So the Lyapounov exponents will be

$$\log_2 \lambda_2 = a \log_2 \lambda_a + (1-a) \log_2 \lambda_b \tag{6.4.46}$$

and

$$\log_2 \lambda_1 = a \log_2 \left(\frac{1}{a} \right) + (1-a) \log_2 \left(\frac{1}{1-a} \right) \ , \tag{6.4.47}$$

where we have taken into account that the probability or asymptotic value of the fraction of time the system has spent in the portion $y < a$ is just a, and the probability the system has been in the portion $y > a$ is $1 - a$. (This is due to the ergodic character of the orbit in the y-direction.)

Putting

$$S(a) = a \log \left(\frac{1}{a} \right) + (1-a) \log \left(\frac{1}{1-a} \right) \tag{6.4.48}$$

and taking into account the previously derived expression

$$d_L = 1 + \frac{\log_2 \lambda_1}{\log(1/\lambda_2)} \ , \tag{6.4.41}$$

we finally obtain the "Lyapounov dimension" of the two-dimensional Baker's map as

$$d_L = 1 + \frac{S(a)}{a \log_2 \left(\frac{1}{\lambda_a} \right) + (1-a) \log_2 \left(\frac{1}{\lambda_b} \right)} \ . \tag{6.4.49}$$

At this point let us remind ourselves that in two dimensions our attractor is the Cartesian product of a Cantor-like set —along the x-direction —and the interval [0,1] along y. So the information dimension will be

$$D_I = 1 + D_I' \ , \tag{6.4.50}$$

where D_I' is the information dimension along the x-axis. To investigate further the relationship of the above result to the information dimension of Baker's system, *Farmer* et al. [6.17] proceed as follows. The information along x, $I(\varepsilon)$, arises essentially as a sum from the contribution of the *two* strips, of height 1 along the y-axis, and widths along the x-axis of λ_a and λ_b respectively. Let us call these contributions $I_a(\varepsilon)$ and $I_b(\varepsilon)$ respectively. The total probability that the system will be in the first strip is a and the total probability that the system will be in the second strip is $1 - a$. Let $n(\varepsilon)$ be the usual number of strips of width ε (resolution length) required to cover the whole attractor. From the scaling properties of the system, it is clear that covering, say, the first strip in the interval $[0, \lambda_a]$, at resolution $\varepsilon \lambda_a$, also requires $n(\varepsilon)$ "cells".

Let us remind ourselves of the definition of the information dimension

$$D_I' = \lim_{\varepsilon \to 0} \frac{I(\varepsilon)}{\log_2(\frac{1}{\varepsilon})} \quad , \qquad \text{where} \tag{6.4.51}$$

$$I(\varepsilon) = \sum_{i=1}^{n(\varepsilon)} P_i \, \log_2\left(\frac{1}{P_i}\right) \tag{6.4.52}$$

and P_i is the probability that the ith N-dimensional "cube" will contain points of the attractor. So, for the strip $[0,\lambda_a]$ of Baker's map, we will have

$$I_a(\varepsilon\lambda_a) = \sum_{i=1}^{n(\varepsilon)} aP_i \, \log \frac{1}{aP_i} = a\left[\log \frac{1}{a} + I(\varepsilon)\right] \quad . \tag{6.4.53}$$

Replacing $\varepsilon\lambda_a$ by ε in the above relationship, we get

$$I_a(\varepsilon) = a \, \log \frac{1}{a} + aI\left(\frac{\varepsilon}{\lambda_a}\right) \quad , \tag{6.4.54}$$

and for the second strip $[1/2,(1/2)+\lambda_b]$ likewise,

$$I_b(\varepsilon) = (1-a)\log \frac{1}{1-a} + (1-a)I\left(\frac{\varepsilon}{\lambda_b}\right) \quad . \tag{6.4.55}$$

Therefore, since $I(\varepsilon) = I_a(\varepsilon) + I_b(\varepsilon)$, we find

$$I(\varepsilon) = aI\left(\frac{\varepsilon}{\lambda_a}\right) + (1-a)I\left(\frac{\varepsilon}{\lambda_b}\right) + S(a) \quad , \tag{6.4.56}$$

where $S(a)$ (6.4.48) has been given in Sect.2.3.2 as the "binary entropy function" representing the information contained in a "head-tail" trial where heads (or tails) have a priori probability a.

Taking into account (6.4.56) in the definition $D_I' = I(\varepsilon)/\log(1/\varepsilon)$ and substituting for $I(\varepsilon)$, $I(\varepsilon/\lambda_a)$, and $I(\varepsilon/\lambda_b)$, we get after some trivial algebra:

$$D_I' = \frac{S(a)}{a \, \log \frac{1}{\lambda_a} + (1-a)\log \frac{1}{\lambda_b}} = d_L - 1 \quad , \tag{6.4.57}$$

and finally

$$D_I = d_L \quad . \tag{6.4.58}$$

For a symmetrical map, e.g., $a = 1/2$ and $\lambda_a = \lambda_b = 1/3$, we get

$$D_I = \frac{1}{\log_2 3} + 1 \quad .$$

For $a = 1/2$, $\lambda_a = \lambda_b = 1/2$, we get

$$D_I = 2$$

— as expected since now the unit square is ergodically covered by the map in both x- and y-directions.

Thus by calculating the Lyapounov exponents of our discrete map or in a continuous attractor, we can get the expression for the information dimension of the attractor giving, in fact, the degree of *compressibility* achieved by the dynamical system.

6.4.3 The Concept of Metric (Kolmogorov-Sinai) Entropy and Its Relation to Information Dimension

So far we have been investigating the amount of information acquired by an observer in making a single, isolated measurement of a dynamical system possessing a strange attractor or, in turn, the information an observer needs in order to define unambiguously one point on the attractor. We now turn to the "dynamical" problem. We assume that the attractor-possessing system "unfolds" before an observer, and we wonder about the new information the observer acquires by making successive measurements in time, that is, by receiving from the system a behavioral repertoire consisting of a discrete time series of "pulses". Obviously, if the attractor is a stable steady state, once the measurement process fixes that state, no further measurements are needed: no new information is to be acquired from a static object.

Likewise, if the attractor is a limit cycle, once the transients associated with the closing of one period are over, further observations merely reproduce redundant information. A similar thing happens to any other periodic attractor, e.g., a (rational) 2-torus. Things change dramatically, however, if the attractor is aperiodic (strange). In such a case the system is unpredictable and the metric entropy presently discussed provides an upper bound on the rate of *acquisition of information* from such an evolving dynamical system. One immediately expects, then, that before the first accumulation point of the control parameter(s) of an attractor-possessing system (where the repertoire of the system consists only of stable periodic trajectories), the metric entropy should be zero and should start "taking-off" just after the appearance of the first aperiodic orbit.

Let us assume that the observer receives a discrete time series which constitutes an ergodic Markov chain possessing n states σ_i (in the form say of n integer numbers), revealing themselves through sequential measurements. The measurements are performed on each of the n elements of the partition of the state space which supports the attractor. If the occurrence of a certain symbol/state depends stochastically on m preceding symbols/states, the process is of mth order. Our first question is: What is the average amount of information per symbol emitted by such a Markov source? Clearly the answer to this should provide an observer familiar with the past history of the source with a measure of the unpredictability of the sequences generated by the source.

The knowledge we have from Chap.4 on elementary information theory (concerning Markov chains of order one) allows us to calculate the entropy of an ergodic Markov source of order m possessing a number of discrete states n. The total number of state sequences one may construct of length m out of n states is n^m. [Alternatively, if we consider each sequence S_m $(\sigma_1,\sigma_2,\ldots,\sigma_m)$ as an m-digit base-n fraction, and arrange all those sequences as points in the interval [0,1], then two neighboring numbers in the unit interval will be separated by a distance n^{-m}, and there will be $1 - n^{-m}$ of them.]

Now each m-digit-long sequence forms a "state" S_m in our m^{th} order Markov chain. The amount of information obtained when the next state is σ_{m+1}, given that the chain is now at the state S_m, is

$$I = -\log_2 P\left(\frac{\sigma_{m+1}}{\sigma_1,\sigma_2,\ldots,\sigma_m}\right) = -\log_2 P(\sigma_{m+1}/S_m) \quad , \tag{6.4.59}$$

where $P(\sigma_{m+1}/S_m)$ is the conditional probability that *given* the "state" S_m, the next (m+1)th symbol will be σ_{m+1}.

To obtain the entropy of the source we have to average the above expression over all possible transitions from S_m to σ_{m+1}, and over all possible n^m "states" S_m. Let $P(S_m)$ be the probability that the particular m-sequence takes place $\left[\sum_{m=1}^{n^m} P(S_m) = 1\right]$. Then the average amount of new information per symbol σ acquired by the observer is

$$\Delta I_m = -\sum_{S_m=0}^{n^m} P(S_m) \sum_{\sigma_{m+1}=0}^{n-1} P(\sigma_{m+1}/S_m)\log_2 P(\sigma_{m+1}/S_m) \quad . \tag{6.4.60}$$

But after Bayes's rule,

$$P(S_m,\sigma_{m+1}) = P(S_m)P(\sigma_{m+1}/S_m) \quad . \tag{6.4.61}$$

where $P(S_m,\sigma_{m+1})$ is the joint probability for S_m and σ_{m+1} to occur *successively*, so we have

$$\log_2 P(S_m,\sigma_{m+1}) = \log_2 P(S_m) + \log_2 P(\sigma_{m+1}/S_m) \quad ,$$

and

$$\log_2 P(\sigma_{m+1}/S_m) = \log_2 P(S_m,\sigma_{m+1}) - \log_2 P(S_m) \quad . \tag{6.4.62}$$

Substituting into (6.4.60), we obtain

$$\Delta I_m = -\sum_{S_m=0}^{n^m} \sum_{\sigma_{m+1}=0}^{n-1} P(S_m,\sigma_{m+1})[\log_2 P(S_m,\sigma_{m+1}) - \log_2 P(S_m)]$$

$$= -\sum_{S_m=0}^{n^m}\left\{\left[\sum_{\sigma_{m+1}=0}^{n-1} P(S_m,\sigma_{m+1})\log_2 P(S_m,\sigma_{m+1})\right]\right.$$

$$- \left[\sum_{\sigma_{m+1}=0}^{n-1} P(S_m, \sigma_{m+1}) \right] \log_2 P(S_m) \right\} \quad . \tag{6.4.63}$$

Since, however,

$$\sum_{\sigma_{m+1}=0}^{n-1} P(S_m, \sigma_{m+1}) = P(S_m) \quad , \tag{6.4.64}$$

we find

$$\Delta I_m = - \sum_{S_m=0}^{n^m} \left\{ \left[\sum_{\sigma_{m+1}=0}^{n-1} P(S_m, \sigma_{m+1}) \log_2 P(S_m, \sigma_{m+1}) \right] - P(S_m) \log_2 P(S_m) \right\} \quad . \tag{6.4.65}$$

Obviously now

$$\sum_{\sigma_{m+1}=0}^{n-1} P(S_m, \sigma_{m+1}) \log_2 P(S_m, \sigma_{m+1}) = P(S_{m+1}) \log_2 P(S_{m+1}) \quad , \tag{6.4.66}$$

so

$$\Delta I_m = I_{m+1} - I_m \quad , \qquad \text{where} \tag{6.4.67}$$

$$I_m = - \sum_{S_m=0}^{n^m} P(S_m) \log_2 P(S_m) \quad . \tag{6.4.68}$$

For an ergodic attractor, ΔI_m is, then, the average amount of new information gained by the observer as the trajectory of the attractor moves from "box" to "box" in the partitioned state space. It is essentially the (exponential) rate at which nearby trajectories diverge locally that is responsible for the generation of new information with each successive measurement (see also Sect.6.5).

Taking the limit as m goes to infinity we get the expression

$$\frac{\Delta I}{\Delta t} = \lim_{m \to \infty} \left(\frac{I_m}{m \Delta t} \right) \quad [\text{bits/s}] \tag{6.4.69}$$

for the rate [bits/s] at which information is generated by the unfolding attractor. The metric entropy is defined from the above expression if we now take the maximum value over all possible partitions β of the state space, namely,

$$\text{metric entropy} = h_\mu = \sup_{(\beta)} \left(\lim_{\substack{m \to \infty \\ \Delta t \to 0}} \frac{I_m}{m \Delta t} \right) \quad . \tag{6.4.70}$$

The metric entropy is then the upper limit (it assumes a perfect observer) of the average amount of new (not implicit in the initial conditions) information gained per second from the dynamical evolution of a strange attractor. As the window of resolution becomes narrower and narrower with increasing iterations, the above rate should increase since now small-scale random (thermal) fluctuations inherent to any large-scale system are not smeared out and are bound to play a macroscopic role. How then can we establish a relationship between the *metric entropy*, on the one hand, and the *information dimension* (and the spectrum of Lyapounov exponents) on the other?

We can, as previously mentioned (see also [6.14]), consider the n^m possible sequences generated by the flow of the attractor, uniformly distributed along the unit interval as a set of m-digit base-n fractions (symbols x_i). Since they are spaced n^{-m} units apart, they are numbered from 0 to $1-n^{-m}$.

Let us now consider the p.d.f. $P_m(x)$ by plotting the probability $P(S_m)$ as a coordinate above the point of the interval corresponding to fraction x. Clearly

$$P_m(x) = P(S_m)n^m \quad , \tag{6.4.71}$$

where $S_m \leqslant x < S_m + n^{-m}$, $S_m = 0$, $n^{-m}, \ldots, 1 - n^{-m}$, and $P_m(x)$ is further specified from the normalization relations

$$\int_0^1 P_m(x)dx = 1 \quad \text{or} \quad \int_0^1 P(S_m)dx = n^{-m} \quad . \tag{6.4.72}$$

In the limit $m \to \infty$, where the Markov chain involved has infinite "memory", the p.d.f. $P_m(x)$ acquires the asymptotic value

$$P_\infty(x) = \lim_{m \to \infty} P_m(x) \quad . \tag{6.4.73}$$

Then $P(S_m)$, the probability of occurrence of the S_m sequence, becomes

$$P(S_m) = \int_{S_m}^{S_m + n^{-m}} P_\infty(x)dx \quad . \tag{6.4.74}$$

This relation means that consideration of symbol sequences of length m amounts to examining the symbol sequence probability density $P_\infty(x)$ at a scale of resolution $\varepsilon = n^{-m}$. This mapping of the set of probabilities $P(S_m)$ onto the unit interval allows the metric entropy h_μ to be expressed in terms of the information dimension. Indeed

$$h_\mu = \lim_{m \to \infty} \left(\frac{I_m}{m}\right) \quad , \tag{6.4.75}$$

for $\Delta t \sim 1$ time unit, or

$$h_\mu = \lim_{m \to \infty} \left(\frac{I_m}{m \log_2 n}\right) \log_2 n \quad . \tag{6.4.76}$$

Since $\varepsilon = n^{-m}$, $\log \varepsilon = -m \log_2 n$, so

$$h_\mu = \lim_{m \to \infty} \left(\frac{I_m}{|\log_2 \varepsilon|}\right) \log_2 n = D_I(P_\infty) \log_2 n \quad , \tag{6.4.77}$$

where $D_I(P_\infty)$ is the information dimension for the asymptotic value taken by the p.d.f., P_∞. Therefore

$$h_\mu = D_I \log_2 n \quad , \tag{6.4.78}$$

where n is the number of *discrete* states/boxes into which we partition the state space —or the number of discrete states we measure on the attractor.

A practical way of achieving a rough calculation of the Kolmogorov-Sinai entropy is by finding the correspondence between a discrete map iterated beyond the chaotic limit and a Markov chain.

Take as an example [6.18] the familiar map

$$X_{t+1} = 4X_t\left(1 - \frac{X_t}{\xi}\right) \tag{6.4.79}$$

where ξ is an integer denoting the partition of the interval. For $\xi = 2$ we have a partition between two symbols, A and B say, in the interval 2 and for $\xi = 4$ we have a partition between four symbols say A, B, C, D (in the interval 4), (for $\xi > 4$ the partition is non-Markovian, see [6.18]).

From geometrical considerations it is quite easy to calculate the elements of the transition probability matrix of the Markovian chain involved. Thus, for $\xi = 2$, for example, the portion of the A = 1 on the horizontal axis which is projected as A on the vertical axis equals $P_{AA} = 1 - (1/\sqrt{2})$; consequently $P_{AB} = 1/\sqrt{2}$; likewise $P_{BA} = 1 - (1/\sqrt{2})$ and $P_{BB} = 1/\sqrt{2}$. In the case $\xi = 4$ the transition matrix is

$P(\sigma_{t+1}/\sigma_t)$	A	B	C	D
A	α	0	0	α
B	β	0	0	β
C	γ	0	0	γ
D	0	1	1	0

where $\alpha = 2 - \sqrt{3}$, $\beta = \sqrt{3} - \sqrt{2}$, and $\gamma = \sqrt{2} - 1$.

From the elements of the transition probability matrix one can calculate the probabilities P(A), P(B) or P(A), P(B), P(C), P(D) from n - 1 of the linear relationships

$$u_K = \sum_{j=1}^{n} u_j P_{jK} \tag{6.4.80}$$

and the normalizing relation

$$\sum_{K=1}^{n} u_K = 1 \tag{6.4.81}$$

where n = 2 or 4.

Therefore, the entropy of the Markov chain involved equals

$$S = \sum_{K=1}^{n} u_K S_K = - \sum_{K=1}^{n} \sum_{i=1}^{n_K} u_K P_{Ki} \log_2 P_{Ki} \quad \text{[bits]} \tag{6.4.82}$$

where (for each case) n_K is the number of states which can be reached in a single hop from state K.

This concludes the section on the parameters characterizing a strange attractor. We have investigated them in some detail since all of them —information dimension D_I, spectrum of Lyapounov exponents $\{\lambda_i\}$, and metric entropy h_μ —are interrelated and are centred around two basic concepts:

a) The *compressibility* displayed by a strange attractor, that is, the degree of reduction of the number of degrees of freedom achieved by the fact that the flow maps onto itself, thereby forming a compact ergodic set in a state space with much higher dimensionality.
b) The concept of *information generation*, which makes (in finite time) the system independent of the initial conditions.

The investigation above is crucial for the main theme of this book: namely the search for mechanisms ensuring the formation of collective properties, thereby allowing the formation of hierarchical systems and the possibility of mutual simulation between them. Indeed compressibility is the sine qua non prerequisite for the formation of such collective properties via the formation of long-range cross-correlations among pertinent macrovariables. The speculation is that biological systems possessing "rhythms" with chaotic properties, i.e., strange attractors, can accomplish reliable information processing in some "economic" fashion. We deliberate on this issue in the next section.

6.5 A Possible Role for Chaos in Reliable Information Processing

This section can be read independently of Sect.6.4.

A problem of fundamental importance in the design of "self"-organizing systems is the "theoretical minimum" amount of hardware complexity (C_H) necessary to drive a given functional repertoire (software complexity, C_S). In general, it is assumed that the curve $C_H = f(C_S)$ is a monotonically rising one, with steepness depending on the specific "wiring mechanism" or the architecture of the given system. This conviction comes traditionally from communication engineering practice, where the act of information processing includes a sequence of an "expansion" and a "contraction" of the dimensionality of the state space, i.e., a proliferation and subsequent compression of the degrees of freedom of the transmitted message.

Indeed, on the transmitting side, the effort towards efficient coding requires the *orthogonality* of the "words" (members) of the sender's repertoire. This is accomplished by expanding the bandwidth W or inflating the transmission time T, thereby increasing the dimensions 2WT of the state space where the individual words (digitized wave forms via the sampling theorem) figure as hypervectors.

At the receiving end a "contraction" is performed, amounting to a set of convolutions between the incoming (noise-contaminated) signal and each member/word of the transmitter's repertoire. Since the individual words are mutually orthogonal, the above operations allow the receiver to detect and correct multiple (albeit a finite number of) errors which occur due to the channel noise.

When we turn to biological systems, we meet a rather confusing state of affairs. Simple parameters such as the weight or the volume of the brain apparently mean very little (the blue whale being a notorious example!). Comparative anatomy does not help either, since this could put the dolphins in a position comparable with humans. On a more sophisticated level, we consider the complexity of the hardware as dependent on a melange of parameters such as the number of interacting components, connectedness, and distribution function of interaction strength.

Nevertheless, on many occasions we witness biological organisms with rather simple nervous systems (hardware)—such as the leech or the cockroach—to be capable of very impressive and complex behavioral repertoires. We begin to suspect that there are systems whose algorithms of information processing do not perhaps follow the principle of man-made artifacts, that is, of increasing hardware complexity in order to achieve great complexity in the behavioral domain.

In this discussion we intend to exploit a new alternative theoretical principle which can satisfy the requirements of a broad functional repertoire with very simple hardware. This principle is based on the fact that information is produced not only by dissipating the degrees of freedom in a system, but also by increasing the resolution in systems with few degrees of freedom.

Certain nonlinear, dissipative systems with just three degrees of freedom can exhibit random behavior which is analogous to that produced by explicit stochastic equations. Instead of creating *new* degrees of freedom, i.e., instead of expanding the bandwidth or the dimensionality of state space, such systems generate iterative self-similar processes (possessing scaling properties) which decrease the resolution or expand the dynamics of trajectories in a low-dimensional state space.[6]

This behavior is the result of a process through which the state trajectories for the nonlinear system enter a low-dimensional region of state space such that nearby trajectories must enter it, but once inside, they diverge from each other. The sensitivity to small differences in initial conditions gives rise to the probabilistic character of an otherwise simple deterministic system.

6.5.1 Theoretical Considerations and General Discussion

Reliable information processing rests upon the existence of a "good" code or language: a set of recursive rules which *generate information* (e.g., aperiodic strings of digits) at a given hierarchical level and subsequently *compress* it at a higher

---

6 More precisely, along a "positive Lyapounov exponent", the system expands trajectories and so it creates variety; along a "negative Lyapounov exponent", the system constrains variety, thereby revealing information.

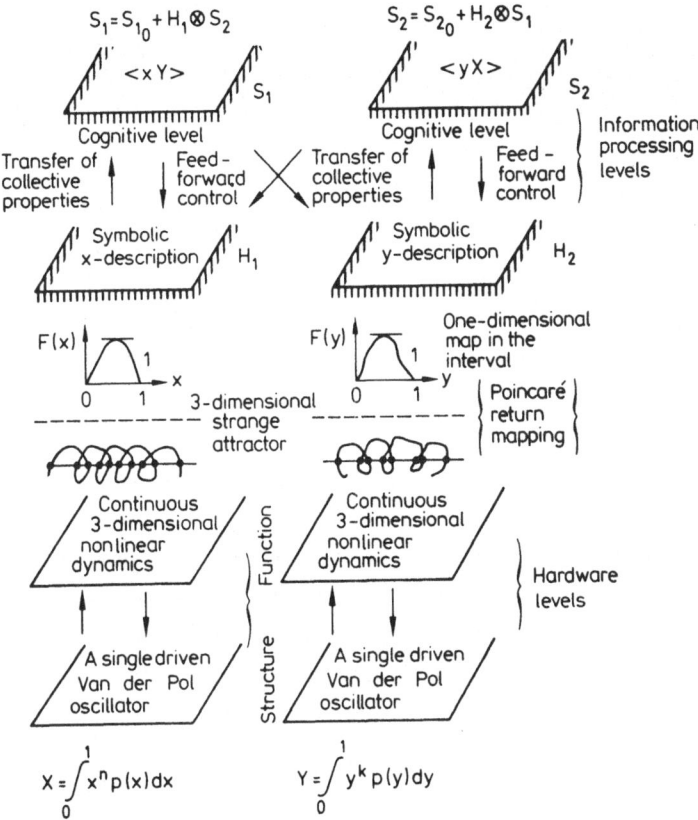

$$S_1 = S_{1_0} + H_1 \otimes S_2 \qquad\qquad S_2 = S_{2_0} + H_2 \otimes S_1$$

$$X = \int_0^1 x^n p(x)\,dx \qquad\qquad Y = \int_0^1 y^k p(y)\,dy$$

Fig.6.30. Layout of a simple communication scheme between two "linguistic" hierarchical systems. The ongoing dynamics at software and hardware levels, respectively, are symbolized by S and H

cognitive level (Fig.6.30). To accomplish this, a language —like good music —should strike at every moment an optimum ratio of variety (stochasticity) and the ability to detect and correct errors (memory). Is there any dynamics available today which might model this dual objective in state space? The answer is, in principle, yes.

We have been investigating recently dynamical systems described by at least three coupled first-order ordinary nonlinear differential equations whose repertoire includes (for different sets of values of the control parameters) multiple steady states, stable periodic orbits (limit cycles), tori, and strange attractors (chaos). For an excellent review see *Shaw* [6.10].

We may consider (subscribing to the scientist's belief that events are deduced only by observation and measurement) that entropy is generated when the volume in state space *expands* through the dynamical evolution of our system (thereby decreasing resolution), and is *compressed* (dissipated, thereby revealing information) when the volume in state space occupied by the flow *contracts* towards a "compact" ergodic flow —the *attractor*. More precisely, for values of the control parameters outside those associated with chaos, information is generated by a physical system via

323

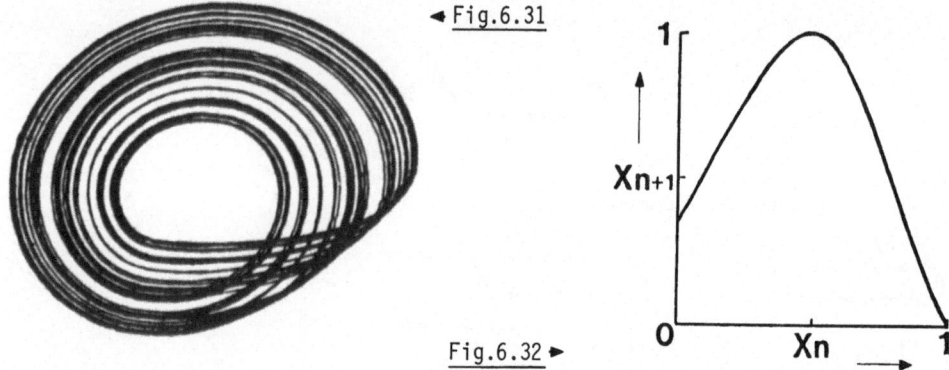

◄ Fig.6.31

Fig.6.32 ►

Fig.6.31. Plane projection of a three-dimensional strange attractor
Fig.6.32. One-dimensional Poincaré map for a three-dimensional strange attractor

cascading *bifurcations* giving rise to broken symmetry. Within the values of the
control parameters which trigger aperiodic trajectories, information is generated
(or dissipated) via *cascading iterations* of the (e.g., one-dimensional) map on the
interval which is constructed as a "Poincaré return map" of the attractor.

The type of "analogue-to-digital conversion" is accomplished by parametrizing
the attractor along a one-dimensional cut, and plotting the position where a trajectory crosses the cut versus the position where it crosses the next time around
the attractor. This stands for a "stroboscopic" pursuit of "level crossings" of the
(three-dimensional) state space flow and gives rise to a Markov chain whose number
of states depends on the partition of the interval [0,1] (Figs.6.31,32).

The change in observable information is generally given by the logarithm to
base 2 of the ratio of states Σ distinguishable before and after some time interval:

$$\Delta I = \log_2 \frac{\Sigma_f}{\Sigma_i} \sim \log_2 \frac{V_f}{V_i} \quad , \qquad (6.5.1)$$

where V_f, V_i are the final and initial volumes in state space. Then the rate of
information creation (or dissipation) is given as

$$\frac{dI}{dt} = \frac{I}{V} \frac{dV}{dt} \sim \frac{1}{\Sigma} \frac{d\Sigma}{dt} \quad . \qquad (6.5.2)$$

The number of distinguishable states $\Sigma(t)$ arising from some initial block volume
in state space need *not*, of course, be directly proportional to the volume change
under the flow.

Now suppose the number of states/blocks in which the system could be found
grows (a) polynomially and (b) exponentially with time.

In the first case,

$$\Sigma(t) \sim t^n \qquad \text{and} \qquad (6.5.3)$$

$$\frac{dI}{dt} \sim \frac{n}{t} \quad : \qquad (6.5.4)$$

the information creation rate of such a system converges to zero as time passes so
the system's behavior is predictable (compressible) for an indefinite time into
the future.

In the second case,

$$\Sigma(t) \sim e^{nt} \quad \text{and} \tag{6.5.5}$$

$$\frac{dI}{dt} \sim n : \tag{6.5.6}$$

such a system is a continuous information source. This information is not implicit
in the initial conditions, whatever they may be, but is generated by the flow itself in state space. This type of information is incompressible and problems associated with handling it may turn transcomputational.[7] We are interested here in nonconservative compact flows, where the phenomenon of attraction (steady state, limit
cycles, or strange attractors) is possible.

In at least three dimensions it is possible to have very simple physical systems,
for example, one forced, negative-resistance oscillator (investigated recently by
Ueda and *Akamatsu* [6.12] and simulated by a van der Pol oscillator driven by a
periodic excitation), displaying flows which, in a compact region of state space,
continuously expand volumes in some dimensions while contracting them in others,
meaning that they generate information for certain variables and compress or dissi-
pate it for others.

For any given one-dimensional map in the interval $y = F(x)$, the probability density $P(x)$ for finding the orbit at x can be estimated via successive iterations from
the recursive relation:

$$P_n(x) = \sum_{i=1}^{2^{n-1}} \frac{P_i(x_i)}{\left| \frac{dF^{(n-1)}}{dx} \right|_{x_i}} , \tag{6.5.7}$$

where $x_i = [F^{(n-1)}]^{-1}(y)$.

Here the first p.d.f. $P(x)$ is arbitrary, say $P_1(x) = \delta(x)$, and the resulting
$P_2(y)$, corresponding to the initial $P_1(x)$ transformed by the action of the map, is
the first-approximation p.d.f. Successive iterations of the map will produce closer
and closer approximations to the correct "equilibrium" value of $P(x)$. If the map
has a stable steady state or a stable periodic orbit, $P(x)$ will converge to a sharp
delta-function spike or a series of delta-like functions on the periodic points.

---

7 If the above system is observed by another cognitive system at a sampling rate
less than n, no prediction of the system is possible. See also Sect.6.5.6 concerning the genesis of "conflict" between two mutually "observed" cognitive systems.
A strategy for "transcending" the high rate of entropy production by the "opponent"
is to form "collective properties" of the observed system; when this trick collapses
too, *intra*-conflict sets in.

The information change ΔI per iteration of a particular point x of the map will be determined as

$$\Delta I = \log_2 \left| \frac{dF}{dx} \right| \tag{6.5.8}$$

and will amount to entropy creation for slopes >1, and to information creation or entropy dissipation for slopes <1. The average information change over the whole interval $0 \leqslant x \leqslant 1$ will be given as

$$\langle I \rangle = \int_0^1 P(x) \log_2 \left| \frac{dF(x)}{dx} \right| dx \quad \text{[bits]} \quad , \tag{6.5.9}$$

or, if $t(x)$ stands for the time between successive passes through the return map, the average information production (or dissipation) rate in time will be

$$\left\langle \frac{dI}{dt} \right\rangle = \int_0^1 \frac{P(x)}{t(x)} \log_2 \left| \frac{dF(x)}{dx} \right| dx \quad \text{[bits/s]} \quad . \tag{6.5.10}$$

Finally, if the initial data from which the process starts are not exactly known, i.e., are determined by an a priori probability density distribution $P_0(x)$ due to observational or internal fluctuations, we get for the informational value of an initial condition the expression

$$s_i = \int_0^1 P_0(x) \log_2 \left(\frac{P_0(x)}{P(x)} \right) dx \quad \text{[bits]} \tag{6.5.11}$$

within the interval; then, the "memory" of the processor measured as the time elapsing until the system becomes causally disconnected from the initial conditions is given as

$$T = \frac{s_i}{\langle dI/dt \rangle} = \frac{\int_0^1 P_0(x) \log_2 \left(\frac{P_0(x)}{P(x)} \right) dx}{\int_0^1 \frac{P(x)}{t(x)} \log_2 \left| \frac{dF(x)}{dx} \right| dx} \quad \text{[s]} \quad . \tag{6.5.12}$$

Beyond this limit T, the system amplifies intrinsic microscopic noise —which, due to the great resolution created by the iterative process, cannot be "smeared out" any longer (see also Sect.6.5.6).

Obviously for values of the control parameters below the initiation of chaos (for which only steady states and periodic cycles exist), the trajectories in state space will only attract, and the corresponding iterations in the Poincaré return map in the interval will monotonically converge. Under such circumstances $\langle I \rangle < 0$, i.e., the system will behave asymptotically as an information sink.

Above the critical value(s) of the control parameter(s) for which continuous stable chaos is established, $\langle I \rangle > 0$ along some variables (information source), and $\langle I \rangle < 0$ along some other variables (information compression or dissipation). So much for the information-production-dissipation interplay.

A dynamical model of natural language should simulate and explain a number of key "idioms" like (a) the ability of the language to form abstraction (collective properties), for instance, cross-correlations among underlying stored patterns and environmental messages, and (b) the perennial paradox of "self"-reference.

At first one might argue that the syndrome of self-reference is related to "circular argumentation". However, this is dynamically attributed to maps folding (iterating) onto themselves, which in turn implies the existence of only steady states and periodic orbits giving rise to information dissipation. The congruence, then, between "circularity" and self-reference appears rather superficial since circularity implies a "halt" in processing while paradox implies "deadlook" in the cognitive machine.

Indeed the paradox of self-reference goes deeper, namely it has to do with the inability of the language to discriminate between statements belonging to hierarchically different cognitive levels (confusing a set for one of its members signals precisely the appearance of the "Russelian" paradox). In short, the paradox of self-reference is associated with the "distress" caused by the inability to compress an irrational number; it is illustrated by the linguistic example: "This sentence means what this sentence means".

In any cognitive process the receiver aims at "breaking" the code of the incoming signal, i.e., "compressing" or abstracting as much as possible the description conveyed by the message, thereby achieving a high degree of predictability. Achieving compressibility is, therefore, tantamount to forming collective properties out of the variables of the signal under study. In the domain of description governed by the solution of the master equation, it is tantamount to compressing the p.d.f. $P(x)$ to achieve the "mean-field" approximation, which involves decoupling and the predominance of the first moment(s) of the p.d.f. over higher moments (variance, etc.). It is remarkable that the p.d.f. $P(x)$ of, for example, the logistic map derived in Sect.6.3.5 has for $a = 4$ a *hyperbolic* shape which makes the median value *least likely* to occur. In such cases, any prediction based on the "law of averages" is bound to give false results.

Remember [6.19] that *near bifurcation points* triggering transitions either among steady states, steady states and limit cycles, or periodic orbits and chaos, especially around period three, the p.d.f. $P(x)$ becomes double-humped, multihumped, or hyperbolic (the median ceases to be the most probable value). All moments of the p.d.f. may become comparable and are coupled, essentially via the nonlinearity of the transitional probabilities of the master equation, from which equation the next higher level emerges through taking successive moments. Thus the "macrodescription" at the level where cognition should take place acquires as many degrees of freedom (moments) as the "microdescription" one hierarchical level below. This is what is meant by saying that the dynamics on two successive hierarchical levels get mixed up: descriptions belonging to two different hierarchical levels become — near bifurcation points — indistinguishable.

In conclusion, when the control parameters allow the existence of only steady states and periodic orbits, away from bifurcations, the mean-field regime is ensured, that is, deliberations at different hierarchical levels — the "microlevel" of variables and the "macrolevel" of moments — are distinct. This is ensured by the small spread of the p.d.f. around the mean.

Yet in such cases stagnation results, since in the close vicinity of the above two kinds of attractors information is neither produced nor dissipated. The "machine" simply halts, as happens, for instance, in the calculation of the digits of a rational number. Continuation of the process literally would lead to circularity. When, on the other hand, the control parameters allow the production of continuous self-sustained chaos, the p.d.f. is broad (it covers dense bands or the whole interval) and it may also, depending on the map, appear quite irregular. This may again signal the breaking of the mean-field regime. Thus in spite of the fact that the strange attractor generates and compresses information at the same time for different variables, continuous chaos is not perhaps the ideal model for a linguistic system.

One should rather settle for an "intermittent" regime where chaos is unstable, so to speak, thereby offering opportunities for spontaneous alternations between expansion and compression of the p.d.f. Fortunately, the repertoire of the above-mentioned simple dynamical systems allows for two such distinct behaviors, namely, (a) metastable chaos and (b) intermittency.

By *metastable chaos* — realized for some windows of the control parameter values — we imply a transitory regime whose decay times have an exponential distribution. After some initial period of irregular oscillation, almost every trajectory eventually settles down to a periodic orbit. Generally, the mean duration of the chaotic regime is not long, typically about 50 iterations, but initial conditions can be found, by trial and error on the map, which remain chaotic much longer.

In other situations, a metastable chaotic regime might decay to a steady state or to one which is also chaotic but which has significantly different behavior, i.e., a different p.d.f., and occupies a different subinterval of the whole attractor. Alternatively, the new regime might also be metastable, eventually decaying to a third regime or possibly back to the original one.

Intermittency — realized for some other windows of the control parameter values of the same dynamical system — refers basically to the physical inability of achieving perfect entrainment or stable, strict locking between, for example, the phase of a dissipative nonlinear relaxation oscillator and the phase of an external periodic excitation.

At the time the limit cycle in the state space is about to close, a "scrambling" process intervenes and destroys the orderly trajectory; it creates for a time interval, which is most probably either too short or too long, a chaotic motion amounting to broadband noise. Then the building up of a limit cycle (a different one) starts all over again. After a while, a new bursting transition to "turbulent chaos" occurs, followed by another limit cycle appearance, and so on, the time sequence of intermittent points being random. The corresponding p.d.f. of the relative times that the system spends in the two regimes turns out to be hyperbolic [6.20].

To sum up, in the neighborhood of steady states and periodic orbits, the "cognitive machine" halts. Circularity sets in, but the p.d.f. is very narrow and the mean-field regime works: abstractions are formed impeccably.

In the vicinity of bifurcations from steady states $\rightleftharpoons$ limit cycles $\rightleftharpoons$ chaos, the "cognitive machine" suffers a "deadlock": The p.d.f. explodes, the mean-field regime breaks down, and abstraction (i.e., cross-correlations) cannot be formed unambiguously. The system at the level of collective properties is turbulent in the sense that it acquires as many degrees of freedom as at the level of single variables.

In self-sustained stable chaos the situation is not necessarily the same, since information is generated along specific directions and compressed along others. Nevertheless, the prospect of a broad and "multiple-spiked" p.d.f. justifies some reluctance in adopting this model for a reliable processor. It seems that the best model for such a processor should be the regime of intermittency and metastable chaos.

Below we intend to discuss this model in a specific example: the electroencephalogram (EEG). Provided that the theory in Sect.6.5.2 makes sense, we can move ahead and propose a set of experiments. Under rather restricted conditions, we can construct from the recorded time series of the EEG a rough picture of the presumed attractor in phase space (for a low dimensionality, see [6.21]).

We begin with the time series obtained by sampling a single coordinate $x(t)$ (from a part of the EEG record) and calculate the delayed values $x_1(t-\tau)$ and $x_2(t-2\tau)$, or for $\tau \to 0$ the time derivatives $\dot{x}(t)$ and $\ddot{x}(t)$, provided that they are independent — something we can check by evaluating the conditional probabilities $P(x/x_1,x_2;t)$. Using these values of the variable and time-delayed variables as independent coordinates x, y, z we may then reconstruct a three-dimensional state-space picture of the sampled system's time evolution.[8]

Having done this, we can construct a set of one-dimensional return maps. From these maps, we deduce the information dimensions of the attractor in terms of the calculated values of the set of "positive-characteristic exponents", i.e., the information production $<I>$ as given by (6.5.9) — along different directions [6.22].

---

[8] In the case where information about the state-space dimensionality is lacking, one proceeds by using an ever-increasing number of variables x, $\dot{x}$, $\ddot{x}$, $\dddot{x}$,..., until saturation in the continuous dynamics is achieved.

Thus in a physiological laboratory with rather sophisticated clinical facilities, an experimental group could perhaps try to investigate and establish relationships between *behavioral states* (congruent to distinct forms of EEG), and *shape* and *dimensionality* of the corresponding regime of the "cognitive processor" responsible for these behaviors.

6.5.2 Application: The Electrical Activity of the Brain–Should It Be Chaotic?

> *"Music is the effort we make to explain to ourselves how our brains work"*
>
> Lewis Thomas
> "The Medusa and the Snail"

Brain-like structures have evolved by performing signal processing, initially by minimizing "tracking errors" on a competitive basis. Such systems are highly complex and, at the same time, notoriously "disordered". The functional trace of the cerebral cortex of the (human) brain is a good example.

The electroencephalogram (EEG) appears particularly fragmented during the execution of mental tasks as well as during the recurrent episodes of rapid-eye-movement (REM) sleep. Stochastically regular or highly synchronized EEGs, on the other hand, characterize a drowsy (relaxing) or epileptic subject, respectively, and indicate — in both cases — a very incompetent information processor.

We suggest that such behavioral changeovers are produced via bifurcations which trigger the thalamocortical nonlinear pacemaking oscillator to switch from an unstable limit cycle to a strange attractor regime or, more correctly, from *intermittency* to *metastable chaos* and vice versa. Our analysis aims at showing that the EEG's characteristics are not accidental but inevitable and even necessary, and therefore functionally significant.

An information processor (analogue or digital) is a cognitive gadget that tracks and identifies the parameters of an unknown signal or "pattern", which is usually contaminated by thermal (equilibrium) noise (white or coloured, additive or multiplicative). In order to accomplish this, the processor has to perform three distinct operations in the following sequence:

a) Produce from "within" a wide variety of (spatial-temporal) patterns (or "templates").

b) Cross-correlate (i.e., "compress") each of those patterns with the incoming one. (See also Sect.7.2.4).

c) Select or filter out, on the basis of some preestablished "hypothesis-testing" or "consensus" criteria, the pattern which forms the greatest cross-correlation with the unknown signal or trigger. (The filtering is usually nonlinear in order to create and enhance contrast, which, by sharpening contours, makes recognition simpler. Selected groups of cerebral neurons — and the Xerox machine — do just that.)

To track a signal, timing is of the essence. [The simplest tracker used in communication engineering practice is the phase-locked loop (PLL).] This means that the existence of self-sustained nonlinear dissipative oscillators (i.e., elements displaying attracting behavior) at the hardware level of the processor is a prerequisite for cognitive operation.

Functionally stable oscillators, in contradistinction to static (switching "on"-"off") devices, offer a number of evolutionary advantages, namely, (a) time keeping, (b) dynamic information storage (dynamic memory), and when triggered by very simple stimuli, possibly (c) an extremely broad spectrum of complex behavioral patterns.

Finally, the oscillators must possess asymptotic stability: one cannot accomplish reception and cognition tasks —which by involving phase locking or compressibility are dissipative (i.e., irreversible) —via Hamiltonian (reversible) working subsystems (although Hamiltonian, "Duffing" oscillators are instrumental in performing switching among patterns); hence the universality of the so-called family of van der Pol oscillators in communication engineering.

Parsimony —which undoubtly possesses survival value —requires that the locally generated dynamical patterns/attractors of the processor should not always be "on". They rather should emerge upon request (i.e., upon triggering from externally impinging stimuli) from a set of available dynamical elements and some basic and rather simple recursive *rules* (algorithms or "maps") for combining those elements.

Below we present a sketch for a dynamical model of a brain processor. Individual neuronal oscillators at the cerebral cortex constitute the above-mentioned set of dynamical elements. The thalamocortical oscillator, on the other hand, is the adaptive agent which performs two distinct operations:

a) It provides pacemaking activity resulting in the formation of internal, synchronized, or coherent (spatial-temporal) neuronal patterns. By making such neuronal groups coherent, the pacemaker helps them elevate themselves above the ambient thermal noise level and also distinguish themselves from coexisting and neighboring neuronal formations, within brief time intervals.
b) It generates *the recursive rules* governing the sequential appearance of these coherent patterns on a time-division-multiplexing scheme.

6.5.3 Experimental Data from EEG Research

In recent years, a very great number of studies ([6.23-26] and references therein) have been devoted to the electrical activity of the (human) brain —activity both spontaneous and evoked —and hypotheses have been advanced concerning the mechanisms and functional significance (if any) of the EEG. It must be stressed from the outset that the brain is *not* a conventional signal processor: due to its enormous redundancy, different departments of the whole system are permeated with "noises" of different kinds, performing overlapping triggering functions at different times [6.27] and having varying degrees of "relevance" in catalyzing specific metabolic processes.

The experimenters' opinions seem to converge nowadays in a number of fundamentals:

a) in pronouncing the EEG the sum total of elemental "slow" neuronal self-sustained oscillatory activities (graded, subthreshold postsynaptic membrane potentials) crowded in rather small volumes of (cerebral) cortical tissue just underneath the specific recording electrode (scalp lead).

b) in seeing as irrelevant for the generation and form of the EEG, spikes traveling along individual neuronal axes, apart from their causing nonlinear coupling among the elemental generators (via presynaptic excitation and neurotransmitter-hormonal discharge into the synaptic cleft).

c) in finding no consistent phase relationships between each of the elemental neuronal oscillators and the gross EEG. In terms of amplitude, the intracellular oscillators are in the range of 50 mV; the EEG in the range of 100 μV. In terms of spectrum, they are more or less identical (narrow-band noise in the interval ~0.3-45 Hz). From the point of view of statistical characteristics, the amplitude p.d.f. of each elemental oscillator is *non*-Gaussian [6.23], while the amplitude p.d.f. for the EEG is Gaussian in the "idle" regime, but becomes more or less skewed during REM sleep and during the execution of (controlled) mental tasks.

This is the "hard" established data. What follows includes what are essentially inferences from the above evidence and further experimental results, still not universally accepted.

A fundamental question is whether the EEG represents activity of synchronized (mutually phase-locked) groups of neuronal oscillators or whether it is produced via summation of statistically unrelated phasors. The experimental probing of the two alternatives has been quite extensive. The current view [6.24] compromises and supports the possibility that the EEG may be produced via sequential and intermittent synchronization of successive, relatively small groups of cortical neurons (not necessarily from the same cerebral area), each of which forms a small (1-10%) subset of the entire neuronal population. Moreover, we are capable of recording activity only from the particular subgroup that happens to be synchronized at the given moment and consequently attains detectable voltage levels (above ~5 μV).

Further, it is indicated that this sequential phase-locking —or more generally, the establishment of long-range space-time coherence among the members of the individual neuronal subgroups [6.25] —is mediated via specific thalamic pacemaking groups of neurons (relaying sensory cues to the cortex.) Thalamocortical and corticothalamic feedback-feedforward loops (the latter ending on nonspecific thalamic nuclei, whose functional role is still elusive) have been experimentally traced and mapped in detail [6.26]. These pacemaking loops are further considered responsible for the scanning or sampling process which allows the sequential formation and breaking of phase-ordered relationships among the neuron members of the individual subsets.

Such a mechanism provides, at the neocortical level, time-division multiplexing of information to be processed. So, although the EEG appears as a continuous stochastic process, it may in fact be considered as a time sequence of discrete evoked potentials, each possessing an average duration of 100-300 ms. Synchronization then aims at enhancing —one at a time —patterns of specific groups of cortical neurons, which then become able to process and store, on a collective basis, specific sensory modalities via (neuronal) pre/postsynaptic membrane ⇌ cytoplasmatic ⇌ nucleic cascade loops.

6.5.4 The Model

We now model the "thalamocortical pacemaker" as a high-amplitude dissipative self-sustained nonlinear (relaxation) oscillator which, in the absence of any environmental input, is intermittently "free running" on an unstable limit cycle with a fundamental ("sampling") frequency ~10 Hz (the "alpha rhythm"). This means that the oscillator has an internal bias which is responsible for this instability.

How such a strong oscillator can synchronize groups of "smaller" oscillators has been sketched in Sect.4.8.2. If our conjecture is correct, we should expect the corresponding EEG to exhibit some pseudo periodicity but also randomness, in the sense that successive amplitude segments of this activity (each segment corresponding to a given stationary state/pattern at the cortical level, formed by the vector sum of the amplitudes of phase-locked neuronal phasors of the pertinent cortical subset) should show no statistical correlation to each other. (The optimal sampling of a random process is at random time intervals: the sampled modalities whose embodiments are the sets of synchronized neurons should thus not overlap.)

This seems indeed to be the case: the amplitude p.d.f. of such an EEG follows a normal distribution, as one could expect from the central limit theorem (or the adherence to the law of averages)[9].

---

9 There appears to be a contradiction here: if the postulated mechanism is indeed intermittency I, the amplitude p.d.f. should not be Gaussian but hyperbolic. However, one should not forget that the processor divides its time not between *one* limit cycle and irregular motion, but among *many* indistinguishable limit-cycle modalities. The experimentally derived p.d.f. therefore represents an ensemble average of individual hyperbolic distributions and it may well be Gaussian.

Consider now what may happen in the regime of REM sleep or the performance of a mental task: information is pumped along the ascending branch of the reticular formation system or through other viaducts of the peripheral nervous system, carrying from within or from outside specific sensory inputs to be identified by the cortex. Our thalamocortical oscillator is now influenced by a fluctuating input, and beyond specific (threshold) values of some control parameter a cascade of bifurcations may set in, as a result of which the intermittent limit cycle is deleted, and the oscillator follows now part of a strange attractor. In other words, it becomes metastably chaotic; its scanning manifestations on the cortical neuronal subgroups turn into a "spasmodic" and non-periodic oscillation.

This hypothesis is corroborated by long-established experimental evidence [6.28]. Increasing arousal of the ascending reticular formation branch leads to polarization of the specifically pacemaking thalamic nuclei, thereby interrupting their sampling function for time intervals concomitant with the degree of arousal, i.e., the intensity of stimulation and therefore the *rate* of information transfer from the environment. Thus, under excitation, the sampling of the cortex of the pacemaking thalamic nuclei switches from intermittently periodic to *sequential*. We witness the establishment of a Markov chain with different corresponding probabilities u_i per stationary synchronized state i (internal pattern), —produced from a transitional probability matrix P_{ij} between successive states/patterns. Therefore, and this is the crucial point, the amount of "holding" time that the cortical processor spends at the specific modality i depends on the length of "disruption" time caused in the scanning pacemaker at i by the information input; this time usually, but not always, increases with the rate or frequency of the incoming trigger.

Evidence supporting this view comes from analyses of the amplitude probability distribution of EEGs indicating changes from Gaussian to skewed forms during performance of mental tasks [6.23]. We are afraid, however, that the interpretation put forth in the above reference, that the skewness in the p.d.f. reflects some "increasing degree of cooperativity among the neuronal generators" is wrong: changes of coupling among neuronal oscillators embodying a pattern may come from different causes; they result *only* in changing (enhancing or degrading) the degree of coherence of the specific pattern to which they contribute but, of course, this has nothing to do with the algorithm of time-division-multiplexing or the policy of switching *among* patterns.

The observed skewness is interpreted by us as indicating a statistical linear correlation ($P_{ij} \neq 0$) or a Markovian characteristic *between* the successive amplitude EEG samples, i.e., between successive synchronized neuronal states, each coding for a single sensory modality. Under such linear interdependence, the central limit theorem does not hold.

Suppose finally that the degrees of arousal of the reticular formation ascending branch, or the average rate of information pumping, increases still further.

We conjecture that in such cases the scanning process of cortical subgroups either gets extinguished (the oscillator is "quenched"), or the scanning rate increases concomitantly with the degree of arousal. In the limit we may consider that *all* the subgroups of cortical neurons are synchronized at the same time, as happens in epileptic convulsions. We do not possess reliable experimental data, and certainly a proper theory is still lacking. Nevertheless, in a recent computing simulation [6.29] hints have been offered that when the *coupling* between the environment and a limit cycle oscillator, already in the chaotic regime, becomes "supercritical", one may witness a *regression* from the chaotic regime back to a limit cycle regime. (The same regression can be achieved by external noise; see Sect.6.5.5).

These hints are relevant to the possible behavior of our thalamocortical pacemaker, which, under strenuous external (photic or acoustic) triggering at its fundamental frequency *or one of its harmonics*, may switch back from the chaotic to a strictly entrained regime with such a high amplitude of oscillations that it can virtually synchronize large numbers of cortical neurons simultaneously.

To sum up, we have tried to appreciate why a cognitive system *may* be chaotic in order to perform effective signal processing. The answer can be offered in the domains of (a) time and (b) frequency.

a) By creating Markovian time-division multiplexing, the system allows separation of sensory modalities/attractors and delegates for each one of these modalities a time-processing interval commensurate with the rate of sensory input.
b) "Turbulent chaos", (nonequilibrium, low-dimensional noise) as a modus operandi of the thalamocortical pacemaker under mild excitation, contains a broad spectrum of temporal (and spatial) frequencies. Thus it can constrain "patches" of postsynaptic functional areas in the cortex and create coherent patterns which can match a wide variety of incoming spatial-temporal patterns.

Specifically *cognition* is manifested at the cortex as a result of a matching process between *pairs* of spatial-temporal patterns, each containing a great number of elemental units (neurons). In each pair, one pattern (the same for all pairs) is the unknown information; it is embodied in incoming triggers, coded either in sequences of pulses from the peripheral nervous system, or, if it comes from other areas of the central nervous system, encoded in strings of macromolecular (neurotransmitter/hormonal) releases from presynaptic endings. The other pattern of the pair is one of the pattern/attractors created by the processor; it constitutes a prestored spatial-temporal "mosaic" embodied in a set of partly synchronized postsynaptic membrane potentials or a spatial-temporal pattern of post-synaptic membrane receptors.

The coupling or cross-correlation between the above two patterns of each pair takes place dynamically via energy exchanges between equal or neighboring frequency pairs (ω_i, ω_j) shared by both spectra, and is estimated quantitatively by the coherence function

$$\text{coh}_{ij}(\omega) = \frac{\varphi_{ij}(\omega)}{\sqrt{\varphi_i(\omega)\varphi_j(\omega)}} \quad , \qquad\qquad (6.5.13)$$

where $\varphi_{ij}(\omega)$ is the cross spectrum and $\varphi_i(\omega)$, $\varphi_j(\omega)$ are the autospectra of the two matched patterns.

The result of the cross-correlation in phase and amplitude determines the "degree of cognition" between the incoming and the preset or the unknown and the expected patterns. What mechanism in the brain decides which coherence function is predominant? It appears that again the reticular formation might be involved: some plausible ways of nonlinear filtering have been discussed in Sect.4.7.6.

In this application we have attributed the phase-locking between individual neuronal oscillators (and therefore the formation of attractors in the cortex) only to the thalamocortical pacemaking activity. Undoubtedly there are other, perhaps equally important, very complicated mechanisms of extending *long-range* spatial-temporal coherence among cortical neurons, mechanisms which have received, with certain exceptions [6.30], very little attention. We just mention the possible role played in that respect by the cortical intercellular electrolyte.

The intercellular fluid appears to contain extensive hydrate networks of complex membrane-bound macromolecules (mucopolysaccharides and mucoproteins). Removal of calcium allows those molecules to bind with water and is concomitant with their spreading out into a loose hydrated network. Restoration of calcium reverses this process, unbinding water and contracting the molecular net, by up to five orders of magnitude in volume.

The existence of very weak EM fields appears to influence strongly Ca^{2+} movement in the intercellular electrolyte, as well as the impedance of the fluid. We may thus envisage new ways of establishing and controlling long-range coherence in the brain and thereby influencing the formation of attractors in the cortex, namely, by random environmental EM fields.

6.5.5 The Dual Role of Intermittency in Information Processing

"Intermittency" models the modus operandi of the "scanner" in the brain, in the absence of an external excitation. It is basically due to the physical inability to achieve —within certain intervals of the control parameters —perfect entrainment, or stable, strict phase-locking (see also Appendix C) between a dissipative nonlinear (relaxation) oscillator and a harmonic stimulus.

As the limit cycle in the state space [phase difference $\varphi(t)$ —instantaneous frequency offset $\dot{\varphi}(t)$] is about to close, a "scrambling" process, as it were, intervenes and destroys the orderly trajectory; it creates for a brief time a chaotic motion in the state space amounting to broad-band noise, and then the building up

of a limit cycle starts all over again. After a while, a new bursting transition
to "turbulent chaos" occurs, followed by another limit cycle appearance, and so on
[6.9]. The time sequence of intermittent points is believed to be random.

Suppose perfect phase-locking *were* possible for certain values of the control
parameters. In such a case, the above sequence would be an *absorbing one*: the scanner would reach a state/modality from which no escape would be possible. The scanning process would then disintegrate and the processor would lock itself at *one*
sensory modality, or it would emit *stereotypically* a single symbol. After a little
while, habituation sets in and information processing stops, bringing the channel
capacity to zero. Furthermore, the channel capacity of such a "brain" would also
deteriorate once intermodular interference (due to overlapping caused by a faulty
scanning mechanism) became critical (for an experimental justification, see [6.31]).

Therefore the mechanism of intermittency, i.e., of slipping cycles near entrainment, acts indeed as the deus ex machina which allows the scanning mechanism to go
on by making the system deal with *one* sensory modality *at a time* and avoiding dynamical fixation on a single modality.

However, intermittency also possesses a so far unsuspected "bad side" which may
prove to be of even more fundamental importance. It concerns the so-called compressibility problem, namely the finding of the "minimal program" or the minium
number of bits or degrees of freedom from which a given sequence of symbols or pattern can be unambiguously reconstructed.

Limitless compressibility would theoretically imply the reduction of the degrees
of freedom of a pattern from $N \gg 1$ to 1. Suppose a given modality is embodied in N
interacting nonlinear oscillators. We may reduce the description and consider the
interacting phases of these oscillators as the degrees of freedom under consideration. Perfect compressibility (full coherence) would amount to establishing among
these phases recursive bounded relationships of the type

$$|\varphi_\kappa - \varphi_\lambda| < \Phi_{\kappa\lambda} \ll 2\pi \quad (\kappa, \lambda \in 1,...,N) \quad , \tag{6.5.14}$$

where all the bounding limits $\Phi_{\lambda\kappa} \to 0$. However, the phenomenon of intermittency may
not allow $\Phi_{\kappa\lambda} = 0$ for any pair κ, λ, so we conclude that the ultimate limit of compressibility is presented by the dynamics of intermittency. Further mathematical
analysis is needed in order to reveal whether or not such a limit constitutes a
universal constant, or whether we may speak of an uncertainty relationship with
the channel capacity of the processor, and the degree of compressibility of a single
(sensory) modality as conjugate parameters.

6.5.6 The Origin of Conflict in Communicating Hierarchical Systems

In two communicating (symbolic) dynamical systems (Fig.6.28), we postulate that
conflict, a by-product of communication, emerges as a dilemma of mutually accommo-
dating excessive entropy production. Each system strives towards "self"-organization
at the expense of the partner.

The hierarchical systems we are dealing with possess *complexity* and *organization*. We have defined complexity as the minimum number of bits required to reproduce a given system. We have defined organization as the ability to *compress* information, which in turn is generated via cascading bifurcations giving rise to broken symmetry or (beyond chaos) via cascading iterations increasing resolution on the map in the interval. "Self"-organization is meant here to imply the function of a cognitive gadget "language" which compresses the complexity of the "opponent" (environment), thereby providing minimal algorithms which reduce and predict the behavior of the other system.

In any "game" of this sort, each participant aims at "breaking" the code of the opponent, i.e., "compressing" as much as possible the description of the partner. Achieving compressibility (or cognition) is therefore tantamount to forming *collective properties* out of the variables of the system under study or, in the domain of a macrodescription governed by the hierarchy of the moments of the master equation, tantamount to compressing the p.d.f. to achieve the mean-field regime (see later in this section).

Most systems relevant to information sciences (e.g., linguistics) contain multiple hierarchical feedback loops, which inevitably make them self-referential. If such systems do not possess "closure" or logical consistency, they are paradoxical. The "truth" — if any — conveyed by such systems is associated with self-consistency and this characteristic implies the existence of stable attractors in the iterative and recursive system we call the self-referential sentence. The suggestion follows that logic — so far a static concept — may be dynamically implemented with evolutionary stability. The incompatibility between *completeness* and *self-consistency* of any closed axiomatic (logical) system manifests itself as an inability to discriminate between statements (strings of symbols) belonging to *different* hierarchical levels.

When paradoxes flare up within a language-possessing system, conflict sharpens; added to the previous *inter*system conflict, we now witness a hierarchically subordinate *inter*system conflict. Intersystem conflict resolution depends now on the parameters determining the intrasystem conflict. The latter, in turn, uses the intersystem communication as an expendable device for advancing its own resolution. The coupling and dynamic evolution of the pair of inter- and intrasystem conflicts has recently been discussed in detail [6.32].

Dynamically speaking, we may say that the origin of conflict boils down to the incompatibility between *compressibility* and *intermittency*: when the p.d.f. cannot be compressed beyond a certain threshold and explodes spontaneously (from a delta-like function to a hyperbolic function — "intermittency"), the mean-field approach breaks down and the system processor at the level of moments becomes "turbulent" in the sense that it acquires as many degrees of freedom as the system at the level of bare variables (one hierarchical level below where the intermittency takes place).

This is what we mean by saying that beyond the Gödelian limit the dynamics of two successive hierarchical levels get "mixed up": from the point of view of a time series, microscopic and macroscopic descriptions become indistinguishable in the sense that *random noise* and *chaos* are statistically similar.

In many instances, hierarchical systems are "almost decomposable", that is, one can study the dynamical activity at a given level taking what is just underneath as a *boundary condition* and what takes place above as a *constant*. This can be done whenever the rate constants differ by orders of magnitude from level to level (the lower level being characterized by larger rate constants or smaller relaxation times).

In linguistic systems, however, this decoupling of levels cannot be accomplished due to the existence of evolutionary feedforward loops mediating the dynamics between successive levels. Specifically, in linguistic systems one uses simultaneously two hierarchical levels where statements and metastatements are interlocked. ("Chicago" is a trisyllabic word *and* a city.)

Let us repeat at this point what we mentioned in Sect.2.3.1 about the circumstances of breaking down of the mean-field approximation. Consider a physical dynamical system reduced to a one-dimensional map. The probability of finding the system at point x at a time t increases due to transition from other points x' of the interval. It decreases due to transitions leaving this point, i.e.,

$$\frac{d}{dt} P(x; t) = \text{rate in} - \text{rate out} = (I) - (0) \quad . \tag{6.5.15}$$

Since the term (I) consists of all transitions from initial points x' to x, it is composed of the sum over the initial points.

Each term of it is the probability of finding the system at a point x' multiplied by the transition probability per unit time for passing from x' to x; thus

$$(I) = \sum_{x'} w(x,x')P(x';t) \quad , \tag{6.5.16}$$

where w(x,x') is the probability of transitions from x' →x.

For the outgoing transitions (0), we have

$$(0) = P(x;t) \sum_{x' \neq x} w(x',x) \tag{6.5.17}$$

where w(x',x) is the probability of transitions from x →x'. (The above transition probabilities are functions of the slope of the map at the corresponding points x, x'.)

The "master equation" then becomes

$$\frac{dP(x;t)}{dt} = \sum_{x'} w(x,x')P(x';t) - P(x;t) \sum_{x' \neq x} w(x',x) \quad . \tag{6.5.18}$$

The "steady state" or "stationary" asymptotic solution of this equation is just what we investigated in (6.5.7). Now the crucial point is that the transition probabilities (generally unknown) are in most cases *nonlinear* functions of x since the slope of the map may be highly nonlinear. [In chemical reactions for instance, these rates are, according to the law of mass action, proportional to the concentrations of the "reacting species", in other words, to the number of ways a pair or a group of "reactant molecules" can be sorted out from the total population. For example, for the reaction $2X \rightarrow X'$, the rate is proportional to $(x/2)(x-1)$, i.e., the equation for the "mean" is $d\langle x\rangle/dt = -K\langle x(x-1)\rangle$.]

Multiplying both sides of (6.5.18) by $x, x^2, \ldots$ and integrating or summing over the interval of x, we get a series of "phenomenological" macroscopic equations with reference to the various moments $\langle x\rangle$, $\langle \delta x^2\rangle$, ... of the p.d.f. $P(x,t)$, namely,

$$\frac{d}{dt} \langle x\rangle = f_1\left\{\langle x\rangle, \langle \delta x^2\rangle, \ldots\right\} .$$

$$\frac{d}{dt} \langle \delta x^2\rangle = f_2\left\{\langle x\rangle, \langle \delta x^2\rangle, \ldots\right\} ,$$

$$\vdots \qquad\qquad \vdots \qquad\qquad .$$

(6.5.19)

Because of the nonlinearity of the operators f_i (due to the nonlinearity of the transition probabilities), we get a very great (infinite) number of coupled non-linear differential equations, with respect to the moments and cross-moments (cross-correlations). Moreover, in the vicinity of bifurcations or for self-sustained chaos filling the whole interval, these moments may become comparable.

We then see very clearly that under such circumstances successive hierarchical levels become indistinguishable, a necessary prerequisite for the appearance of the paradox of self-reference and the "deadlock" of a linguistic processor.

6.6 Comments on the Effects of Internal Fluctuations and External Noise on the Stability Properties of Dynamical Systems

So far we have considered deterministic nonlinear dynamical systems where the stochasticity manifested by some of them has been attributed to *intrinsic noise* amplified by the cascading bifurcations and iterations of a strange attractor. When *internal* fluctuations cannot be ignored and excluded from average behavior (see the previous section) and are therefore taken into account, the dynamical system can no longer be modeled by a set of coupled nonlinear differential equations with deterministic values of variables and/or parameters.

One should rather start from a master-equation or Fokker-Planck equation formalism and express the solution of the resulting p.d.f. in terms of a "stochastic" potential function — in contrast to the "deterministic" potential function which was used for the simple examples treated in Chap.2 (Sects.2.2.3 onwards). When one does just that and investigates the stochastic analogue of, for example, the simple nonlinear overdamped oscillator, something very interesting occurs (for details, see [6.33,34]): the stochastic potential does *not* coincide with the deterministic potential, which, as we recall, has the quadratic form

$$V(x) = \frac{1}{2} \alpha x^2 + \frac{1}{4} \beta x^4 \quad , \tag{6.5.20}$$

but contains in addition an *odd* cubic term which makes the potential asymmetrical beyond bifurcation, its two valleys having unequal depths. As a result, beyond bifurcation the system has a *choice* between two different and *not* equally probable alternatives.

In fact, this possibility of a *preferential* move to one among more than one successor states provided by bifurcation —a problem we have been unable to answer so far as the explicit calculation of transition probabilities is concerned — is illustrated also in the double-hump form that the p.d.f. acquires as a solution of the master equation beyond the critical point, where the two humps are of *unequal* height.

Let us now briefly comment on the role of *external* noise superimposed on the system of the deterministic nonlinear equations. Since the equations are nonlinear, the external noise will have an additive *and* a multiplicative effect, namely it will influence both the amplitude and the phase of the system's variables. *Horsthemke* et al. [6.35] have proved that the external noise may either delete bifurcations existing in the deterministic dynamical system or, more importantly, provoke new ones not included in the repertoire of the deterministic system. Thus a system may be *driven* to differentiation and "self"-organization, not so much by its intrinsic potentialities, but by destabilization of singularities which either did not exist at all in the absence of external noise or were not prone to destabilization from the action of the intrinsic fluctuations of the system alone.

External noise in chaos was first examined by *Mayer-Kress* and *Haken* [6.36] as well as by *Crutchfield* et al. [6.37]. The main results of applying external Gaussian noise to the logistic map can be summarized as follows:

a) The external fluctuations remove detailed structure by smearing the sharp peaks of the p.d.f. of the map.

b) Period orbits broaden into bands similar to chaotic attractors.

c) Fluctuations increase the degree of randomness of chaos while destroying the periodic windows encountered in the chaotic regime. They affect the local stability properties of the attractors by deleting existing bifurcations and adding new ones, at the same time leaving their global stability relatively unchanged. In some cases, however, fluctuations are instrumental in completely destroying the *chaotic* regime and making the system *regress* to a limit cycle [6.38].

d) Under the influence of external noise, the bifurcation diagram of the map is "blurred", renormalized, and scaled in the sense that it comes out as a convolution of the deterministic bifurcation diagram with a Gaussian probability distribution with respect to the control parameter of the map.

7. Epilogue: Relevance of Chaos to Biology and Related Fields

7.1 Computational Complexity

Although a strange attractor as an *object* of *observation* frustrates the (naive) observer, due to his practical inability to deduce reliable predictions from a series of measurements he performs on the system, a strange attractor *as a cognitive device possessed by the observer* plays exactly the opposite role, namely it gives rise to the compressibility of seemingly chaotic sequences of observed phenomena.

In the simple case of the Lorenz attractor, for example, the fractal dimensionality as calculated by *Mori* [7.1] is $D_F \sim 2.06$. The compressibility then has a minimum value $N - D_F = 3 - 2.06 = 0.94$ bits, so any initial condition under the action of the attractor is compressed by almost 30%.

It is perhaps the possession of such "cognitive devices" by the software of the brains of higher mammals (almost exclusively humans) which makes possible social cohesion, science, technology, and economic and cultural development in an evolving world. Since the most elemental and basic intellectual activity is *computation*, it is perhaps relevant to ask *how* humans compute and whether the possible existence of chaotic attractors in their brain function can be held responsible for the way decision making takes place. The prompt answer is, of course, that we do not have the slightest idea. Still, we suspect that the human way of computing can be not only serial (sequential) but perhaps parallel and hierarchical also. Let us elaborate on this point by referring to what Bremermann developed some years ago as the physical limitation of serial computation, known as Bremermann's limit. (For an updated version, see *Bremermann* [7.2].)

The idea is that any signal-processing activity involving information storage and retrieval should be based on some "material" background (hardware) such as punched cards, magnetic tapes, and so on. Suppose we have a physical body of mass m which is used for processing, with 100% efficiency. The most natural way to store information which can subsequently be retrieved *unambiguously* is to "deposit" each bit on each energy level of the material body. How many bits (n) can be stored in this way in a mass m? Obviously,

$$n = \frac{E_{max}}{\Delta E} \quad , \tag{7.1.1}$$

where E_{max} is the maximum amount of energy one can extract from the physical body, and ΔE is the resolution with which one energy level can be observed.

But $E_{max} = mc^2$, and from the uncertainty principle $\Delta E \cdot \Delta t \sim h$, where Δt is the observation time and h Planck's constant. Therefore the maximum possible number of bits our processor can manipulate in time Δt is

$$n = m \frac{c^2}{h} \Delta t \quad [bits] \quad \text{or} \tag{7.1.2}$$

$$n_0 = \frac{c^2}{h} \quad [bits/gs] \quad . \tag{7.1.3}$$

This number is

$$n_0 \sim 2 \times 10^{47} \quad [bits/gs] \quad .$$

Suppose now that the whole physical universe is a perfect information processor working on this job since the moment of the big bang. How many bits (n') could the universe have processed serially during its lifetime (so far)? The age of the universe is roughly $\tau \sim 10^{17}$ s, and its total estimated mass $m \sim 10^{58}$ g. Then

$$n' = 2 \times 10^{47} \times 10^{17} \times 10^{58} \sim 10^{122} \quad [bits] \quad .$$

Is this number big or small? Well, in order to appreciate what humans do in this respect, let us consider an information "box" (Fig.7.1) with Λ inputs and r outputs and ask about the rate of information flow in such a box. We take as an example a big international airport where the input stands for say 1000 requests in dyadic form ("to land or not to land?") per day and the output with $r = 2$ stands for "Yes", "No". The traffic is handled by ~ 20 humans —air traffic controllers. How much information flows into the box per day?

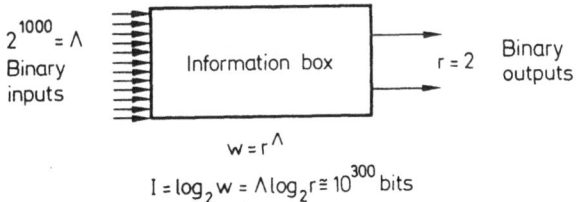

$$w = r^\Lambda$$
$$I = \log_2 w = \Lambda \log_2 r \cong 10^{300} \text{ bits}$$

The total number of complexions is $w = r^\Lambda$, where $r = 2$ and $\Lambda = 2^{1000}$; the amount of information is then

$$\log_2 w = \log_2 r^\Lambda \quad \text{or}$$

$$\log_2 2^{2^{1000}} = 2^{1000} \sim 10^{300} \text{ bits!}$$

(Dividing this amount by the number of seconds in 24 hours multiplied by the number of air traffic controllers does not help much!) How do humans (or machines designed *by* humans) handle in 24 hours an amount of information *180 orders of magnitude* greater than the amount of information that a perfect serial computer possessing the size and the age of the universe can process? The answer is, through "short cuts", that is, through instabilities giving rise to cascading bifurcations, beyond each of which *the hierarchically arising new pattern(s)* may provide the solutions! *Such patterns could take eons to emerge if one had to produce them serially.* Bifurcations, then, and the emerging higher hierarchical structures may provide the deus ex machina for solving by "intuition" complex problems. Still, there are particular problems, the so-called NP-complete problems (NP standing for nondeterministic polynomial time), which stubbornly refuse even an algorithm for their solution.[1]

Take for example the classical problem of the "traveling salesman". A salesman has to visit N cities, passing at least once by each one of them, and return to the point of departure. How should he schedule his visits in order to minimize the total distance? Regrettably, there is no algorithm yet available for this problem. The salesman has to try before his departure *all* possible combinations which are N! or, if N is large, N^N after applying Stirling's formula. We say that the number of calculations needed increases *exponentially* with the complexity of the problem. In tractable problems the number of calculations grows only as a *polynomial* function of the size of the problem. The difference between exponential and polynomial solutions in terms of computing time may be huge. For example, if $N = 30$, 3^N computations take ~6.5 years of conventional digital computer time, but N^3 computations take just ~0.027 s. If $N = 60$, 3^N computations take ~10^{15} years, whereas N^3 computations take a mere ~0.216 seconds. Should the design of chaotic processors in the future alleviate the difficulty? Biological organisms most probably possess such processors (DNA and the cerebral cortex are good candidates). Still, we do not yet *know* how they work and so we cannot design artifacts mimicking or surpassing them.

7.2 Towards a Dynamic Theory of Language

7.2.1 The Nature of the Problem

One important issue that modern communication theory does *not* deal with is the physical substratum of information. We have advanced here the thesis that information is generated by the ever-increasing *complexity* of a "self"-organizing hierarchical system, which evolves via cascading bifurcations giving rise to broken symmetry. We call "language" the process which *reveals* that information, namely the cognitive gadget which *compresses* the complexity generated by broken symmetry, thereby providing "minimal-length" algorithms for triggering an "internal representation" or the replication of the physical system involved. This compressibility has an ob-

---

1 "Analog computers" using physical bodies in the vicinity of "phase transitions" could provide the machinery needed here.

vious "survival value" since it allows the possessor of language to reduce and predict a rapidly changing environment.

Let us consider a macroscopic physical system at its lowest hierarchical level, namely the level of *equilibrium thermal* ("molecular") *mixing*. At that level of disorder, the system possesses the maximum number of degrees of freedom or the fullest possible symmetry (if by degree of symmetry we imply the number of equivalent alternative descriptions). However, even an isolated system in the process of *approaching* equilibrium possesses *less* symmetry than the symmetry of the microscopic laws of motion of its constituent interacting parts. For an external observer the system is therefore "almost" structureless, and so it affords the briefest description of one bit length: There is *no* time-reversal symmetry, and this is reflected by the irreversible approach to equilibrium. Otherwise, once the unique stable steady state of thermal equilibrium is reached, "nothing happens": the symmetry becomes maximum, the number of degrees of freedom likewise, and the complexity is minimum since the number of bits required to specify a system at thermodynamic equilibrium is the smallest possible: To reproduce such a system we may start from almost any initial condition.

Imagine now that the system comes in contact with the environment; it may again settle to equilibrium when new conditions for minimization of its *free energy* are met; or, as a consequence of internal and/or external microperturbations (which are inevitable due to the enormous number of the constituent components of the system and its environment), the one and only stable steady state of equilibrium may become deleted. Beyond a certain critical distance from equilibrium (which the system passes triggered by fluctuations), at least two new, alternative stationary configurations (containing their own symmetrical branches) emerge. Assume that the system settles in one of them by a more or less random "choice".

In this metastable state the system now forms, say, a spatial static pattern (i.e., a crystal) which, although it may appear quite symmetrical, has a number of degrees of freedom far below that corresponding to the symmetry of the laws of motion of its constituent molecules. Nevertheless, we do not yet speak of "broken symmetry" but rather of "repealed" symmetry [7.3], since the probability of "tunneling" to the symmetrical metastable state remains finite (although it rapidly decreases with the scale of the system). If, however, by coupling the system with an external field (scalar or polarized vector), we perturb the bifurcation diagram so that all alternative mirror symmetrical configurations but one are "repressed" (as in the case of biological macromolecules) [7.4], we do witness a genuine case of "broken" symmetry (up-down, left-right), and the asymmetrical configuration at the corresponding steady state is stable for a wide range of the external parameters. We say that the *complexity* of the system has increased, meaning that the length of instructions or the minimum amount of bits an external observer requires for the replication of the system has increased.

At the same time the statistical order of the system may increase *or* decrease, which means that the entropy may go up or drop. Thus, in the regime of broken symmetry the system may appear *more or less ordered* but is definitely *more complex*.

We can imagine now a succession (cascade) of symmetry breaking, bifurcation "episodes" stimulated by the shifting of the numerical values of the control parameters beyond discrete thresholds; then an external observer will witness the emergence of new hierarchical levels in the system concerned, that is, the formation of new spatial-temporal patterns of fluctuating statistical order (entropy) but steadily increasing complexity. In order to provide minimal-length algorithms, the observer will eventually need to "compress" or to "abstract" more as he tries to describe the system at higher and higher hierarchical levels.

The basic issue under discussion is whether the observer need be external to the system. Is it possible that the evolving system, beyond a certain stage of development, creates levels of "symbolic activity", thereby providing for its own internal descriptions and its own internal controls? If such is the case, then the system becomes "self"-organized.

7.2.2 Structural and Functional Hierarchical Levels

In the common parlance of "general system theory", languages are tools that (biological) organisms use to *probe*, *model* (i.e., provide internal descriptions), and *control* their "environment". The "environment" may include semiautonomous parts of the organisms themselves, according to the frame of reference; in fact, what essentially distinguishes "organisms" from mere "ecosystems" is the establishment of information transactions among various parts of the system or the ability to "map", i.e., to devise one-to-one or multi-to-one correspondences between patterns which are formed within the organism.

The essential purpose of a language is recursively to reduce or constrain the ever-increasing complexity of an evolving natural system — thereby revealing "information" which is subsequently used for control and optimization tasks. The information itself is generated via cascading bifurcation processes in the natural system, inasmuch as these processes lead to broken symmetries. Broken symmetries increase the complexity of the system by sequentially reducing the number of degrees of freedom of the new pattern(s) (via the triggering of long-range spatial-temporal correlations). The greater the reduction of the degrees of freedom, i.e., the greater the number of constraints a pattern incorporates, the greater the length of the minimum program necessary for the "complete" description of the system [7.5]. The sequence of broken symmetries constitutes, we could say, the "history" of the system. Thus the minimal algorithm is the most laconic "storyteller". (In the same sense cosmology can be compressed to the account of selective hierarchical inhibitions since the time of the big bang.) In view of such a grand scope, it would be rather unimaginative to call language simply "a set of aperiodic sequences of symbols, stochastically interdependent through given rules of grammar and syntax". Languages *themselves* are to be considered hierarchical self-organizing systems whose action and evolution depend on the *coupling* (via feedforward-feedback loops mediated by some sort of highly specific or contextual "enzymatic" activity) between physical (energetic) and symbolic (informational) dynamical processes.

The central problem here has to do with the way symbolic interactions "emerge" from energetic interactions. In other words, the problem is how to devise an *interface* between the structure ("hardware") of our system —where the dynamics is mediated by energetic interactions —and their function ("software") —where the dynamics is mediated by strings of interdependent symbols.

We could say that at the structural level(s) our systems behave like "thermal engines", while at the functional level(s) they behave like (auto)programmed "computers". However, whereas the energetic interactions at the structural level(s) are basically classified and well understood (at the scale of interest they are governed by classical electromagnetism and quantum mechanics), the nature of the symbolic interactions at the functional level(s) appears elusive. It is still not possible to say whether the "grammatical" and the "syntactical" rules which determine the interactions between symbols forming "words" and words forming "sentences" respectively (ranging from the simplest language, e.g., the genetic code, all the way up to the human language) come as "epiphenomena", i.e., as mappings of collective properties (convolutions of energetic interactions with environmental fluctuations which take place at the hardware-software interface), or whether they possess also some intrinsic, nonreducible property.

For instance, we do not understand how the "symbols" and the discrete grammatical and syntactical rules of the human language(s) are related to the series of the discrete (neuronal spikes, synaptic processes) as well as the continuous nonlinear electrochemical interactions (graded potentials, EEG) between groups of neurons in various cerebral tissues of the corresponding "hardware" (the human brain).

Specifically, there are two aspects which puzzle us as far as the mechanisms of structural-functional interface are concerned. Taking the example of the genetic "code", we wonder about (a) the *arbitrariness* and (b) the *randomness* in the succession of the bases of DNA, i.e., the combinations of symbols $(A \rightleftharpoons T)$ and $(C \rightleftharpoons G)$, which largely determine the morphogenetic procedure for the possessing organism. When we say that the succession of the bases is arbitrary, we mean that we have no way so far to deduce it from the nature of molecular forces acting between the individual bases.

When, on the other hand, we speak of randomness, we mean that even if we are given a string of say one million of such bases, we have no way of *predicting* the next one. This means that the succession of symbols of the genetic code appears to be "incompressible". We cannot, for the time being, devise an algorithm which would produce a succession of N steps of the DNA sequences and possess fewer than N steps.

The problem becomes infinitely more perplexing when natural (human) languages are under consideration, since we have not the slightest idea about the possible relationship between the human genetic code and the (human) natural language(s). It is roughly estimated [7.6] that the stored information in the human DNA falls short, by at least two orders of magnitude, of the amount required for the *complete* specification of the circuitry of the human cortex. Therefore a substantial portion of the structure (interneuronal synaptic connections) as well as the function of the human cortex should be attributed to the stochastic interplay between the developing human neonate and its environment.

In conclusion, the problem of trivial reduction of the language to its underlying hardware does not appear an easy one; even if it were so, it would lead to the classical (and rather trivial) "egg-chicken" paradox. We believe that part of the difficulty is that most people dealing with the related problem of "pattern recognition" adopt a "static" perspective. In what follows, we suggest an "evolutionary" model within the current paradigm of hierarchical self-organizing systems. In this paradigm, structures are most naturally represented by numbers (digits, symbols); inversely, symbols, grammar, and syntax are represented by dynamical rules governing the interactions among nonlinear competing (lumped as well as distributed) oscillatory "modes".

7.2.3 An Evolutionary Linguistic Model: Digits and Patterns

Self-organizing systems can be thought of conceptually as open-ended "pyramids" of hierarchically ascending "platforms" classified in two categories: hardware (H), which is energetic and structural, and software (S), which is informational and functional. Of course, the category of a level changes as we move "upward", i.e., a certain level is "functional" for what is underneath and "structural" for what is above. A given hardware-software pair communicates as follows: From H →S we have transfer of collective properties of the dynamics at H, weighted over environmental fluctuations. From S →H we have feedforward controls constraining the dynamics at H. The dynamics at the successive structural levels are characterized by fluctuating *order* (low entropy). Successive functional levels, on the other hand, are characterized by ever-increasing *complexity* (high organization and redundancy) and usually *high* entropy. The coexistence of high redundancy with high entropy, i.e., the coexistence of *organization* and *disorder*, is not only inevitable and perfectly compatible, it is also necessary in most cases, where the number of the components of the system during transient or metastable periods of its lifetime (i.e., during morphogenesis) rapidly increases with time. Details are given elsewhere (see Appendix A), but for the sake of the argument it is enough to remember the relationship between redundancy $R(t)$ and entropy $S(t)$:

$$R(t) = 1 - \frac{S(t)}{\log_2 W(t)} \quad , \tag{7.2.1}$$

where $W(t)$ equals the maximum number of complexions at steady state equilibrium. For a hardware composed of, say, switching "on"-"off" elements, $W(t) \sim 2^{N(t)}$, where $N(t)$ is the number of the components, organization obviously does take place whenever

$$\frac{\partial R}{\partial t} \geq 0 \quad \text{or}$$

$$\frac{d}{dt} [\ln S(t)] \leq \frac{\dot{N}}{N} \quad , \qquad N \gg 1 \quad , \tag{7.2.2}$$

i.e., whenever the instantaneous disorder of the system increases more slowly than the rate of activation of new components according to the above limit. This means that by simply "throwing" new components into the system, we do not improve organization. The new components have to establish among themselves and the preexisting ones ordered ("phase-locked") relationships —either under the influence of a central "pacemaker" or via self-organization (see also Appendix A), so that (7.2.2) remains fulfilled. [Presumably, the dynamics of a strange attractor could implement the (7.2.2) criterion.]

To understand the intrinsic coupling between "material" patterns and symbols we have to appreciate that the action of symbols (whatever they are) is always mediated via a temporal or "pulsing", triggering "order", which determines the context. In engineering digital communication systems for example, timing is of the essence. In such systems information is represented by sequences of digits (highly stable pulses) selected from a source of "symbols". The digits are grouped into code words —only certain combinations comprising "legitimate" words —on the basis of a given *grammar*. These words are in turn grouped into frames —the meaning of the word depending upon its position in the frame —on the basis of a given *syntax*. To extract information from a sequence like this, the observer (receiver) must know when each new symbol, each word, and each new frame begins. The establishment of the necessary *timing* and *synchronization* between transmitter and receiver is therefore of basic importance.

Thus the existence of self-sustained oscillatory patterns at the structural "hardware" level(s) is an evolutionary prerequisite for the emergence of symbols/ digits performing meaningful, i.e., contextual, activity. Symbols (as we shall see below) perform two functions. On one hand, they *trigger* the emergence of patterns, while on the other, they *characterize* or "label" these patterns.

Let us try to outline how these activities may be dynamically conceived. Consider a physical system of N variables: we can describe the evolution of the system in continuous time by a set of N coupled nonlinear differential equations of the autonomous form

$$\frac{dx_i}{dt} = f_i(x_1, x_2, \ldots, x_N; \mu) \quad , \tag{7.2.3}$$

where the nonlinear operators on the right may include space derivatives. We refer to (7.2.3) as a "reaction-diffusion" model. The symbol μ stands for a parameter (or a set of parameters) mediating the coupling between the variables $x_1, \ldots, x_N$ and the environment; μ is also called the "control" or "external parameter", denoting the open character of the system and its free interchange of energy with the environment.

The steady states of the system are the real solutions of

$$\frac{dx_i}{dt} = 0 \quad ;$$

their stability is controlled by the parameter μ. Suppose that the system resides at one of the available steady states which, let us say, is stable. Then, as a consequence of the coupling between internal-external microfluctuations, the external parameter μ is shifted beyond a certain critical value μ_c, for which the hitherto stable steady state becomes unstable. This means that from all available eigenvalues of the (linearized) interaction matrix of the sytem, at least one acquires a positive real part λ_0 —thereby making the corresponding mode or eigenfunction of the system unstable.

For simplicity, let us assume for the moment that only one mode (out of the many which can be simultaneously excited in the system) becomes unstable. Let us further assume that this excited mode is characterized by the minimum damping constant

$$\frac{1}{|\lambda_0|}$$

or that its amplitude $\xi_0(t)$ is the most slowly evolving in comparison with the other (stable) modes. Under this condition, one can apply the "slaving principle" [7.7] to deduce approximately that the excited slow mode eliminates adiabatically the fast stable modes and determines the character of the emerging spatial-temporal pattern beyond the instability. After Haken, we call this dominant eigenfunction the "order parameter".

We see therefore that at least in the above simple case a pattern is completely specified by the numerical values (or two digits/symbols) of *two* parameters: the critical (threshold) value μ_c of the control parameter —responsible for triggering the bifurcation and determining the history or the context— and the quasi-static amplitude of the "order parameter" ξ_0, which determines the behavior of the new pattern. The emergence of only *one* order parameter allows a tremendous reduction in the number of degrees of freedom of the system at the next higher level. The entropy drops since the partition at the higher levels amounts to a delta-like function, but we do not witness yet the emergence of function or complexity.

The order parameters themselves obey a new set of differential equations characterizing the goings-on at the new hierarchical level; they have nothing to do with

the previous lower-level dynamics. We might say, so much for the relationship between *symbol* and *pattern*, where are grammar and syntax? Unfortunately, things are
not always (in fact they seldom are) so simple as to allow the straightforward
application of the "slaving principle" and the "adiabatic" elimination of fast stable
modes.

   In general, beyond the threshold of instability *many* modes are simultaneously
excited, with comparable damping constants and comparable (time-varying) amplitudes.
These modes are self-sustained nonlinear oscillations (lumped or distributed) which
compete. In a lattice of at least two dimensions this irregular beating may go on
in a stable, self-sustained manner; the pattern assumes the form of a *multimode*
nonlinear lumped or distributed oscillator [7.8].

   Now, depending on the set of the threshold values of the coupling coeffcients
(control parameters) (μ), the set of the (rms) amplitude values of the excited
competing modes (ξ), and the set of the imaginary parts of the eigenvalues (the
threshold values of the frequencies Ω) of the excited competing modes, many extremely interesting dynamical regimes can be observed in such a *multimode* nonlinear
oscillator. Let us just mention some characteristic ones and give references for a
fuller account [7.9].

a) Total Entrainment

It is possible that if the frequencies Ω of the competing modes (which can be deduced from ξ and μ) are close enough and the amplitudes ξ_i of some of these modes
are above a certain threshold, spontaneous fusion (entrainment) of the individual
frequencies may take place. The whole system oscillates at one fundamental composite frequency, which may range (depending on the strength and the topology of
coupling) from the arithmetic mean

$$\frac{\sum_{i=1}^{K} \Omega_i}{K}$$

to the geometrical mean

$$\left(\sum_{i=1}^{K} \Omega_i \right)^{1/K} \quad ;$$

or it may equal the (weighted) quadratic average

$$\left(\frac{\sum_{i=1}^{K} \xi_i^2 \Omega_i^2}{\sum_{i=1}^{K} \xi_i^2} \right)^{\frac{1}{2}} \quad ,$$

where K is the number of the excited modes. This is certainly self-organization, reminding us, however, more of "crystallization" mediated by a simple first- or second-order phase transition.

The result is a regime of high order (low entropy) but still "shallow" complexity. We simply witness the emergence of one "hypersymbol", i.e., again one order parameter, and that is all. Some beating may occur between harmonics among the fused nonlinear oscillators, but we assume that in the above ideal case this amounts to some noisy "hiss" or "blurring" of the hypersymbol.

b) Part Entrainment, Part "Jittery" Phase Locking

This is a more realistic case. Again, for different combinations of the numerical values of the sets of μ and ξ, we may witness the compartmentalization of the multi-mode oscillator in say, Λ groups. Members of each group are strongly coupled so they become mutually entrained; we end up with Λ symbols (Λ "local" order parameters).

The individual groups are weakly coupled and so they establish among themselves "jittery" phase locking, i.e., a set of flexibly bounded time-ordered recursive relationships which may be considered as standing for some primitive "rules of grammar". We thus witness the formation of many "words", namely,

$$\Lambda^1 + \Lambda^2 + \Lambda^3 + \ldots + \Lambda^\Lambda = \Lambda(\Lambda^\Lambda - 1)/(\Lambda - 1) \quad ,$$

of length from 1 bit up to Λ bits ($\Lambda \gg 1$).

c) Chaos

In general, between the excited modes we have irregular beating among pairs of fundamental frequencies as well as among all possible harmonics and subharmonics. The competing modes of the system, as a result of such interactions, may be inhibited (quenched); excited, with birth and death rates depending again on the sets $\{\mu\}$ and $\{\xi\}$; partly entrained in groups; partly phase-locked among some groups; or, last, free running.

The resulting activity, which gives rise to aperiodic self-sustained spatial and temporal oscillations, is called "turbulent chaos" (in fact just three beating incommensurate frequencies are enough). While steady states give points in the state space and limit cycles give closed, stable (periodic) trajectories, dissipative chaos, on the other hand, is characterized by one or more "strange attractors", i.e., finite areas or bounded volumes (compact sets) which attract all nearby trajectories, and within which, trajectories do not cross each other, but rather converge *and* diverge exponentially. We witness, therefore, the emergence of quasi-*stochastic* patterns as evolutionary sequences arising from absolutely deterministic nonlinear differential equations which describe classical electromagnetic and quantum mechanical interactions!

Above all, and beyond symbol formation and grammatical rules, language is characterized by highly *aperiodic* strings of *words*; it is therefore tempting to speculate that some "mild" turbulent chaos (amounting at least to narrow-band noise) at a hardware level is necessary for the functional manifestation of language. Specifically, we can model "syntax", at least in a number of simple cases, as the rules followed by strings of an aperiodic Markov chain whose number of states, elements of the transitional probability matrix, as well as the (time-varying) percentage of state occupancy, come as consequences of antagonistic competitive processes ("games") played inductively between interacting self-sustained modes. The rules of competition which determine the "logic" of the inductively played games, and from which the transition matrix results, are determined again by the sets $\{\mu\}$ and $\{\xi\}$.

More formally, we can search for the "syntactical rules" of chaos by looking at the sequences of crossings that the moving trajectory realizes along a Poincaré mapping, i.e., a cross section of the state-space manifold with a plane. Iterating such a map for different values of the control parameters gives rise to an uncountable number of periodic and aperiodic trajectories (or strings of 0's and 1's) which stand for syntax itself. (Not all possible sequences are allowed; the "filtering" mechanism, setting the syntactical rules, is provided by the shape of the map.)

We sum up this section: Each pattern at a hardware level is associated with a number of digits —we could classify them into two categories, "pre-digits" and "post"-digits. Pre-digits compose the set of the successive critical (threshold) numerical values of the control parameter, which triggered the corresponding sequence of bifurcations whose final product of broken symmetry is the pattern under consideration. The more abstract the pattern is, the longer the sequence of pre-digits needed to prescribe (and preserve) its evolutionary history. We may say that the pre-digit set is the evolutionary "Zip code" of the pattern. The sequence of pre-digits appears to be truly random, i.e., incompressible. If the set of pre-digits constitutes the *address* of the pattern, the set of post-digits constitutes its "I.D. card" or rather its "fingerprints". This contains (for the pattern concerned) the set of the amplitudes of the groups of "local" order parameters.

From these two sets of threshold values $\{\mu\}$, $\{\xi\}$, the letters, grammars, and syntax of the pattern concerned are determined unambiguously, although prediction is severely limited (see Sects.6.5-6.5.6). So the pyramid-forming pairs of structural-functional levels of a self-organizing system appear to accommodate themselves between the levels of microscopic equilibrium thermal mixing, where "nothing happens" (full symmetry, minimal complexity), and the macroscopic nonequilibrium, fully developed turbulent chaos where almost "everything is possible" (broken symmetry, maximum complexity).

At a level where homeostasis is the goal, the system must be highly *ordered* and possess relatively little complexity. The heart is a good example: it goes functionally around a limit cycle and becomes intermittently chaotic only when fibrillation occurs.

On the other hand, at a level where reliable signal processing is the goal, the system must be disordered and highly *complex*. The cerebral cortex of the human brain is a good example. The EEG is practically chaotic during the execution of mental

tasks and during the episodes of REM sleep. A highly ordered, synchronized EEG indicates an epileptic brain and constitutes a very incompetent information processor.

7.2.4 Unresolved Problems: Communication Between Two Hierarchical Systems

Language determines the cognitive pathways of a biological organism into the world.
It generates and puts forth in ways still poorly understood and impossible to implement analytically (or with any contemporary machine) many alternative aperiodic
strings of symbols ("hypotheses") which the organism correlates with a given string
of environmental cues (which it perceives, processes, and codes on the basis of still
unknown decision-making criteria) and finally interprets as the most likely "model"
the string which forms with the incoming sequence the greatest cross-correlation.

How do languages evolve? They probably do so by making "errors", i.e., by adding
or deleting symbols or by modifying the stochastic interdependence between symbols
(or by modifying the elements of the transition probability matrix in the Markovian
paradigm). In short, they evolve by randomly modifying existing chaotic maps.

These errors are random but they are also indispensable. A "non-erring" language
stagnates since it cannot cope with the devising of "compressed" descriptions of a
rapidly evolving environment. (Suppression of the activity of the immune system —
which has many similarities with natural languages — is a striking example.)

The survival value of an "erring language" can be best illustrated in the communication between two competing organisms (forming say a symbiotic "host"-"parasite"
pair). Each partner involved in this game tries to "break" the code of the opponent
first, to provide a maximum compression of the opponent's behavioral repertoire in
order to forecast and control it. At the same time, he tries to "fool" the opponent
by increasing the complexity or randomizing at his own behavioral level.

The first requirement stimulates the emergence of many alternative aperiodic
strings of states and leads to chaos at the cognitive level (C) of the organism
involved, while the second leads to the establishment of chaos at his behavioral
level (B). The two requirements are in conflict and the process is self-defeating.
Indeed, if one partner "over"-randomizes at the B level, the opponent is unable to
cognize his behavior and so he stops sending triggers. The communication (*symbiosis*)
gets disturbed and this may result in overt pathogenicity. Thus the possessor of
such an erratic behavior is in danger of jeopardizing his own homeostasis. If, on
the other hand, the partner "over"-randomizes at the C level, the opponent's code
is quickly "cracked" and the partner involved stops questioning further; again the
symbiosis ends. Also the partner-possessor of such an erratic cognitive pattern is
in danger of not being able to detect and correct errors during the decoding procedure. The problem for each organism is therefore to compromise between homeostasis
and reliable reception from the partner in order to maintain symbiosis and to allow

time for *learning* to take place, with the concomitant emergence of new, more abstract, cognitive levels.

Thus communication is a transient process which, via an intersystem game, tries to build up a "questionnaire" which will eventually crack the code of the opponent and allow reliable prediction of his future behavior. Therefore what we call "natural selection" has to do with the differential advantages acquired by the possessor of a language that has the ability to detect and correct decoding errors by converting minimal amounts of energy to maximum information, within given time intervals.

How fast can a language afford to evolve in order to overcome quickly evolving "opponents"? There are intrinsic "hardware" limits. For each organism these limits are determined by cellular (neuronal) biorhythms which control the manufacture of proteins and other macromolecules (hormones, neurotransmitters), whose complexes mediate the hardware of the memory bank of the organism.

When environmental pressures prompt an organism to accelerate the pace of **metabolic signal processing beyond a certain limit, the organism is either eliminated** (due to the error accumulation resulting from manufacturing distorted metabolites) or jumps to a higher cognitive level by performing *selective* detection and processing (meta-coding) in order to minimize "rate-distortion" on the basis of some decision-making criteria.

In the last analysis, each organism is consumed in an exploratory processing of the triggers it is able to detect from the environment. Each organism incorporates memory mechanisms, i.e., modes of coded descriptions of subsets of environmental states. Whenever these mechanisms (whose hardware consists of nets of hard nonlinear oscillators or active "switching" elements) are activated by environmental cues, they cause responses which may be classified in two distinct categories.

a) Those which result from the cross-correlation between already-processed environmental signals and internally stored patterns above a certain threshold of quality ("information").
b) Those which do not recognize the input ("noise").

In the first case, the result of cross-correlation (cognition) gives rise to a succession of metabolic steps (signal processing), via which the response of the organism is mediated and the storing of the experience is consolidated in the memory bank. Any future theory of language has to elucidate further the nature of storing, retrieval, and renewal processes in such a dynamic memory. Persistence of memory constitutes another mystery.

Normally, natural language is *recursive*, i.e., it approaches its goal via successive iterations either *deductively* or *inductively* (via learning). Abnormally, however, language may stagnate and consequently display a circular modus operandi or even a "dead end". (Absorbing Markov chains are good models of circularity.)

Limited recursion is established (together with an inability to form higher cognitive hierarchical levels) when bifurcations leading to broken symmetries simply cannot go on —or when early stabilization occurs (over-homeostasis).

Thus the organism cannot differentiate cognitively beyond a certain stage. Communication with the environment may provide the necessary stimulus for secularly shifting key control parameters beyond their critical values, thereby allowing the bifurcations which may give rise to chaotic syntax formation or push the organism back to the recursive algorithm. The spontaneous and continuing *folding* of polypeptide chains into three-dimensional structural proteins provides such an example of cascading broken symmetry which —via enzymatic feedforward action —dispels stagnation or circularity in the genetic language.

Nevertheless, the mystery "Who makes the understanding?" lingers on. The human mind possesses the unique capability of "mapping" the external (as well as part of the organism's internal) world —that is, it "compresses" long and complex strings of impinging environmental stimuli ("observations") —and then uses these "minimal length algorithms" to simulate physical phenomena —thereby revealing the "laws of nature".

We may theorize that this process of "self"-organization and category formation is implemented via a set of coexisting (strange) attractors in the cognizant aparatus —each one of which attracts (and therefore compresses) whole subsets of "initial conditions" the sum total of which constitutes the set of external stimuli (Fig.7.2).

This set of initial conditions forms the "basin" of the attractors and the processes of partition and category formation in the mind involve the topology of the separatrices among the individual subsets of the basin.

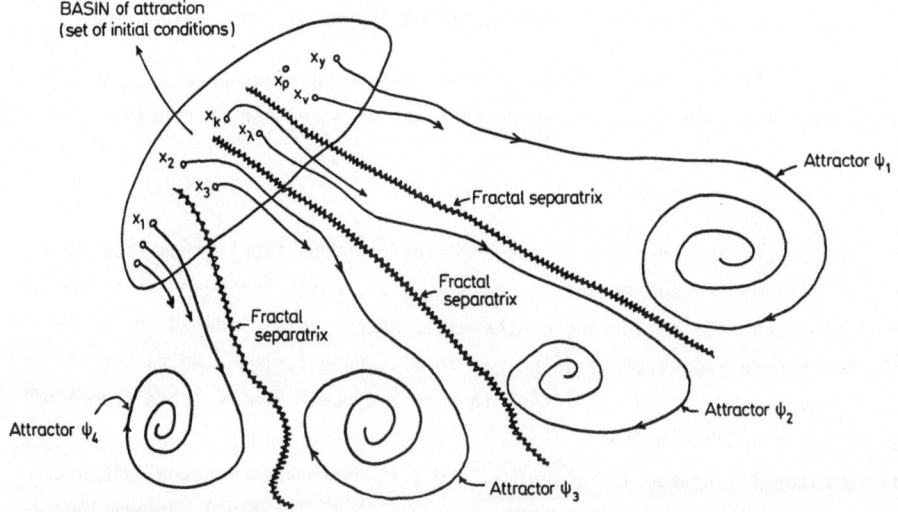

Fig.7.2. Sketch of a cognitive channel. The x's and ψ's belong to different hierarchical levels

Let us then try to sketch the *dynamical equivalent* of an information channel mediating the mapping between a set of messages of the "transmitter" and a set of "regimes" of the "receiving" or rather the cognitive apparatus —playing the role of memory bank. (Let us remind ourselves again that an information channel according to Shannon's "static" information theory provides a one-to-one mapping between the transmitter and the receiver and is in no way associated with the act of "cognition". The sole role of a classical information channel is the mere copying of the transmitted patterns, *as they are*, at the receiving end with the greatest possible reliability.)

We propose that the set of messages should be *identified* as a set of initial conditions or the basin of the attractors in an N-dimensional state space where N stands for the number of variables mediating the dynamical phenomenon under consideration. The receiver, or more generally the cognitive apparatus, is identified as the set of the attractors of much lower dimensionality embedded as invariant compact subsets in the same N-dimensional state space.

Falling into the attractor via (short or long) transients in state space corresponds to the act of progressive signal abstraction (or compression). The issue is: what about the dynamical isomorph of the "channel"? or, how is the set of initial conditions, that is, the messages, *partitioned among different categories*, that is, among more than one coexisting attractors (steady states, limit cycles, tori, or strange attractors)? One has to realize here that the mapping between messages and attractors (or "templates") is *not* one-to-one: whole subsets (of initial conditions) of the basin converge towards one and only one attractor. Alternatively some points of the basin may undergo a perpetual (or too long) transient trajectory without converging anywhere, or they may even be repelled to infinity. How can we tell which subset of the basin will land on which attractor?

The problem in this general form is, of course, of formidable difficulty. A very elementary case (two simple attractors, namely two steady states) has recently been addressed by *Grebogi* et al. [7.10]. They considered a two-dimensional non-invertible map (the "channel") and found a *fractal separatrix* between the two subsets of the basin.

The essential outcome of the above work is that the precision with which a given initial condition is known may play an extraordinary sensitive role in the prediction of the final attractor upon which this initial state will eventually land. This means that an extraordinarily high accuracy is sometimes necessary in our measurements for the reliable prediction of the pertinent attractor ("compressor") into which the observed phenomenon will be categorized: clearly then, the set of attractors play in our model the role of dynamically stored patterns in the brain with which whole sets of externally impinging stimuli have to be "convoluted". [This process may not be accomplished in one single hierarchical step; in a second step the attractors (if numerous at the lower level) will form a hyperbasin for a new hierarchy of fewer hyperattractors and so on.]

Knowing the state space dimensionality of the impinging messages *and* the (fractal) dimensionalities of the corresponding attractors one can evaluate in each case the degree of (average) *compressibility* which a given N-dimensional stimulus is subjected to. Furthermore, investigating the topology of the separatrices of the basin one can calculate the uncertainty (or the information gained thereof) associated with the category to which the given stimulus belongs. Together, *the entropy of partitioning* and *the degree of compressibility* give the *two* essential macroparameters characterizing the *cognitive process at a given hierarchical level*.[2] Once *the mapping* (that is the "linguistic scheme" adopted by the individual) is known, both the above parameters can in principle be calculated.

Finally we would like to argue that the number of subsets of the basin in a given N-dimensional state space which are *not* attracted at all is the *rule* rather than the exception. Consider the following example [7.11]. One is asked to sort out a particular arrangement of a deck of cards. Since each arrangement is equiprobable, its probability being ~1/52! it follows that $\log_2(52!)$ "yes-no" decisions are required to reconstruct a particular order from a shuffled deck. Now $\log_2(52!)$ ~ $\log_2 10^{68}$ ~200 bits; so one would think that the specific order (or string) can be deduced in about 200 "yes-no" guesses. The real issue is however that in this case we do not possess the "linguistic tool" to classify the card arrangements so as to be able to make judicious dichotomies for "yes-no" answers: our knowledge of card arrangements is confined to a very small subset of "compressible" sequences that is of arrangements which are reproducible with fewer instructions than their own length. To reconstruct any other arrangement among the incompressible ones one has to *name* the position of *each* card (or symbol) separately; it turns out then that the difficulty in *categorizing* springs from the deficiency in *compressing* or devising a set of (strang) attractors.

We have appreciated already in Sect.2.3.7b the fact that the percentage of compressible sequences in a string, say of 0's and 1's of length N, is extremely small. Considering the 2^N a priori equiprobable sequences of length N (the members of the basin) the number of compressible sequences to, say, K bits is just

$$\Sigma = 2^1 + 2^2 + \ldots + 2^{N-K-1} = 2^{N-K} - 2 \qquad (7.2.4)$$

and the percentage of compressible sequences is

$$\eta = \frac{\Sigma}{2^N} \sim 2^{-K} \ . \qquad (7.2.5)$$

Therefore, although in principle there is a numerous set of attractors at each cognitive level, the percentage of those which really deserve the name, that is,

---

2 According to our model in Sect.6.5.2 the EEG is the pacemaking agent which permutates the attractors on a time division multiplexing basis.

possess an information dimensionality $c = N - K$ bits $[K \sim O(N)]$, is just $\eta = 2^{-K}$, which is insignificantly small.

The moral is that the human mind can simulate a very small subset of natural phenomena. Their overwhelming majority literally "pass through" our heads undetected. Even so, how is it that our templates/attractors *really* have as their basin the set of *externally* impinging stimuli? To accept this one has to a priori digest the hypothesis that *there exists* a "homology" between *some* patterns of the external world and the categories of the software of the human brain. But if the human brain is just an unsuspected evolutionary by-product of the "laws of nature" how is it that its software can simulate the very laws which, acting upon matter, triggered its formation?

7.3 Concluding Remarks

Possible application of chaotic dynamics to more "exotic" fields like economics, sociology, and psychology, although very fashionable today (especially among non-physicists), is in our opinion either premature (in the case of economics and sociology) or, beyond a certain point, just futile (in the case of psychology). In the case of economics and sociology, many more data are needed. The case of psychology is more subtle.

Let us comment briefly on the last issue. Psychologists are trying to elicit the "inner" dynamics of the psyche by dealing presumably with (human) behavior, but behavior constitutes only the tip of the iceberg in the whole spectrum of human problems. Behavior itself can to some extent be treated in the realm of game theory. Game theory among more than *two* agents or persons (presumably a fair "game" for chaotic dynamics) is plagued, however, with new and difficult concepts, for instance, "coalitions" between two persons against a third one. These concepts can no longer be expressed mathematically and modeled in serious ways. *Individual* behavior itself may, of course, be quite erratic, and again there is a temptation to try to frame it within the chaotic paradigm. If we consider behavior as a one-dimensional "mapping" of an ongoing complex internal chaotic dynamics, what are the chances of reliably inferring this internal dynamics of the mind from the observed behavior? The answer is nil: there is no one-to-one mapping between mind and behavior. A lot of things going on in the mind of the individual are not expressed at all — seemingly obstructed by some "refractoriness" or an inhibition barrier — due not only to possible lack of "talent" and the feeling that some thoughts are not socially "appropriate" but simply because man conducts an internal dialogue with himself, the outcome of which remains mostly inconclusive; it constitutes an ongoing conflict between different hierarchical levels.

Indeed intrapersonal conflicts (e.g., contradictory deliberations between emotional hierarchical levels) are ad hoc "non-constant-sum" debates. In such conflicts the ("neurotic") individual is often considered as waging a "civil war", as it were, by making himself oscillate between two extremes (leading essentially to the same outcome):

a) a tendency to activate as many "dormant" potentialities as possible, via trespassing self-imposed constraints ("values") at the expense of becoming incoherent, and

b) another tendency of compulsively staying "self-consistent" at the expense of stagnation.

Hence the dilemma. This occasional incompatibility between man's ability to reason deductively and his rather chaotic emotional deliberations has been somehow related to the conflict (or less-than-perfect integration) between his giant (left-hemisphere) neocortex and his emotionally driven (acquired and innate) value systems, residing, speaking in terms of hardware, in the "paleomammalian" and "reptilian" parts, respectively, of his brain. These are the limbic system [7.12] and the lower parts of the brain comprising the olfactostriatum, the corpus striatum, and the globus pallidus.

Under "ideal" conditions man's "instinctively" driven decision-making system should cooperate with his neocortex (primarily oriented towards the external environment) in comparing alternative courses of action and selecting one that seems best in terms of a built-in system of values. However, from an evolutionary point of view, the innate human values are just inherited decision-making or rather behavior-turnover criteria that were built into the animal during the long process of evolution. Thus insistence on the above "ideal" conditions (reminiscent of the "lost paradise") would not differentiate man from other species endowed with fixed and very effective "subroutines" wired in their "reptilian" brain. Human behavior is influenced par excellence not only by these inherited primary decision criteria, but also by *beliefs* that are rapidly acquired via the use of language during the lifetime of the individual as a result of "inductively playing" with the symbolic or man-made environment.

These latter criteria, "imprinted" in the limbic system, are what should perhaps most properly be called "social values" (collective properties), i.e., learned objectives determining "character" or the capacity to act. Still, in man there is a recurrent discrepancy between these latter "software" instructions and "hardware" inbuilt primary impulsive drives —a discrepancy that leads quite often, especially in a rapidly changing symbolic environment, to "deadlock" or to what more plainly has been categorized as human "neurosis".

Thus, in our opinion, real psychological problems never lead to reliable "observables": they constitute in a sense, unsettled theological disputes that the individual carries along with himself throughout a lifetime. Such issues are not negotiated —as for example Freud and his followers advocated —through rationalization or communication: no more than guilt is alleviated through confession. Aside from a thin patina of knowledge and affective commitment, the individual man as a "self"-cognitive system and as a feeling being is essentially alone in a universe which is neither friendly, nor hostile, but something much more sobering: indifferent.

A real scientist strives for knowledge and feels guilty if there is any piece of new information around that he has not already digested. This type of striving being is not exactly longing for immortality but rather for *familiarity* with the physical world so that extinction will not take him unawares. When man dies he perhaps ceases having thoughts as well. This, at least, was the prospect which most terrified the mighty J. von Neumann, "the perfect intellect", on his deathbed. So much dynamics going on without having the opportunity to share it even through the passivity of a disembodied spirit! How will chaos look a hundred years from now? What will be the concomitant progress in optical computing and signal processing — which will not only perhaps allow man to "tame" simple chaotic dynamics but also use them for the development of self-programming computers, taxing and extending not only our perceptive abilities but also our very imagination? We end up with the timid hope that the incorporeal soul which has expressed itself so far dimly through an imperfect carbon biochemistry will be able, when interfaced with silicon advanced architecture, to hold onto itself with greater confidence.

Είπεν αυτοίς ο Ιησούς·
ει τυφλοί ήτε, ουκ αν
είχετε αμαρτίαν· νυν δε
λέγετε ότι βλέπομεν·
η ουν αμαρτία υμών μένει.

Ιωάννης Κεφ. 9, παρ. 41

Appendix

A. A View of the Role of External Noise at a Neuronal Hierarchical Level

A.1 Introduction to the Problem

In this appendix we would like to give a more detailed account of our view concerning the way the neuronal genome might interact with a post-synaptic membrane aggregate, thereby leading to the renewal of the set of structural proteins covering such a site. Again let us start by pointing out that any self-organizing system is a hierarchical structure[1]. It simultaneously undergoes a variety of distinguishable activities so that different types of description (i.e., different sets of variables and parameters) are appropriate to the study of these several activities. Which hierarchical level we deal with depends therefore on which *aspect(s)* of the behavior of the system we are interested in, i.e., on the way in which we interact with the system. At each hierarchical level we introduce a phase-space description involving type and number of variables, parameters, and number of states pertaining to that particular level.

Progressive organization *can* take place at a given hierarchical level where the system propelled *by a set of triggers* (to be defined presently) moves to steady states where the number of (flexibly bounded) ordered relationships between the variables increases with time.

The motion and transactions that the system undergoes in phase space at a fixed hierarchical level (such as sequential passage from steady state to steady state, "time of rest", or degree of stability at each state as well as flight time between successive steady or stationary states) depend on (a) information transmission (afferent signals) coming from the level immediately below; (b) information coming

---

1 Graphically, a hierarchical structure should be represented as an "inverted tree" with branches spreading laterally as one moves downwards. For the sake of simplicity, we consider here a single branch, so instead of having at a given hierarchical level many independent, equal-height "platforms" with communication channels converging *in groups* (and getting multi-cross-correlated) at higher platforms, we depict and subsequently treat in the text a single sequence (branch). "Oblique" contributions concerning information arriving at a given hierarchical level from more than the one level underneath are labeled as "total environment".

from the external environment (i.e., via the peripheral nervous system); and (c) feedback control (efferent signals) coming from the level immediately above.

Each hierarchical platform can then be considered in a very simplified form as a dynamical *constraint* where a "storage and integration" or cross-correlation procedure takes place continuously. This means that the signals transmitted sequentially from "below" $x_\lambda(t)$ (each one corresponding to a different state of the system at the level immediately below) *and* the ones arriving from the external environment $w_\lambda(t)$ are time averaged, upon arrival at the receiving platform, with the result that some of the variables x_λ are completely eliminated or "washed out" at the higher level while the others are nonlinearly correlated or matched:

$$<x_\lambda^\nu(t)w_\lambda^k(t + \tau)> \quad ,$$

where $0 <\tau <T$, and T is the transmission time interval either from the lower level or from the external environment. Events taking place simultaneously at different hierarchical levels have therefore *correlative* rather than *causal* correspondence. The mapping taking place at the interface between successive levels is *not* one-to-one.

The variables y_i at the "receiving" level under consideration, Y (displaying the autonomous innate behavior of the system at that level Y), are also much more constrained or "abstract" than these at the "transmitting" lower level X. Thus the higher hierarchical levels are more organized or redundant and, in a description sense, "simpler" than the lower ones. In this appendix we are concerned with the time evolution and self-organization of an advanced biological system at one or two specific higher hierarchical levels.

About the noise in the system we make the following assumptions: although (strong) noise causes both amplitude and phase (jitter) modulation, we are considering here low-level (quasi-stationary) additive noise acting only on the amplitude of the variables involved.

After the above preliminary remarks let us proceed. We consider an open nonconservative system Σ_i (described at a specific hierarchical level, say Z) consisting of a number $n_z(t)$ of nonlinear interacting components ("reactants"). These components (variables) can be imagined as self-sustained oscillators (either of a Hamiltonian or nonconservative type) lumped or distributed (normal modes). Each active component of the system is represented by a random phasor (i.e., a nonharmonic oscillation in a pool of weak additive noise):

$$z_k(t) = A_k(t)e^{j\varphi_k'(t)} \quad , \tag{A.1.1}$$

where $\varphi_k'(t) =\omega_k t +\omega_k(t)$, characterized by a continuous joint probability density distribution function (p.d.f.): $\rho_k(A_k,\varphi_k;t)$.

Dynamically, the system Σ at the level Z is modeled by a set of $n_z(t)$ coupled nonlinear differential (rate) equations which, in the absence of time-varying differentiation ($\nabla^2 z_i =0$), read:

$$\frac{dz_1}{dt} = f_1\left(z_1, z_2, \ldots, z_k, \ldots, z_{n_z(t)}\right) \, ,$$

$$\cdots$$

$$\frac{dz_k}{dt} = f_k\left(z_1, z_2, \ldots, z_k, \ldots, z_{n_z(t)}\right) \, , \qquad (A.1.2)$$

$$\cdots$$

$$\frac{dz_{n_z(t)}}{dt} = f_{n_z(t)}\left(z_1, z_2, \ldots, z_k, \ldots, z_{n_z(t)}\right) \, ,$$

and unless otherwise stated, it is assumed that

$$\left|\frac{dn_z(t)}{dt}\right| \ll \left|\frac{\partial A_k}{\partial t}\right| \, , \quad \left|\frac{\partial \varphi_k}{\partial t}\right| \, .$$

By expressing the dynamical behavior of the system at the level Z as in (A.1.2), we essentially imply that all activity in it originated as noise which has been more or less *constrained* at a rather early age of the organism through an interplay between the environment and the dynamic activity of the hierarchical levels below Z. The noise has not been wholly constrained, however: some of it (which we call additive noise here) escapes control and gives rise to an element of arbitrariness or flexibility, or even (it depends on how we look at it) "originality", in the system's behavior at the level Z.

In order to follow the behavior of Z in time, it is important to derive in such a situation not the analytical expressions of the coupled stochastic processes z_k, but rather their joint probability density distribution

$$\rho(z_1, z_2, \ldots, z_{n(t)}; t) \, .$$

This could be achieved in principle by deriving from (A.1.2) and subsequently solving, under given initial and boundary conditions, the corresponding $[n(t) + 1]$-dimensional Fokker-Planck equation [A.1] which (in the limit of weak additive noise in the neighborhood of a steady state) describes the diffusion in time of the joint p.d.f. $\rho(z_1, z_2, \ldots, z_{n(t)}; t)$.

From this distribution one can derive the "instantaneous" (conditional) entropy of the system

$$S(t) = c_1 + c_2 \int_{\upsilon} \rho\left(z_1, z_2, \ldots, z_{n(t)}; t\right)$$

$$\times \log_2 \rho\left(z_1, z_2, \ldots, z_{n(t)}; t\right) d\upsilon \, , \qquad (A.1.3)$$

where $d\upsilon = dz_1, dz_2, \ldots, dz_{n(t)}$; the integration is extended over the volume υ in phase space; $Z, z_1, z_2, \ldots, z_{n(t)}$ become eventually statistically correlated as a result of the organizing process going on; and c_1, c_2 are constants.

We have to point out immediately that in most biological systems the variables and parameters at a given hierarchical level are in general unknown; so, depending on the relevance of the variables chosen for the model, the calculated entropy (which is a property of the state, i.e., a probability density distribution with respect to a specific partition) may or may not represent the degree of disorganization of the "real" Z. The relevance of the introduced model can only be inferred from the overt behavior of the organism at a level higher than Z.

We now proceed in the formulation of our problem as follows: our open system Σ at the level Z interacts energetically with the environment Σ' (both external and internal). Consequently, the differential of the entropy of the system Z, dS(t) can be written [A.2] as

$$dS(t) = dS_i + dS_e \ , \tag{A.1.4}$$

where dS_i is the entropy production associated with the irreversible processes inside Z ($dS_i > 0$ always) and dS_e is the entropy flow associated with the energetic exchanges with the "total" environment Σ'. The sign of the term dS_e is undetermined. In the trivial case $dS_e > 0$, i.e., disorder or "ataxia" entering the system Z from the outside, we see an acceleration of the process of disorder moving the system toward the (thermodynamic) equilibrium faster. In the case $dS_e < 0$, order or "negentropy" enters from the environment. It is understood, of course, that it is the system at Z itself which "decides" about the sign of dS_e. In other words, information transfer takes place between a series of events in Σ' and another series of events in Σ at the level Z, if there is some degree of correlative correspondence between them. Two possibilities exist now: either $|dS_e| < dS_i$ —as a result, the degradation of the system at Z will slow down and perhaps level off in time —or $|dS_e| > dS_i$. In this latter case, dS(t) < 0, i.e., the entropy of the system at Z will decrease at the expense of the external environment and/or other hierarchical levels of the system.

In this discussion we start with a *given* partially organized hierarchical level Z of an organism Σ and examine the condition under which we can make it *more coherent* (i.e., increase the number of ordered relationships between appropriate dynamical variables, e.g., φ_k) by absorbing information from the environment and the lower than Z levels, in the presence of additive fluctuations existing between the components of the system at Z.

Under these circumstances and for a fixed number of active components n_z, it appears that noise will cause deterioration, i.e., it will eventually overtake whatever organizing process is continuously acting on the phases of the oscillators. Therefore the condition $|dS_e| > dS_i$ cannot be fulfilled in such a case as above. After all, the Fokker-Planck equation —or any master equation for that matter— is a *diffusion* equation; thus what is predicted is a spreading rather than a "shrinking" of the p.d.f. $\rho(\varphi_1, \varphi_2, \ldots, \varphi_n; t)$ with time. Consequently the entropy of the system at Z, S(t) will increase for any (small) deviation from a steady state. Nevertheless, the organization at Z which can properly be defined in terms of its *redundancy* R [expressing intercomponent $(\varphi_k, \varphi_\lambda)$ correlation] *may* increase even if the entropy $S(t)$ goes up, provided that the maximum entropy of the system increases *faster* than $S(t)$ with time. Since

$$R = 1 - \frac{S(t)}{S_{max}} \quad \text{and} \tag{A.1.5}$$

$$S_{max} = c_3 + c_4 \log_2 \lambda(t) \quad , \tag{A.1.6}$$

where λ is the number of complexions at equilibrium,

$$\lambda = \frac{n!}{(\frac{n}{2})!(\frac{n}{2})!} \quad ,$$

(we consider bistable "off"-"on" elements), it is obvious that

$$\frac{\partial R}{\partial t} > 0 \quad , \qquad \text{if}$$

$$\frac{d}{dt}(\ln S_{max}) > \frac{d}{dt}[\ln S(t)] \quad , \qquad \text{or if}$$

$$\frac{d}{dt}[\ln S(t)] \leq \frac{\dot{\lambda}(t)/\lambda(t)}{\log_2 \lambda(t)} \quad , \qquad \text{or}$$

$$\frac{d}{dt}(\ln S) \leq \frac{\dot{\lambda}(t)}{\lambda(t)} \quad , \qquad \text{for} \quad \lambda(t) \gg 1 \quad . \tag{A.1.7}$$

Of course, for an invariant number of active components $n = \text{constant}$ at Z, $\lambda = \text{con-}$ stant, too, and the system can never comply with the above requirement. The only possibility for organization is that *new* degrees of freedom enter the organizing process under an instantaneous overall entropy growing more slowly than $\log_2 \lambda(t)$. It is noted that λ is an increasing function of n.

By what mechanism could such a thing possibly be realized? We elaborated on specific examples in Sects.2.2.3-6; let us discuss for the moment two possible cases where this process *might* occur.

1) During the embryogenesis of a biological organism the redundancy as expressed above *must* increase, although it is known that during this particular period of development the internal entropy production of the system ΔS_i increases very quickly. This has indeed been shown by *Trincher* [A.3] who gives experimental evidence that the entropy production dS_i/dt per unit mass and unit time (in cal/h) increases during the first embryogenetic period of the differentiating organism, then passes through a maximum, and finally decreases and levels off to a steady regime with the attainment of the organism's adult state.

In this case, the increase of redundancy is accomplished, we believe, through a "fast" increase of $n(t)$, i.e., cell division and multiplication. On the other hand, during the adult life of the organism, cells keep dividing but do not normally multiply. The values of $\partial R/\partial t$ fluctuate slightly about the region of "zero plus-minus" while the system exists in a dynamical stationary state, e.g., "homeostasis". (In cases where $\partial R/\partial t > 0$, $R \to 1$, the abnormally tight organization of "petrification" would follow, making the system unable to display adaptive behavior; for $\partial R/\partial t < 0$, $R \rightarrow 0$, a "death process" would be initiated.)

2) A second, most striking case concerns the electrical activity of the brain as related to changing physiological states: intracellular recordings have revealed spontaneous wavelike activity with amplitudes in the range 5-20 mV in many cerebral structures [A.4-6]. These intracellular recordings show that even in the absence of propagated nerve impulses, rhythmic noiselike slow potential changes occur due to (subthreshold) fluctuations of the cell's membrane potential.

Power density distributions in these records closely resemble spectra from similar analyses of gross electroencephalographic records (EEG) from the same regions: this means that the EEG can be considered as the (scalar) projection of a vector superposition of n(t) phasors with central frequencies ω_λ in the neighborhood of ~10 Hz —each representing the individual (dendritic) neuron's fluctuating postsynaptic membrane potential. Now, as has been pointed out many times by workers in this field [A.5], the spontaneous electrical activity of the brain still remains one of the most elusive aspects of cerebral functions. It is a (nonstationary, beyond time intervals of a few seconds) continuous wave activity of variable amplitude and frequency with inconstant phase relationships ($\varphi_\lambda - \varphi_k$ = unbounded) between its constituents (phasors) z_λ, z_k; its overall character is quite similar to random noise {although increasing skewness of the p.d.f. of the EEG composite phasor during performance of mental tasks [A.4,5] indicates progressive (linear) statistical coupling between the individual neuronal groups}. However the cognitive and behavioral levels of hierarchy (just above the neuronal level) are under such conditions rather orderly, so, unless the electrical activity of the brain is not related to behavior, there is a possibility that unlike man-made communication systems, in which noise is an encumbrance, the brain is a system which may use noise as an inevitable and "beneficial" factor.

It is known that neuron cells do not divide in the brain beyond a rather early age of the organism so one could not justify any increase of brain organization on that basis. There is, however, the possibility (and this is our hypothesis) of a number of "phasors" in the system's Z hardware level being "dormant" for given periods of time, but being excited whenever the fulfilment of the condition $\partial R/\partial t > 0$ at the level involved or *at a higher level* requires action. This necessity is presumably "appreciated" by the neuronal (Z) level after a suitable efferent feedback signaling comes down from the cognitive level just above. We should point out once again that just activating above-threshold, previously "dormant" nonlinear oscillators is not enough; the procedure has to take place under a slower increase of entropy S(t), i.e., the newly activated degrees of freedom must be subsequently subjected to a "phase-locking process" similar to that already operating on the "older" active variables of the system at Z. In the following section, we examine in detail the possibility of achieving amelioration of the redundancy in the system Σ at the hierarchical level Z by allowing the number of self-sustained oscillators $n_z(t)$ to vary with time, as a result of switching "on" and "off" caused by (weak) noise catalytic action. With the aid of such a model, we may approach an understanding of how (mild) noisy activity in a biological organism at a given level (the neuronal one) may result in orderly behavior at a higher (cognitive) level.

A.2 Organization Through Weak Stationary-Amplitude Noise

We consider a system Z consisting of $n(t)$ random phasors z_λ with fixed central frequencies ω_λ very close together, and amplitudes $A_\lambda(t)$ and phases $\varphi_\lambda(t)$ all random processes. Let us briefly consider first what may happen dynamically for $n = \text{constant}$, in the absence of noise, i.e., in cases where z_λ are deterministic nonlinear coupled oscillators with independently varying amplitudes and phases between individual oscillators. Since the frequencies ω_λ are very close together, the system can find itself in either (a) the regime of frequency entrainment [A.7] where all oscillators operate in synchrony — thereby implying functional homogenization of the system — or (b) the regime of (flexibly bounded) phase locking $|\varphi_{\lambda+1} - \varphi_\lambda| < \Phi_\lambda$ $(\lambda = 1, \ldots, n)$ — which implies a sequential process or a differential mode of function (and suggests the possibility of dynamical information storage via the above sequential phase relations).

Now, in the general case with z_λ random phasors, we are faced with the problem of estimating the redundancy of the system at the level Z as a function of time. Assuming that the amplitudes $A_\lambda(t)$ of the phasors are statistically independent of the phases $\varphi_\lambda(t)$ and also vary slowly with time, we can express the instantaneous entropy of the system as

$$S(t) \sim \int_{-\infty}^{\infty} \int_{-\infty}^{\infty} \cdots \int_{-\infty}^{\infty} \rho(\varphi_1, \varphi_2, \ldots, \varphi_\lambda, \ldots, \varphi_{n(t)}; t)$$

$$\times \log_2 \rho(\varphi_1, \varphi_2, \ldots, \varphi_\lambda, \ldots, \varphi_{n(t)}; t) d\varphi_1 d\varphi_2 \cdots d\varphi_{n(t)} \quad , \tag{A.2.1}$$

where $\rho(\varphi_1, \varphi_2, \ldots, \varphi_{n(t)}; t)$ is the joint p.d.f. for the phases of the system. In general, the above expression is of academic interest due to the impossibility of calculating the joint p.d.f. In cases where, however, the probability density distribution of the resultant vector $\rho(R, \theta)$ can be deduced explicitly — during time intervals of quasi-stationary behavior — we can heuristically replace $S(t)$ by

$$\int_{-\infty}^{\infty} \rho(\theta) \log_2 \rho(\theta) d\theta \quad , \tag{A.2.2}$$

and the redundancy of the system takes the form

$$R(t) = 1 - \frac{\int_{-\infty}^{\infty} \rho(\theta)\log_2\rho(\theta)d\theta}{\log_2\lambda(t)} \quad , \quad \text{where} \tag{A.2.3}$$

$$\rho(\theta) = \int_0^{\infty} \rho(R,\theta)dR \quad . \tag{A.2.4}$$

We shall now assume that each phasor z among n independent units of the system is a *hard* nonlinear oscillator in a pool of ambient noise and obeys a dynamical stochastic equation of the form

$$\ddot{z} - 2\varepsilon(1 - 4\alpha z^2 + 8\beta z^4)\dot{z} + \omega_0^2 z = \xi(t) \tag{A.2.5}$$

where $\xi(t)$ is a stationary random process with zero mean value and power spectral density $N(W/Hz)$.

We are interested in the probabilities of temporal excitation and quenching (interruptions) of self-sustained oscillators like the one above contained in the system, under the influence of surrounding noise $\xi(t)$. A study of the system (A.2.5) has been carried out by *Stratonovich* [A.8], who showed that under the influence of noise an initially non-excited or "dormant" hard oscillator can be asynchronously activated [A.9], in spite of a *stable* steady state at equilibrium (i.e., at rest). Conversely, the noise may have the opposite effect of asynchronously quenching existing stable oscillations, thereby pushing the "live" oscillator back into its stable, non-excited state. If the (mean) time that the individual oscillator of the ensemble Z is in the (stable) *excited* state is greater than the (mean) time that the oscillator is in the *dormant* (stable) state, then n(t) will increase, and so there is a possibility of increasing the redundancy of the system at the studied level strictly "by noise".

The rates of excitation, "birthrate" K and quenching, "deathrate" K_1 have been calculated by *Stratonovich* [A.8] as

$$K = 2\varepsilon R \sqrt{\frac{\varepsilon u}{2\pi N}} e^{-f(R)/N} \quad \text{and} \tag{A.2.6}$$

$$K_1 = \frac{\varepsilon R}{\pi R_1} \sqrt{uu_1} e^{[-f(R)+f(R_1)]/N} \quad , \quad \text{where} \tag{A.2.7}$$

$$R = \left(\frac{\alpha - \sqrt{\alpha^2 - 4\beta}}{2\beta}\right)^{\frac{1}{2}} \quad , \tag{A.2.8}$$

$$R_1 = \left(\frac{\alpha + \sqrt{\alpha^2 - 4\beta}}{2\beta}\right)^{\frac{1}{2}} \quad , \tag{A.2.9}$$

$$u = -(1 - 3\alpha R^2 + 5\beta R^4) \quad , \tag{A.2.10}$$

$$u_1 = (1 - 3\alpha R_1^2 + 5\beta R_1^4) \quad , \tag{A.2.11}$$

$$f(R) = \varepsilon\left(\frac{R^2}{2} - \frac{\alpha R^4}{4} + \beta \frac{R^6}{6}\right) \quad , \tag{A.2.12}$$

$$f(R_1) = \varepsilon\left(\frac{R_1^2}{2} - \frac{\alpha R_1^4}{4} + \beta \frac{R_1^6}{6}\right) \quad , \tag{A.2.13}$$

and N is the spectral density of the surrounding noise in W/Hz. Therefore, the ratio of (mean) times which the oscillator(s) spend(s) in the non-excited and the excited state is given by

$$\frac{K_1}{K} = \frac{e^{f(R_1)/N}}{\sqrt{2\pi}R_1}\sqrt{\frac{u_1 N}{\varepsilon}} \quad . \tag{A.2.14}$$

Substituting the above parameters we find

$$\frac{K_1}{K} = \frac{(\alpha^2 - 4\beta)^{\frac{1}{4}}}{2\sqrt{\pi}\varepsilon}\sqrt{N}\ e^{c/N} \quad , \qquad \text{where} \tag{A.2.15}$$

$$c = \frac{\varepsilon}{12\beta}\left(\alpha + \sqrt{\alpha^2 - 4\beta}\right)\left[1 - \frac{\sqrt{\alpha^2 - 4\beta}}{4\beta}\left(\alpha + \sqrt{\alpha^2 - 4\beta}\right)\right] \quad . \tag{A.2.16}$$

For α, β of the same order of magnitude (strong, hard nonlinearity), we may have either $c > 0$ or $c < 0$; in the special case of $\alpha > \beta\sqrt{2}$, $c \sim \alpha\varepsilon/12\beta$. Thus, the above results indicate that for given parameters α, β, ε, the mean time during which the oscillator(s) is (are) in the non-excited state versus the corresponding time that the oscillator is "alive" varies as $\sqrt{N}\ e^{1/N}$. We can always choose the parameters α, β, ε so that for a given (moderate) level of noise N, $K_1/K < 1$, i.e., "birthrate"/ "deathrate" > 1. Thus the oscillator or oscillators spend *more* time in the excited state and consequently the number of excited oscillators within a given time interval Δt increases. The number of oscillators which are "on" at a given moment we call $n(t)$ and $n_1(t)$ is the corresponding number of oscillators that are quenched (or rather, quiescent). We assume an overall constant number of oscillators in the system, i.e., $n + n_1 = n_0 = $ constant. Each oscillator of the species n that "dies" becomes a member of the species n_1 and vice versa. The rate equations for n and n_1 assume the form:

$$\frac{dn}{dt} = K_1 n_1 - Kn \qquad \text{and}$$

$$\tag{A.2.17}$$

$$\frac{dn_1}{dt} = Kn - K_1 n_1 \quad ,$$

from which the closure condition follows as $n_1 + n = $ constant. The steady states of (A.2.17) are those points for which dn/dt and dn_1/dt vanish simultaneously. Every point on the line $K_1 n_1 - Kn = 0$ on the plane (n, n_1) will have this property, so for a fixed overall number of oscillators n_0, the steady-state point will be at the intersection of the straight line $n + n_1 = n_0$ and $K_1 n_1 - Kn = 0$.

Since $K_1 + K > 0$, the steady state is stable. If n_{eq} and $n_{1_{eq}}$ are the steady-state numbers of excited and non-excited oscillators, we will have

$$n_{eq} + n_{1_{eq}} = n_0 \quad \text{and} \tag{A.2.18}$$

$$Kn_{eq} - K_1 n_{1_{eq}} = 0 \quad ; \tag{A.2.19}$$

calling

$$\frac{K}{K_1} = \mu \quad , \tag{A.2.20}$$

we find

$$n_{eq} = \frac{\mu n_0}{\mu + 1} \quad , \tag{A.2.21}$$

$$n_{1_{eq}} = \frac{n_0}{\mu + 1} \quad . \tag{A.2.22}$$

For the cases of interest where $\mu > 1$, $n_{eq} > n_{1_{eq}}$, and we see that the steady-state numbers of excited versus "dormant" oscillators do not depend on the individual rate constants but only on their ratio μ. We also note that these steady-state populations depend on the overall number of oscillators in the system. So the maximum entropy of the system at the steady state will be

$$S_{max} \sim \log_2 \lambda_{eq} \quad , \qquad \text{where} \tag{A.2.23}$$

$$\lambda_{eq} = \frac{n_{eq}!}{\left(\frac{n_{eq}}{2}\right)! \left(\frac{n_{eq}}{2}\right)!} \quad . \tag{A.2.24}$$

We therefore deduce that a "mildly" intense noisy environment (of additive stationary noise) may indeed exercise "beneficial" effects (that is, act as a more or less "nutritious" environment where oscillators literally "feed on noise"), by increasing the redundancy of the system under the condition:

$$\frac{d}{dt} [\ln S(t)] \leqslant \frac{\dot{\lambda}}{\lambda} \quad . \tag{A.2.25}$$

It remains to say something about the origin of the necessary organizing process (or pacemaker) which will exploit the redundant degrees of freedom and lead the system to the condition (A.2.25). We are specifically referring to the case of cerebral organization. At present we think that such a pacemaker (the thalamocortical pacemaker) has indeed been detected in the brain (Sect.6.5.2).

A.3 Relevance of the Model to Neuronal and Cognitive Organization

In Sect.A.2 we proposed a model involving a switching "on"-"off" process in an ensemble of hard nonlinear oscillators asynchronously activated or quenched by additive noise (and subsequently constrained by the thalamocortical pacemaker). It is time now to examine the relevance of this model to the dynamical processes that take place at the neuronal level. Do switching "on" and "off" (synaptic and/or genetic) processes exist in cortical or thalamic centers, as adaptive responses to changing physiological and environmental conditions? Let us first try isomorphically to identify the fluctuations that take place at the level of the single neuron with the hard nonlinear oscillators of the model, or at least point out some analogies between them. The available experimental evidence [A.10] favors a scheme in which individual synapses act as unitary generators of wave activity and in which the aperiodic oscillations seen in intracellular recording already represent the outcome of summation of individual synaptic contributions. By their nature, the excitable postsynaptic portions of the neural membranes are unstable and, under certain conditions, there is a possibility of sustained oscillations.

The action of synaptic currents is to modify the resting potential. Based on estimates of the resting potential in long dendritic arborization of central neurons, the suggestion has been made [A.10] that these branches, on account of their high axial resistance and large surface area, may be at a substantially lower resting potential than the cell body of the same neuron. The lower resting potential perhaps predisposes these elements to oscillatory behavior, and synapses located on these branches might act either to induce or to quench ongoing oscillations. In this model, then, synapses may not be the prime source of wave activity but they play a regulatory role, insofar as synaptic action is needed either to excite or to quench dendritic potential waves.

In spite of the above suggestions, we are still reluctant to identify the hard nonlinear oscillators (of our model) excited or quenched by noise ("catalytic" action) with individual neuron's postsynaptic membrane potential fluctuations. We are rather inclined to consider these potential oscillations as more or less corresponding to the additive noise in our model, and search elsewhere *within the neuron* for the analogue of "the bistable hard nonlinear oscillator".

Cumulative experiments extensively reviewed by *Hydén* [A.11] seem to point the way [A.12]. Hydén and his group found that learning experiments with rats, for example, were consistently accompanied by a relative mRNA increase per neuron in the learning cortex of 60-100%. This indicates a stimulation of the neuron's genome, i.e., this mRNA response can be interpreted as reflecting an activation of hitherto silent gene areas in brain cells. We suggest below that such an activation of (generally repressed) genetic loops (switches) can be caused by ionic fluctuations penetrating the genome of the neurons, and may be due either to a direct external stimulation (e.g., microwave radiation [A.13]) or to coupling with the continuous noisy electrical wave activity of the postsynaptic membrane potentials (also triggered by hormones and neurotransmitters).

Indeed, let us examine briefly a single genetic "Jacob-Monod" loop or "genetic switch" (Fig.A.1). In this simplified sketch, messenger RNA is made in the nucleus.

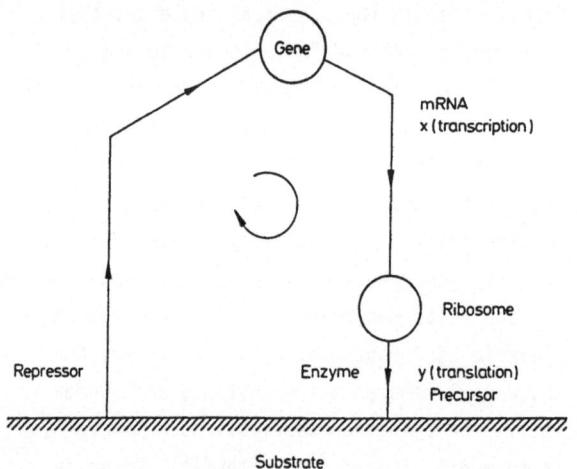

Fig.A.1. A genetic feedback loop x, y. The level of nucleic acids and the level of proteins are to be considered as two successive hierarchical levels

Its instantaneous concentration at any moment is x(t). At the ribosome the message is translated and enzyme molecules synthesized with corresponding instantaneous concentration y(t). This enzyme catalyses the reaction from an inactive precursor to a repressor molecule. The repressor molecule (protein) in turn reacts with the gene so that when a repressor is attached to the gene no more mRNA is made and the activity stops.

Goodwin [A.14] tried to show that under certain conditions the above feedback loop can be an oscillatory system with the instantaneous concentrations x and y fluctuating in a nonlinear fashion around their steady-state values. One especially restrictive assumption made by Goodwin was that the enzyme concentration y decays at a constant rate (independent of the concentration). Also Goodwin's original scheme did not take into account any time delay between x and y. Later *Goodwin* [A.15] modified his model equations to include a time lag and decay terms dependent on the concentrations of x and y, and, after some analogue computing experiments, he concluded that the system oscillates in a self-sustained way for a restricted range of parameter values. Nevertheless, *Griffith*, who critically examined this model later [A.16], came to the conclusion that undamped oscillations in x and y can never occur for *any* realistic values of the parameters. As a result, it appears that the simple genetic loop as depicted in Fig.A.1, when deterministically considered, cannot behave but as a switch —mostly repressed, i.e., in the "off" position.

More recently, however, *Tiwari* et al. [A.16] simulated the Jacob-Monod system with a digital computer, allowing inclusion of stochastic variations of the variables x, y. They found that the result of stochastic influences (noise) in the system was a maintained oscillation: "A single feedback repression loop does not constitute a precise 'clock' system, but a number of such loops operating independently, could constitute a reasonable precise clock", they concluded. It appears, therefore, that in the genetic control circuits, stochastic elements induced from outside are necessary prerequisites of excitation and production of continuous undamped oscillations —in an otherwise "dormant" hard switching element. In our theory, these stochastic elements can be identified with ionic fluctuations coming from the continuous post-synaptic membrane activity penetrating the genome and massively exciting a number of hitherto repressed genes. These subsequently oscillate regularly during the time interval Δt for which the stochastic influence has the proper characteristics.

From computer analysis of EEG patterns during learning processes, there is some evidence [A.18] relating EEG frequency distribution and overall power in the spectrum just after the presentation of learning material with the quality of later recall. In these experiments, more powerful, faster, and longer-lasting EEG activation after presentation of the learning material was systematically associated with increasingly better quality of later recall, and inversely very low levels of activation were associated with failure of storage. Therefore the level of EEG activation relates to efficiency of storage, low levels being incompatible with retrievable storage.

In accordance with our model we suggest that the EEG must indeed be particularly active during the period of memory consolidation in order to drive effectively as many loops in individual neurons as possible (simultaneously in many cerebral areas) from the dormant to the excited self-sustained oscillating state. Only through the resulting massive mRNA increase and the correlated production of new specific proteins —part of which thereafter move outside the cytoplasm and "coat" the corresponding dendritic membranes —is it possible to obtain an efficient consolidation of memory. The presence of the new proteins and the corresponding conformational changes at the postsynaptic positions modify, in turn, the degree of excitatory or inhibitory character of the synaptic sites involved. This is done by influencing the electrokinetic potential of the membrane (or zeta potential) and, therefore, the electrophoretic velocity at the individual synaptic clefts [A.19].

Thus there may exist a continuous feedback coupling between the postsynaptic noisy membrane potentials and the "underneath" genetic activity in the genome of the neuron(s). It is not known, of course, what types of genes are activated within given time intervals Δt_i. These intervals must be long enough (at least of the order of 100 minutes) to allow protein manufacture and memory consolidation.

We suggest that continuity in cognition requires a rather strong similarity between the types of individual proteins sequentially coating the dendritic and somatic postsynaptic areas; otherwise, "gaps" or "cognition failure" may occur. As a matter of fact, "cognition" manifested at hierarchical levels above the neuronal one arises as a matching process between at least two spatial-temporal random processes: one associated with a set of incoming triggers (a sequence of pulses from the peripheral nervous system or some part of the central nervous system, e.g., in the form of neurotransmitter releases), and the other, a prestored spatial-temporal random process correlative with a set of postsynaptic membrane potentials, or a spatial-temporal pattern of postsynaptic receptors.

The coupling or cross-correlation between the above two random processes takes place via energy exchanges between homologous frequency (ω_i, ω_j) pairs shared by both spectra and is estimated by the coherence function

$$\text{coh}(\omega) = \frac{\Phi_{ij}(\omega)}{\sqrt{\Phi_i(\omega)\Phi_j(\omega)}} \quad , \tag{A.3.1}$$

where $\Phi_{ij}(\omega)$ is the cross spectrum and Φ_i, Φ_j are the autospectra of the corresponding processes. The result of this cross-coupling in phase and amplitude determines the "degree of cognition" between the *incoming* and the *preset* or the "unknown" and the "expected" patterns.

A typical case of cognitive failure is represented in schizophrenia where preliminary EEG analyses [A.20] showed no conclusive differentiation in EEG types of activity between normal and schizophrenic persons. We would like tentatively to put forward here the hypothesis that cognitive failure may be due to the excitation of types of genes never activated before in the neuronal genome or "infantile" genes, i.e., genes once coded for synthesis of the foetal components that have been since then "quenched" permanently. The activation of such types of otherwise normal genes, functioning nevertheless in abnormal patterns, results in the production and coating of the dendritic neuronal surfaces with corresponding "archaic" types of proteins.

These types of proteins working "out of joint" in time may block and brake the normal continuation of the cognitive process. It has been recently conjectured [A.21] that perhaps improper activation (via, e.g., hormonal action) of foetal genes in adult organisms may cause epigenetic cancer in dividing cells. Epigenetic cancer can, of course, be considered as a concomitant of a cognitive failure at some organic hierarchical level(s). The prospects of such an analogy between mechanisms of epigenetic cancer and schizophrenia seem tempting for further research.

Finally, let us point out that our hypothesis about the nature of schizophrenia forwarded above by no means contradicts results deduced from accumulated experimental findings about the tranquilizing effect exercised by phenothiazines, butyrophenone, and other "antischizophrenic" drugs. These drugs presumably block dopamine postsynaptic receptor sites. (This neurotransmitter is subsequently taken up by the presynaptic ending which initially released it or breaks down into the synaptic cleft.) One could therefore naturally speculate that the antischizophrenic action of these drugs is mediated via a blockade of dopamine receptor sites in the brain. The dopamine receptor sites blocked as above may not, however, be directly related to schizophrenic abnormality, but may represent only an indirect effect "several steps removed from the primary cause of the disease" [A.21]. This deduction comes from the fact that these drugs do not cure schizophrenia but only facilitate remission of the disease. By decreasing the activity of the pertinent neurotransmitter (dopamine) and thereby decoupling sensory perceptions from the "internal" set of postsynaptic membrane receptors, the drugs, we believe, simply block the formation of "distorted" cognitive patterns, i.e., the formation of corresponding cross-correlation. The primary reason for the distortion of these patterns according to our model is the existence of genetically produced "improper proteins" coating the receptors' postsynaptic sites.

It is up to experimentalists to clarify the issue by showing or disproving the existence and the amount of such hypothesized types of "infantile" proteins, e.g., in the serum of schizophrenics. Such proteins might also manifest antigenic properties.

B. On the Difficulty of Treating the Transaction Between Two Hierarchical Levels with Continuous Nonlinear Dynamics

In Sect.4.7 we gave —in a specific example —a detailed account of the communication process between *two* hierarchical systems modeled by controlled Markov chains. In the present appendix, we wish to justify the *partitioning* of the state space at each hierarchical level into discrete elements by showing the intrinsic difficulties associated with a treatment based on continuous dynamics. Let us adopt, then, a continuous state-space description involving a number of dynamical variables and a set of parameters pertaining to the particular hierarchical level.

B.1 The Level Q of Partner I

The dynamics at level Q are described by the Itô stochastic differential equation

$$dq_1 = \varphi_1(q_1; v_1; t)dt + \Gamma_1(q_1; v_1; t)d\mu_1 \quad , \tag{B.1.1}$$

where

q_1 is the state vector for the level Q,

t is the time,

v_1 is the control vector to be specified later,

$\mu_1(t)$ is a Wiener-Lévy process with autocorrelation

$$E\left[\mu_1^T(t_1)\mu_1(t_2)\right] = M \min(t_1, t_2)$$

where M is a positive definite covariance matrix,

$\varphi_1(q_1; v_1; t)$ is a vector function describing the dynamical structure of level Q parametrized on v_1, which incorporates all the underlying game parameters,

$\Gamma_1(q_1; v_1; t)$ is a matrix function parametrized as above.

The evolution of the joint probability density function (p.d.f.) of the state vector components is described by the Fokker-Planck-Kolmogorov equation.

Similarly for the Q' level of partner II we have the Itô stochastic differential equation

$$d\mathbf{q}_2 = \varphi_2(\mathbf{q}_2; \mathbf{v}_2; t)dt + \Gamma_2(\mathbf{q}_2; \mathbf{v}_2; t)d\mu_2 \quad , \tag{B.1.2}$$

where $\mathbf{q}_2$ is the state vector for the level Q'.

B.2 Homeostasis and Cross-Correlations

In the phase space of level Q (or Q') we single out a region H_1 (or H_2) characterized as homeostatic: it is the region within which the activities at the level Q
(Q') should evolve preferentially but not exclusively in order to achieve the intrinsic regulation of the system at the level concerned. However, in order to ensure adaptability to the partner, the organism should experience high cross-correlations between the incoming set of triggers $\mathbf{s}_2(t)$ induced by the higher level W'
of the partner and the state vector $\mathbf{q}_1(t)$. For any given set of triggers emanating
from the partner, the above requirement of high cross-correlation implies a rather
extensive use of the trajectory at the level Q (Q'). However, such an extensive
"wandering" in the state space jeopardizes the requirement for homeostasis.

In general, therefore, the two basic deliberations for each partner, namely the
homeostatic necessity and the tendency to continuous adaptability to the partner's
cues, seem to be in conflict. The higher levels W (W') act as stochastic controllers
aiming for a compromise in the above conflict.

Let $u_1(t)$ be the probability that at time t the state $\mathbf{q}_1(t)$ lies within the homeostatic region H_1, i.e.,

$$u_1(t) = P(\mathbf{q}_1(t) \in H_1) = \int_{H_1} P(\mathbf{q}_1; t)d\mathbf{q}_1 \quad , \tag{B.2.1}$$

where $P(\mathbf{q}_1; t)$ is the joint p.d.f. of the components of the state vector $\mathbf{q}_1(t)$.

The cross-correlation between signals from the level W' and the state of level
Q is given by

$$r(t,\tau) = E\left[\mathbf{s}_2^T(t + \tau)\mathbf{q}_1(t)\right] \quad , \tag{B.2.2}$$

where T denotes the transpose of a vector and τ is a delay parameter, fixed throughout the interaction of the two systems, e.g., $\tau = 0$. Note that the vectors $\mathbf{s}_2$ and
$\mathbf{q}_1$ have the same dimensionality. What are afferently reported from the level Q to
the level W are time averages, approximating the above. In other words, let T be
the duration of the time window

$$I = \{t': t - T \leqslant t' < t\} \quad ; \qquad \text{then}$$

homeostasis $u(t)$ = percentage of time that $q_1(t') \in H_1$

(homeostatic region) for $t' \in I$.

Cross-correlation $r(t,\tau) = \int_{t-\tau-T}^{t-\tau} s_2^T(\lambda + \tau) q_1(\lambda) d\lambda$,

where we select either $\tau = 0$, or τ such that $E[r(t,\tau)]$ is maximum.

B.3 The Level W of Partner I

The dynamics at level W are described by the following stochastic differential equations:

$$dw_1 = f_1(w_1;u,r)dt + G_1(w_1;u,r)d\xi_1 \quad , \tag{B.3.1}$$

$$s_1 = h_1(w_1;u,r) \quad , \qquad \text{where} \tag{B.3.2}$$

$w_1(t)$ is the state vector at level W,

$\xi_1(t)$ is a Wiener-Lévy process accounting for the intrinsic spontaneity of the system at the level W,

$s_1(t)$ is the observable behavior which acts as a trigger for the partner.

B.4 The Controller

The feedforward control mechanism for each system has as inputs the homeostasis and correlation afferent signals u and r, as well as the state vector w_1 of the higher level W. The state equations for the controller are thus

$$\dot{v}_1 = g_1(v_1, w_1, u, r) \quad ,$$

where g_1 is a vector function to be specified so that the objective function (joint figure of merit) F is maximized. Here F is defined as follows:

$$\mathcal{F} = \mathcal{F}_1 \mathcal{F}_2 \quad \text{(multiplication criterion)} \quad ,$$

where

$$\mathcal{F}_1 = \lambda_1 E[u_1] + (1 - \lambda_1)E[r_1] \quad ,$$

$$\mathcal{F}_2 = \lambda_2 E[u_2] + (1 - \lambda_2)E[r_2] \quad ,$$

for $\lambda_1, \lambda_2 \in (0,1)$, or

$$\mathcal{F} = \sigma_1 \mathcal{F}_1 + \sigma_2 \mathcal{F}_2 \quad \text{(additive criterion)}.$$

The characteristic of a multiplicative figure of merit is that it strongly favors *parity* between the partners, while the additive criterion puts under the same denominator the equal partners as well as the fully dominating partner.

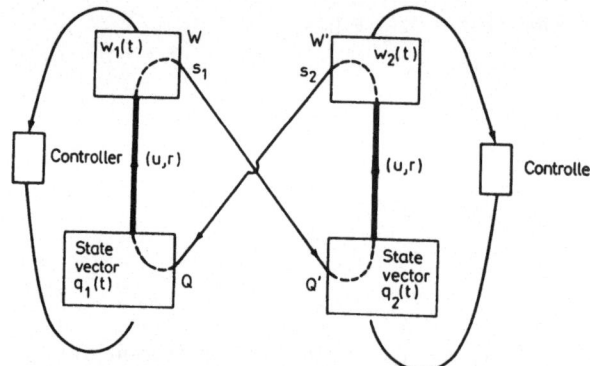

Fig.B.1. State variables model
for two communicating hierarchical systems

For reasons of simplicity, in the work reported in Sect.4.7 we quantized the
state space at each hierarchical level and admitted only a finite number of states
which stand for the salient subtraits at the level involved. Thus the stochastic
nonlinear differential equations corresponding to the continous state description
have now been replaced at all hierarchical levels by discrete-time Markov chains
characterized by appropriate transition matrices P_{ij} fully describing the transitions between the possible states of the systems at the levels involved (Fig.B.1).

C. Noisy Entrainment of a Slightly Nonlinear Relaxation Oscillator by an External Harmonic Excitation

C.1 General Description of the Model

The most characteristic parameter of a self-sustained (nonlinear) dissipative oscillation is the amount of energy exchange between the oscillating system and the
environment. If the interaction between the oscillator and the external stimulus is
small (as in a "pendulumlike" oscillator), in each period of the oscillation the
energy content can change only by a small amount and the total change lasts many
periods. On the other hand, if the interaction is large (as in a relaxation oscillator) the total change of the energy content occurs within a few periods.

In this appendix we will adopt as a model of self-sustained oscillations (simulating, for example, a biological rhythm) the simple van der Pol oscillator

$$\frac{d^2x}{dt^2} - \omega_0\varepsilon(1 - \beta x^2)\frac{dx}{dt} + \omega_0^2 x = \omega_0^2 E \cos\omega t \quad , \tag{C.1.1}$$

driven by a harmonic *zeitgeber* of (constant) amplitude E and frequency $\omega \neq 0$ (we
have $\Delta\omega = \omega - \omega_0$). The coefficient ε indicates the amount of energy exchange. For
$\varepsilon \ll 1$ the oscillation belongs to the "pendulum type", and for $\varepsilon \gg 1$ the oscillation
belongs to the "relaxation type".

In pure pendulum oscillations, the frequency ω_0 is independent of all external influences, but the range of oscillation varies with varying environmental conditions. In pure relaxation oscillations, on the other hand, the range of oscillation is independent of all external influences while the frequency changes by a large amount with varying external conditions. From the known properties of biological rhythms, we may conclude that self-sustained biological clocks must occupy an intermediate position between the two extreme types of "pendulum" and "relaxation" oscillation.

In what follows we will take $\varepsilon = 0.1$ and $\beta = 0.1$.

C.2 A Method for the Study of Entrainment

For a study of entrainment processes in nonlinear oscillators subjected to an external harmonic excitation, we will introduce in this discussion a method based on the study of trajectories in the $d\varphi/dt$, φ plane, where $\varphi(t)$ is the instantaneous phase difference between the *zeitgeber* and the oscillator. This method affords an abstraction and concentrates on the main features of the problem under consideration.

Merely by inspection of these trajectories in the $\dot{\varphi}$, φ plane, we can immediately recognize the achievement of entrainment (strict or "loose") or the lack thereof. Out of the variety of trajectories of that sort, we can, in general, single out four characteristic types.

C.2.1 Strict Entrainment

This can also be seen as suppression of autoperiodic oscillation (Fig.C.1a). In this case, starting from an arbitrary initial condition $\dot{\varphi}(0)$, $\varphi(0)$, the oscillator eventually synchronizes with the excitation and a constant phase difference Φ_0 is established. In a nonlinear dissipative system, this forced "harmonization" of anharmonic oscillations amounts to "quenching" and can be achieved for rather high values of E; this case is not of physical interest here.

C.2.2 Loose or "Jittery" Entrainment

Here from a given initial condition the phase difference never becomes time independent but rather remains bounded, and a limit cycle is reached in the $(\dot{\varphi},\varphi)$ plane (Fig.C.1b). This limit cycle corresponds to a loose entrainment, which means that the oscillator, although locked in at the frequency ω, becomes phase modulated or "jittery". This effect has been systematically observed in many biological rhythms (including the circadian clocks) under the influence of the natural environment.

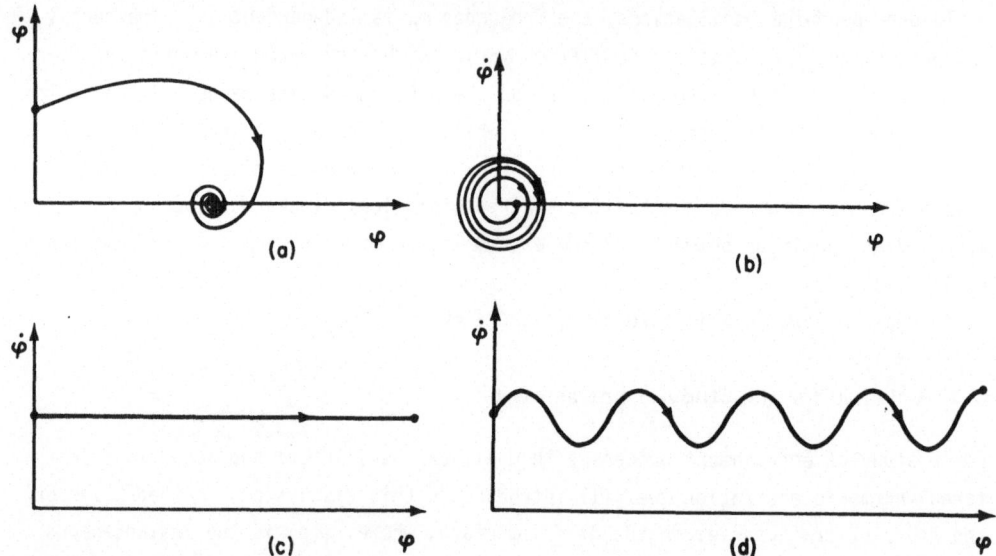

Fig.C.1a-d. Typical trajectories in the $(\varphi,\dot\varphi)$ plane. **a** strict entrainment; **b** loose ("jittery") entrainment; **c** pure free-running oscillation; **d** jittery free-running oscillation

C.2.3 Pure "Free-Running" Oscillation

Here the oscillator continues running at its eigenfrequency. Entrainment is never reached and the phase difference varies linearly with time (Fig.C.1c).

C.2.4 Free-Running Oscillation

In this case entrainment is never achieved, the phase difference changes secularly with time, and the instantaneous frequency offset vibrates about a constant mean value (Fig.C.1d).

C.3 Mathematical Treatment and Computer Simulation

We now transform (C.1.1) in order to determine the differential equations of motion in the $(\dot\varphi,\varphi)$ plane. We put the solution of (C.1.1) in the form

$$x(t) = A(t)\sin[\omega t - \varphi(t)] = A(t)\sin\psi(t) \quad , \tag{C.3.1}$$

where $\psi(t) = \omega t - \varphi(t)$. This represents a simultaneously amplitude- and phase-modulated wave form. Here $\varphi(t)$ is the instantaneous phase difference between the *zeitgeber* and the oscillator, and $A(t)$ is the time-varying amplitude of the oscillation. We have

$$\frac{dx}{dt} = \frac{dA}{dt}\sin\psi + A\left(\omega - \frac{d\varphi}{dt}\right)\cos\psi \tag{C.3.2}$$

and

$$\frac{d^2x}{dt^2} = \left[\frac{d^2A}{dt^2} + 2\omega A \frac{d\varphi}{dt} - A\omega^2 - A\left(\frac{d\varphi}{dt}\right)^2\right]\sin\psi$$

$$+ \left(2\omega \frac{dA}{dt} - 2 \frac{dA}{dt}\frac{d\varphi}{dt} - A \frac{d^2\varphi}{dt^2}\right)\cos\psi \quad . \tag{C.3.3}$$

Substituting into (C.1.1) we obtain

$$M(t)\sin\psi + N(t)\cos\psi = 0 \quad , \tag{C.3.4}$$

where

$$M(t) = \frac{d^2A}{dt^2} - \varepsilon f \frac{dA}{dt} + (\omega_0^2 - \omega^2)A + 2\omega A \frac{d\varphi}{dt} - A\left(\frac{d\varphi}{dt}\right)^2 + \omega_0^2 E \sin\varphi \quad , \tag{C.3.5}$$

$$N(t) = - A \frac{d^2\varphi}{dt^2} - 2 \frac{dA}{dt}\frac{d\varphi}{dt} + \varepsilon f A \frac{d\varphi}{dt} + 2\omega \frac{dA}{dt} - \varepsilon f \omega A - \omega_0^2 E \cos\varphi \quad , \tag{C.3.6}$$

and

$$f = (1 - \beta x^2)\omega_0 = \omega_0\{1 - \beta A^2 \sin^2[\omega t - \varphi(t)]\} \quad . \tag{C.3.7}$$

We determine $A(t)$ and $\varphi(t)$ so that the quadratic components vanish simultaneously, i.e., $M(t) = 0$ and $N(t) = 0$, in which case (C.3.4) is fulfilled.

The coupled nonlinear differential equations for $A(t)$ and $\varphi(t)$ are therefore

$$\frac{d^2A}{dt^2} - \varepsilon f \frac{dA}{dt} + (\omega_0^2 - \omega^2)A + 2\omega A \frac{d\varphi}{dt} - A\left(\frac{d\varphi}{dt}\right)^2 = -\omega_0^2 E \sin\varphi \quad , \tag{C.3.8}$$

$$- A \frac{d^2\varphi}{dt^2} - 2 \frac{dA}{dt}\frac{d\varphi}{dt} + \varepsilon f A \frac{d\varphi}{dt} + 2\omega \frac{dA}{dt} - \varepsilon f \omega A = \omega_0^2 E \cos\varphi \quad . \tag{C.3.9}$$

This is a four-dimensional time-varying nonlinear system which may be simulated on an analogue or digital computer. Before we proceed with the simulation, we write the system equations in canonical form. We choose the following state variables:

$$q_1(t) = \varphi(t) = \text{phase error} \quad ,$$

$$q_2(t) = \frac{d\varphi(t)}{dt} = \dot{\varphi}(t) = \text{instantaneous angular frequency offset} \quad ,$$

$$q_3(t) = A(t) = \text{amplitude} \quad ,$$

$$q_4(t) = \frac{dA(t)}{dt} = \text{rate of change of the amplitude} \quad .$$

The canonical form of the equations is found to be

$$\frac{dq_1}{dt} = q_2 \quad , \tag{C.3.10}$$

$$\frac{dq_2}{dt} = \left(\frac{2q_4}{q_3} - \varepsilon f\right)(\omega - q_2) - \omega_0^2 \frac{E \cos q_1}{q_3} \quad , \tag{C.3.11}$$

$$\frac{dq_3}{dt} = q_4 \quad , \tag{C.3.12}$$

$$\frac{dq_4}{dt} = \epsilon f q_4 + q_3 \left[(\omega - q_2)^2 - \omega_0^2 \right] - \omega_0^2 E \sin q_1 \quad , \tag{C.3.13}$$

where

$$f = \omega_0 \left[1 - \beta q_3^2 \sin^2 (\omega t - q_1) \right] \quad . \tag{C.3.14}$$

The only explicitly time-varying term in this system is f. The variation of f is at an angular frequency 2ω when phase-locking occurs, i.e., when $q_1(t)$ is bounded. Explicitly,

$$f = \omega_0 \left(1 - \frac{1}{2} \beta q_3^2 \right) + \frac{1}{2} \omega_0 \beta q_3^2 \cos(2\omega t - 2q_1) \quad . \tag{C.3.15}$$

The state equations are thus of the nonautonomous form

$$\dot{q} = y(q; t) \quad , \tag{C.3.16}$$

where q is the state vector

$$q = (q_1, q_2, q_3, q_4)^T \tag{C.3.17}$$

and y is a nonlinear vector function,

$$y(q; t) = [Y_1(q_1, q_2, q_3, q_4; t), \quad Y_2(q_1, q_2, q_3, q_4; t) \quad ,$$
$$Y_3(q_1, q_2, q_3, q_4; t), \quad Y_4(q_1, q_2, q_3, q_4; t)]^T \quad . \tag{C.3.18}$$

If the nonautonomous system (C.3.16) starts at some initial state $q(0)$, it traverses a trajectory in the four-dimensional state space. Here we are merely interested in the projection of this trajectory on the (q_1, q_2), i.e., $(\varphi, \dot{\varphi})$, plane. This is because the most direct method for the study of entrainment of frequency is to concentrate on the behavior of the *phase error*. The amplitude variations are of secondary importance in such a study. The elimination of state variables $q_3 = A$ and $q_4 = \dot{A}$ cannot be performed analytically in the differential equations (C.3.10-13) in order to get directly the differential equations of motion of the projection in the $(\varphi, \dot{\varphi})$ plane. Equations (C.3.10-13) could be viewed as a parametric representation of this motion. In the digital computer simulation of this system, starting at some initial state $q(0)$, we concentrate on the corresponding values of $\varphi(t)$ and $\dot{\varphi}(t)$ at the same instant t. For the numerical integration with a digital computer of the system (C.3.10-13), we use a Runge-Kutta method.

C.4 Behavior of the Oscillator Under an Applied Harmonic Excitation (Entrainment)

We assume that at time $t = 0$ an external electromagnetic excitation field is applied to the van der Pol oscillator. The normalized magnitude of the field is E, and its angular frequency ω is taken to be different from the angular eigenfrequency

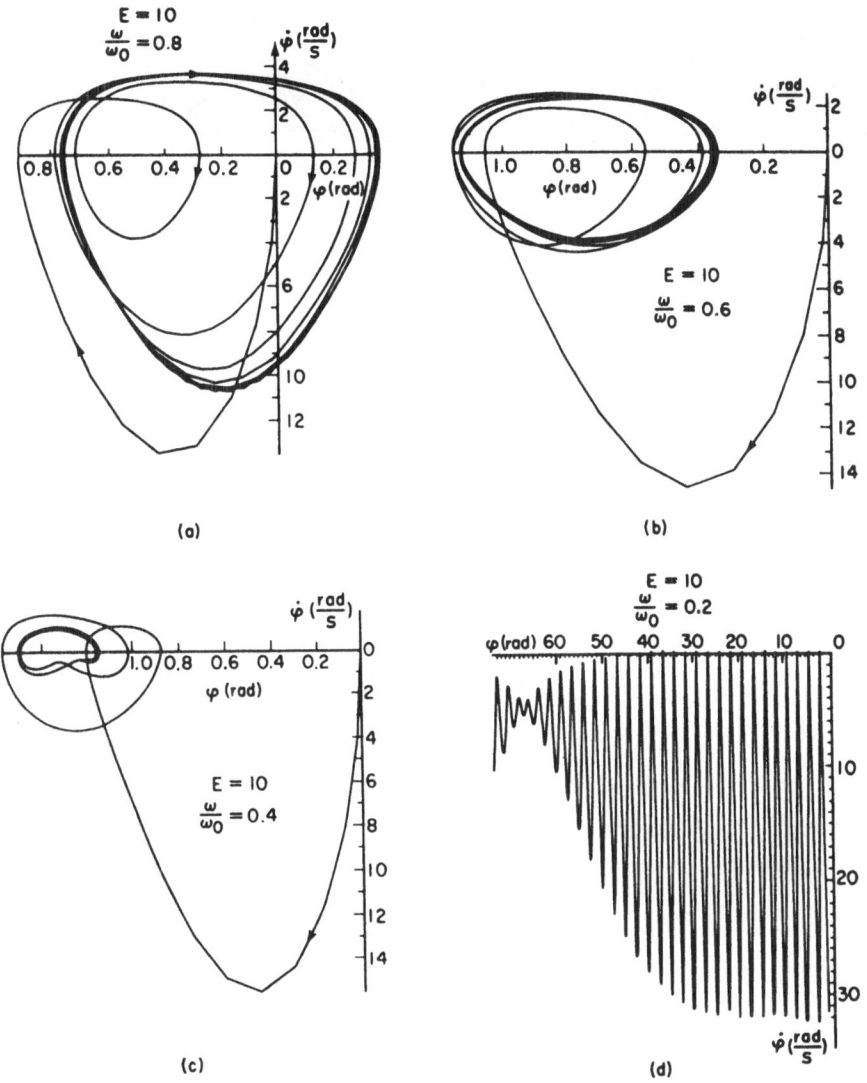

Fig.C.2a-d. $(\varphi,\dot{\varphi})$ trajectories for $E = 10$, $\omega/\omega_0 < 1$. In **a**, **b**, **c**, **d**, ω/ω_0 is 0.8, 0.6, 0.4, and 0.2, respectively

ω_0 of the oscillator. Thus we have a suddenly applied input mistuning $\Delta\omega = \omega - \omega_0$, or a normalized input mistuning $\gamma = (\omega/\omega_0) - 1$.

In the computer simulation, we took two different values of the normalized field amplitude, namely $E = 10$ and $E = 1$, and 23 values of the angular frequency ω for each E, i.e., $\omega/\omega_0 = 0.1$ to 2 with steps of 0.1, and then 2.5, 3, and 10, or normalized input mistuning $\gamma = 0$, ± 0.1, ± 0.2, ..., ± 0.9, 1, 1.5, 2.9. We thus had 46 program runs. From the computer plotter we obtained the corresponding $(\varphi,\dot{\varphi})$-plane trajectories and graphs $x(t)$. We display here only a selected group of the drawings.

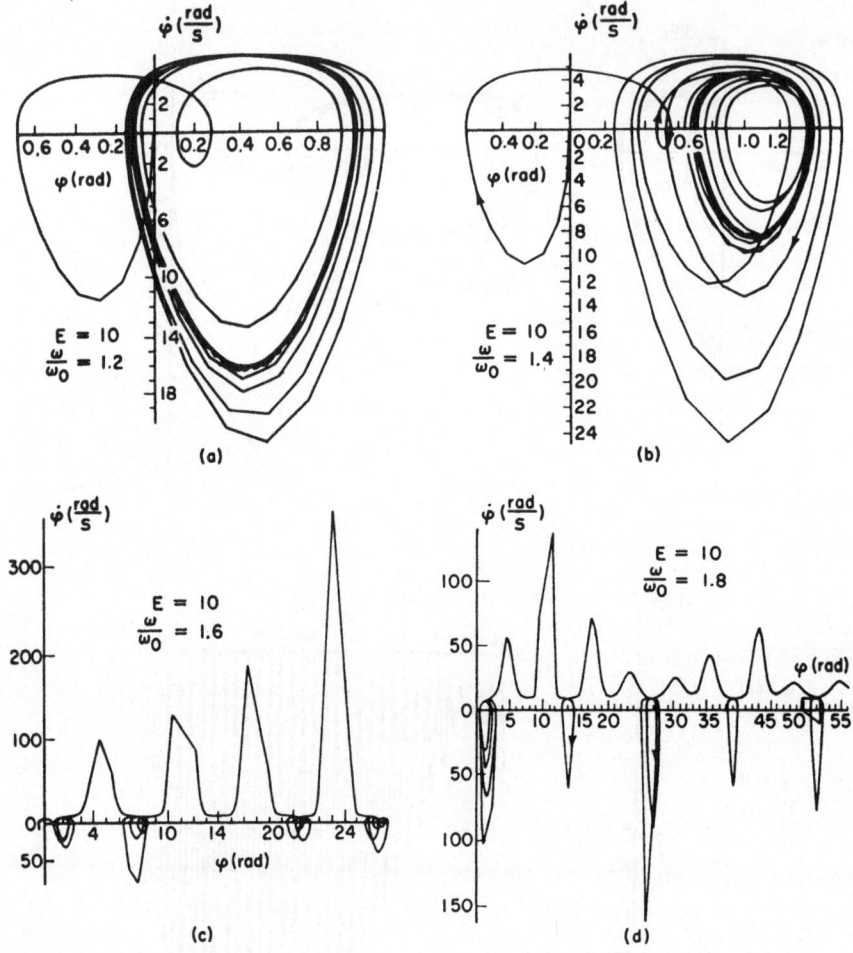

Fig.C.3a-d. $(\varphi, \dot{\varphi})$ trajectories for $E = 10$, $1 < \omega/\omega_0 < 2$, In a, b, c, d, ω/ω_0 is 1.2, 1.4, 1.6, and 1.8, respectively

Figures C.2-4 correspond to the case $E = 10$.[1] Figure C.2 depicts the $(\varphi, \dot{\varphi})$ trajectories for negative input mistuning, i.e., for $\omega/\omega_0 < 1$. We observe that for $0.4 < \omega/\omega_0 < 1$, entrainment takes place, since the trajectories tend to limit cycles. This is a loose kind of entrainment, i.e., the frequency is locked but there is a phase jitter (phase modulation). The phase error φ remains bounded, e.g., for $\omega/\omega_0 = 0.6$ the phase error is $-1.2 < \varphi < -0.35$. For $\omega/\omega_0 < 0.2$ we observe free-running, i.e., entrainment breaks.

Thus we observe that in our case of $\varepsilon \ll 1$ (quasi-pendulum oscillations), there is a maximum value $0 < \gamma_e < 1$ (where γ_e is a function of E, β and ω_0) such that for $-\gamma_e < \gamma < 0$ or $(1 - \gamma_e)\omega_0 < \omega < \omega_0$, entrainment takes place, whereas for $\gamma < -\gamma_e$, i.e.,

---

1 For $E = 1$ similar results hold, the difference being that as E decreases, the pulling region shrinks.

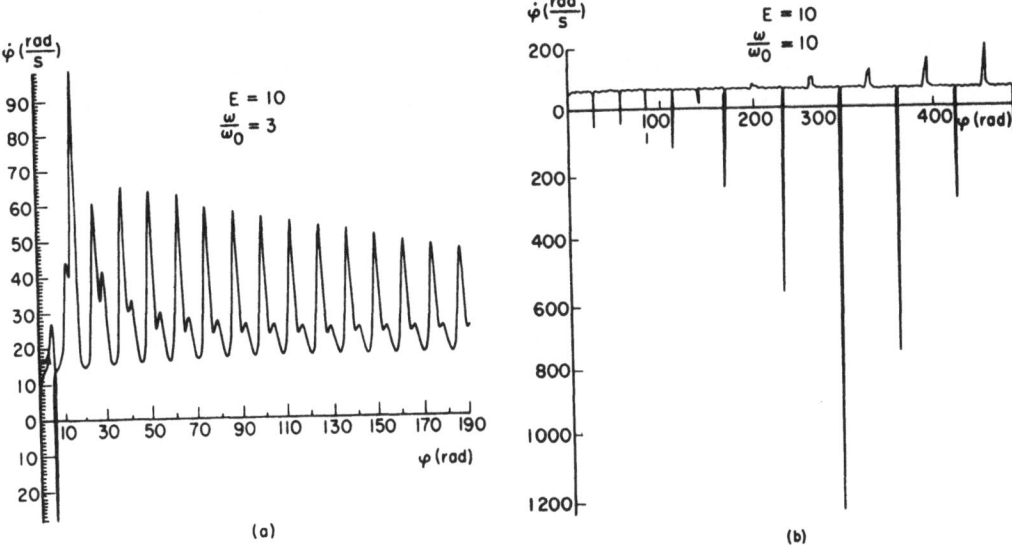

Fig.C.4a,b. $(\varphi,\dot{\varphi})$ trajectories for $E = 10$, $\omega/\omega_0 \geqslant 3$. In **a** and **b** ω/ω_0 is 3 and 10, respectively

$\omega < (1 - \gamma_e)\omega_0$, free-running occurs. [We point out, however, that for values $\varepsilon \gg 1$ (transition from pendulum to relaxation oscillations), one could also observe subharmonic lock-in.]

In Figs.C.3,4 we show the phase diagrams for $1 < \omega/\omega_0 < 2$ and $\omega/\omega_0 \geqslant 3$, respectively. For $1 = \omega/\omega_0 \leqslant 1.4$, entrainment takes place (Figs.C.3a,b), whereas for $\omega/\omega_0 < 1.6$, free-running is established. There is an upper limit γ_u such that for $0 < \gamma < \gamma_u$, entrainment takes place, whereas for $\gamma > \gamma_u$, free-running is established.

Comparing Figs.C.2 and C.3 for opposite values of γ, we observe asymmetry as far as the shape, dimensions, and limit cycle are concerned. The entrainment region

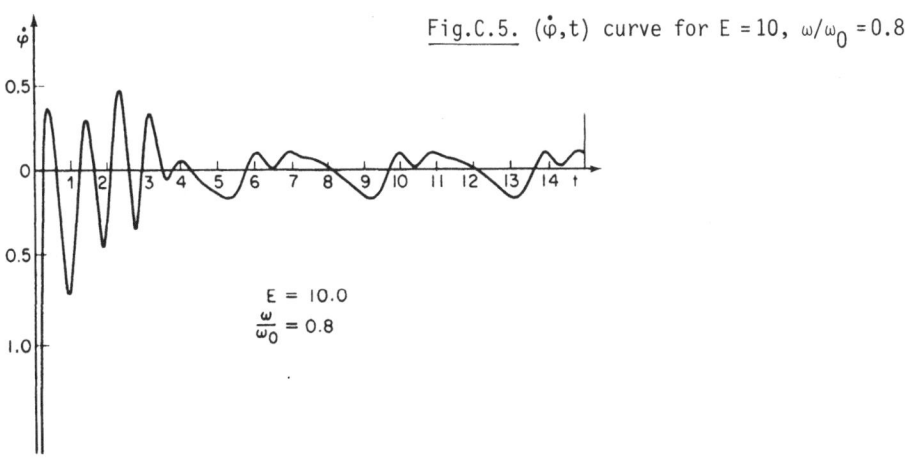

Fig.C.5. $(\dot{\varphi},t)$ curve for $E = 10$, $\omega/\omega_0 = 0.8$

$-\gamma_e < \gamma < \gamma_u$ is *not* symmetrical around zero ($\gamma_e > \gamma_u$). An almost "pure" form of free-running is observed as the mistuning ω/ω_0 increases (Fig.C.4).

Finally, the behavior $\dot{\varphi} = f(t)$ was calculated in order to estimate the locking time. From Fig.C.5 we observe that this "entrainment time" is of the order of a few autoperiods $T_0 = 2\pi/\omega_0$ for $E = 10$ and $\omega/\omega_0 = 0.8$.

References*

Chapter 2

2.1 J. Narlikar: The Structure of the Universe (Oxford University Press, Oxford 1977)
2.2 J. Dyson: Sci. Am. 226, 184 (September 1971)
2.3 H. Haken: Synergetics: An Introduction, 3rd ed., Springer Ser.Synergetics, Vol.1 (Springer, Berlin, Heidelberg 1983)
2.4 R. Rosen: Bull. Math. Biol. 38, 87 (1976)
2.5 W. Hahn: Theory and Application of Lyapounov's Direct Method (Prentice-Hall, Englewood Cliffs, NJ 1963)
2.6 N.N. Bogoliubov, Y.A. Mitropolski: Asymptotic Methods in the Theory of Non-Linear Oscillators (Mir, Moscow 1966)
2.7 C. Bennet: Int. J. Theor. Phys. 21, 905 (1982)
2.8 P.C.W. Davies: The Physics of Time Asymmetry (University of California Press, Berkeley, CA 1977)
2.9 R.E. McMurtrie: J. Theor. Biol. 50, 1 (1975)
2.10 G. Chaitin: Sci. Am. 230, 47 (May 1975)

Chapter 3

3.1 J.A. Wheeler, R. Feynman: Rev. Mod. Phys. 17, 157 (1945)
3.2 L. Brekhovskikh: Waves in Layered Media (Academic, New York 1960)
3.3 D. Gabor: "Light and Information", in Progress in Optics 1, 503 (North Holland, Amsterdam 1961)
3.4 D. Gabor: Philos. Mag. 41, 1161 (1950)
3.5 G.T. di Francia: Opt. Acta 2, 5 (1955)

Chapter 4

4.1 R.W. Hamming: Coding and Information Theory (Prentice-Hall, Englewood Cliffs, NY 1980)
4.2 P. Beckmann: Probability in Communication Engineering (Harcourt Brace and World, New York 1967)
4.3 G.D. Forney: Concatenated Codes (MIT Press, Cambridge, MA 1966)

---

*We provide only - with due apologies for omitting perhaps important material - references which have a direct bearing on the author's mind in the design of these lectures.

4.4   D.A. Huffman: Proc. IRE 40, 1098 (1952)
4.5   L. Gatlin: Information Theory and the Living System (Columbia University Press, New York 1972)
4.6   V.I. Siforov: "On the Theory of Radio Communication Channels with Randlomly Varying Parameters", in Proc. 13th Int. URSI Conf., ed. by S. Silver (North-Holland, Amsterdam 1963) pp. 164-194
4.7   G.T. di Francia: Opt. Acta 2, 5 (1955)
4.8   D.S. Lebedev, L.B. Levitin: Inf. Control 9, 1 (1966)
4.9   H.A. Johnson, K.D. Knudsen: Nature (London) 206, 930 (1965)
4.10 B.R. Gaines: "On the Complexity of Causal Models", IEEE Trans. SMC-6, 56 (1976)
4.11 J.S. Nicolis, M. Benrubi: J. Theor. Biol. 59, 76 (1976)
4.12 E.S. Gershon, R.H. Belmaker, S.S. Kety, M. Rosenbaum: The Impact of Biology in Modern Psychiatry (Plenum, New York 1977)
4.13 S.Wolf, B.B.Berle: The Biology of the Schizophrenic Process (Plenum, New York 1976)
4.14 O.W. Hill: Modern Trends in Psychosomatic Medicine, Vol.3 (Butterworth, London 1976)
4.15 P. Watzlawick, J.H. Beavin, D.D. Jackson: Pragmatics of Human Communication (Norton, New York 1967)
4.16 G. Bateson, D.D. Jackson, J. Haley, J. Weakland : Behav. Sci.1, 196 (1956)
4.17 C.E. Sluzki, D.C. Ransom: Double Bind: The Foundation of Communicational Approach to the Family (Grune & Stratton, New York 1976)
4.18 J.S. Nicolis, E.N. Protonotarios, I. Voulodemou: "Control Markov Chain Models for Biological Hierarchies", in Applied General Systems Research. Recent Developments and Trends (Plenum, New York 1977); also in J. Theor. Biol. 68, 563 (1977)
4.19 M. Verzeano: "Pacemakers, Synchronization, and Epilepsy", in Synchronization of EEG Activity in Epilepsies, ed. by H. Petsche, M.A. B. Brazier (Springer, Berlin, Heidelberg 1972) p. 154
4.20 M.E. Scheibel, D.B. Scheibel: "Patterns of Organization in Specific and Non-Specific Thalamic Fields", in The Thalamus, ed. by D.P. Purpura, M.D. Yahr (Academic, New York 1966)
4.21 M. Verzeano et al.: "Evoked Responses and Network Dynamics", in The Natural Control of Behavior, ed. by R.E. Whalen, R.F. Thoson, M.Verzano, S. Weinberger (Academic, New York 1970)
4.22 G. Moruzzi, H.W. Magoun: Clin. Neurophysiol. 1, 455-458 (1949)
4.23 W. Kilmer, W.S. McCulloch: "The Reticular Formation Command and Control System", in Information Processing in the Nervous System, ed.by G. Leibovic (Springer, Berlin, Heidelberg 1972)
4.24 M.E. Scheibel, D.B. Scheibel: "Anatomical Basis of Attention Mechanisms in Vertebrate Brains", in The Neurosciences, A Study Program, ed. by G. Quatron et al. (Rockefeller University Press, New York 1967)
4.25 A. Turing: Proc. Roy. Soc. London, Ser. B 237, 37 (1952)
4.26 G. Nicolis, I. Prigogine: Self-Organization in Non-Equilibrium Systems (Wiley, New York 1977)
4.27 H. Haken: Synergetics: An Introduction, 3rd ed., Springer Ser. Synergetics, Vol.1 (Springer, Berlin, Heidelberg 1983)
4.28 J.S. Nicolis, E.N. Protonotarios: Int. J. Bio-Med. Comput. 10, 417 (1979)

Chapter 5

5.1   A. Rapoport: Two-Person Game Theory, the Essential Ideas (University of Michigan Press, Ann Arbor 1966)
5.2   J.S. Nicolis, J.M. Argitis, D. Carabalis: Kybernetes 12, 9-20 (1983)
5.3   H. Bremermann: J. Theor. Biol. 87, 671 (1980)
5.4   J. Maynard-Smith: J. Theor. Biol. 47, 209 (1974)

5.5   J. Maynard-Smith, G.R. Price: Nature (London) 246, 15 (1973)
5.6   A. Rapoport: General Systems Yearbook 20, 49 (1975)
5.7   J.C. Frauenthal: Mathematical Modeling in Epidemiology (Springer, Berlin, Heidelberg 1980)

Chapter 6

6.1   A.N. Kolmogorov: "General Theory of Dynamical Systems and Classical Mechanisms", in Proc. 1954 Intern. Cong. Mathematics (North-Holland, Amsterdam 1957) pp. 315-333
6.2   E. Fermi, I.R. Pasta, S.M. Ulam: Los Alamos National Laboratory Report No. LA-1940 (1955)
6.3   M. Henon, C. Heiles: Astron. J. 69, 73 (1964)
6.4   K.J. Whiteman: Rep. Prog. Phys. 40, 1033 (1977)
6.5   S. Smale: Bull. Am. Math. Soc. 73, 747 (1967)
6.6   Y. Ueda: J. Stat. Phys. 20, 181 (1979)
6.7   Y. Ueda, N. Akamatsu: IEEE Trans. CS-28, 217 (1981)
6.8   J.S. Nicolis, G. Galanos, E.N. Protonotarios: Int. J. Control 18, 1009 (1973)
6.9   M.J. Feigenbaum: Los Alamos Science (Summer 1980) p.4
6.10 D.R. Hofstadter: Sci. Am. 236, 16 (Nov. 1981)
6.11 E.N. Lorenz: Tellus 16, 1 (1964)
6.12 P. Manneville, Y. Pomeau: Physica 1D, 219 (1980)
6.13 R. Shaw: Z. Naturforsch. 36a, 80 (1980)
6.14 J.D. Farmer: Z. Naturforsch. 37a, 1304 (1982)
6.15 J.S. Nicolis, E.N. Protonotarios: Int. J. Bio-Med. Comput. 10, 417 (1979)
6.16 P. Grassberger: "On the Hausdorff Dimension of Fractal Attractors", Preprint WVB 80-33 (Oct. 1980)
6.17 J.D. Farmer, E. Ott, J.A. Yorke: Physica 7D, 153 (1983)
6.18 G. Nicolis, R. Lefever: Phys. Lett. 62A, 469 (1977)
6.19 G. Nicolis, I. Prigogine: Self-Organization in Non-Equilibrium Systems (Wiley, New York 1977)
6.20 J.E. Hirsch, B.A. Huberman, D.J. Scalapino: Phys. Rev. A25, 519 (1982)
6.21 N.H. Packard, J.P. Crutchfield, J.D. Farmer, R.S. Shaw: Phys. Rev. Lett. 45, 712 (1980)
6.22 H. Mori: Prog. Theor. Phys. 63, 1044 (1980)
6.23 R. Elul: Science 164, 328 (1969)
6.24 R. Elul: "Randomness and Synchrony in the Generation of EEG", in Synchronization of EEG Activity in Epilepsies, ed. by H. Petsche, M. A.B. Brazier (Springer, Berlin, Heidelberg 1972) pp.59-77
6.25 W.R. Adey: Int. J. Neurosci. 3, 271 (1972)
6.26 M. Verzeano: "Pacemakers, Synchronization, and Epilepsy", in Synchronization of EEG Activity in Epilepsies, ed. by H. Petsche, M.A. B. Brazier (Springer, Berlin, Heidelberg 1972) p.154
6.27 J.S. Nicolis, M. Benrubi: J. Theor. Biol. 59, 76 (1976)
6.28 G. Moruzzi, H.W. Magoun: Clin. Neurophysiol. 1, 455 (1949)
6.29 R.M. May: Intern. Conf. on Non-Linear Dynamics, New York, 1979, Ann. N.Y. Ac. Sci. 357, 267-281 (1981)
6.30 W.R. Adey: In The Neurosciences: A Study Program, ed. by F.O.Schmitt (Rockefeller University Press, New York 1970) p.181
6.31 S.J. Hutt, H. Fairweather: Electroencephalogr. Clin. Neurophysiol. 39, 43 (1975)
6.32 J.S. Nicolis, E.N. Protonotarios: Int. J. Bio-Med. Comput. 10, 417 (1979)
6.33 G. Nicolis, R. Lefever: Phys. Lett. 62A, 469 (1977)
6.34 G. Nicolis, J.N. Turner: Ann. N.Y. Ac. Sci. 316, 251 (1979)

6.35   W. Horsthemke, M.M. Mansour: Z. Phys. $\underline{24}$, 307 (1976)
       W. Horsthemke, M.M. Mansour, L. Brenig: Z. Phys. $\underline{28}$, 135 (1977)
       L. Arnold, W. Horsthemke, R. Lefever: Z. Phys. $\underline{29}$, 367 (1978)
6.36   G. Mayer-Kress, H. Haken: J. Stat. Phys. $\underline{26}$, 149 (1981)
6.37   J.P. Crutchfield, J.D. Farmer, B.A. Huberman: "Fluctuations and
       Simple Chaotic Dynamics", Phys. Rep. $\underline{92}$, 45 (1982)
6.38   K. Matsumoto, I. Tsuda: "Noise Induced Periodicity", J. Stat. Phys.
       $\underline{31}$, 87 (1983)

Chapter 7

7.1   H. Mori: Prog. Theor. Phys. $\underline{63}$, 1044 (1980)
7.2   H.J. Bremermann: Int. J. Theor. Phys. $\underline{21}$, 203 (1982)
7.3   P.W. Anderson: Science $\underline{177}$, 393 (1972)
7.4   G. Nicolis, I. Prigogine: Proc. Nat. Ac. Sci. USA $\underline{78}$, 659 (1981)
7.5   F. Papentin: Naturwissenschaften $\underline{67}$, 174 (1980)
7.6   H.J. Bremermann: "Quantitative Aspects of Goal-Seeking Self-Organiz-
       ing Systems", in Progress in Theoretical Biology, Vol.1,ed. by R.
       Rosen, F. Smell (Academic, New York 1967) pp.59-77
7.7   H. Haken: Synergetics: An Introduction, 3rd ed., Springer Ser. Syn-
       ergetics, Vol.1 (Springer, Berlin, Heidelberg 1983)
7.8   A.C. Scott: IEEE Trans. CT-$\underline{17}$, 55 (1970); IEEE Trans. SMC-$\underline{1}$, 267
       (1971)
7.9   R.D. Parmentier: IEEE Trans. CT-$\underline{19}$, 142 (1972)
       T. Endo, T. Ohta: IEEE Trans. CS-$\underline{27}$, 277 (1980)
       T. Endo, S. Mori: IEEE Trans. CS-$\underline{23}$, 100 (1976)
7.10   C. Grebogi, E. Ott, J. Yorke: Phys. Rev. Lett. $\underline{50}$, 935 (1983)
7.11   A. Rapoport: Behav. Sci. $\underline{1}$, 303 (1956)
7.12   P. McLean: Psychother. Psychosom. $\underline{28}$, 207 (1977)

Appendix

A.1   W.C. Lindsey: Synchronization Systems in Communication and Control
       (Prentice Hall, Englewood Cliffs,NJ 1972)
A.2   P. Glansdorff, I. Prigogine: Thermodynamic Theory of Structure,
       Stability and Fluctuations (Wiley, New York 1971)
A.3   K.S. Trincher: Biology and Information (Consultants Bureau,
       New York 1965)
A.4   R. Elul: "Statistical Mechanisms in Generation of the EEG", in Pro-
       gress in Biomedical Engineering, ed. by L. Fogel (Spartan Books,
       Washington, DC 1966) pp.131-150
A.5   R.Elul: "Brain Waves: Intracellular Recording and Statistical Ana-
       lysis Help Clarify Their Physiological Significance", in Data Ac-
       quisition and Processing in Biology and Medicine, ed. by O. Enslein
       (Pergamon, Oxford 1968) pp.93-114
A.6   W.R. Adey: In The Mind: Biological Approaches to Its Function, ed.by
       W. Corning, J. Balaban (Wiley, New York 1968) pp.69-99
A.7   J.S. Nicolis, G. Galanos, E.N. Protonotarios: Int. J. Control $\underline{18}$,
       1009 (1973)
A.8   R.L. Stratonovich: In Non-Linear Transformation of Stochastic Pro-
       cesses, ed. by P.I. Kuzretsov, R.L. Stratonovich, V.I. Tikhonov
       (Pergamon, London 1965)
A.9   N. Minorski: Non-Linear Oscillations (Van Nostrand, Princeton, 1962)
A.10   R. Elul: "Scanning of Cortical Neurons in the EEG:A Possible Mechan-
       ism of Attention and Consciousness", in Information Processing in
       Dendrites, ed. by A. Scheibel, M. Scheibel (Brain Information Ser-
       vice, University of California, Los Angeles 1969)

A.11 H. Hydén: In <u>Beyond Reductionism</u>, ed. by A. Koestler, W. Smythies (Hutchinson, London 1969)

A.12 J.S. Nicolis, E.N. Protonotarios, E. Lianos: Biol. Cybern. <u>17</u>, 183 (1975)

A.13 R.J. Spiegel, W.T. Jones: Bull. Math. Biol. <u>35</u>, 591 (1973)

A.14 B.C. Goodwin: <u>Temporal Organization in Cells</u> (Academic, London 1963)

A.15 B.C. Goodwin: Adv. Enzyme Regul. <u>3</u>, 425 (1965)

A.16 J.S. Griffith: J. Theor. Biol. <u>20</u>, 202 (1968)

A.17 J. Tiwari, A. Fraser, R. Beckman: J. Theor. Biol. <u>45</u>, 311 (1974)

A.18 D. Lehman, M. Koukkou: Electroencephalogr. Clin. Neurophysiol. <u>37</u>, 73 (1974)

A.19 R. Elul: Nature London <u>210</u>, 1127 (1966)

A.20 D. Giannitrapani, L. Dayton: Electroencephalogr. Clin. Neurophysiol. <u>36</u>, 377 (1974)

A.21 T.H. Maugh: Science <u>184</u>, 147 (1974)

A.22 H. Snyder: <u>Madness and the Brain</u> (McGraw-Hill, New York 1974)

Subject Index

Analog-to-digital conversion 123
Animal conflicts 237,248
Attractors 5

Baker's transformation 312
Bifurcation 2
"Big-bang" 62
Boltzmann's constant 49
Brains's reticular formation 204
Bremmerman's limit 342
Broken symmetry 18

Cantor set (symmetric, asymmetric
  black-white, gray) 275
Channel capacity 130
Chaos (dissipative) 2,276
Chaos (Hamiltonian) 268
Coding theory 144
Cognitive channel 356
Coherence 22
Collisions 51
Communication between hierarchical
  systems 181
Competing species 217
Complexity 1,7
Compressibility 5,303
Computational complexity 342
Constant-sum games 218
Continuous systems 377
Cosmology 62

Dimensionality of an attractor 301
Dipole moment 77
Discrete systems 276
Dispersive medium 90
Dominant strategy 218

Electroencephalogram (E.E.G.) 330
Entrainment (strict, jittery) 381
Entropy 46
Entropy of electromagnetic radiation
  112
Entropy of Markov chains 150
Ergodicity 262
Error correction codes 144
Evanescent waves 98
Evolutionary stable strategy (E.S.S.)
  251
Expanding universe 67

Fluctuations 340,369
Fokker-Planck equations 172
Fractal objects 284
Fractal separatrix 357

Game of "Chicken" 233,234
Game theory 217
Gravitation 12

H-theorem 52
Hierarchical levels 1,5